International Association of Geodesy Symposia

Wolfgang Torge, Series Editor

Springer

Berlin
Heidelberg
New York
Barcelona
Budapest
Hong Kong
London
Milan
Paris
Santa Clara
Singapore
Tokyo

International Association of Geodesy Symposia

Wolfgang Torge, Series Editor

GPS Trends in Precise Terrestrial, Airborne, and Spaceborne Applications

Symposium No. 115

Boulder, CO, USA, July 3–4, 1995

Convened and Edited by
Gerhard Beutler
Günter W. Hein
William G. Melbourne
Günter Seeber

Gerhard Beutler
Astronomisches Institut
Universität Bern
Sidlerstraße 5
CH-3012 Bern, Switzerland

Günter W. Hein
Hochschule der Bundeswehr
Werner-Heisenberg-Weg 39
D-85579 Neubiberg, Germany

William G. Melbourne
JPL, 4800 Oak Grove Drive
Pasadena, CA 91109, USA

Günter Seeber
Institut für Erdmessung
Universität Hannover
Nienburger Straße 6
D-30167 Hannover, Germany

Series Editor
Wolfgang Torge
Institut für Erdmessung
Universität Hannover
Nienburger Straße 6
D-30167 Hannover, Germany

```
         Library of Congress Cataloging-in-Publication Data


GPS trends in precise terrestrial, airborne, and spaceborne
  applications : Boulder, Colorado, USA, July 3-4, 1995 / convened and
  edited by Gerhard Beutler ... [et al.].
      p.   cm. -- (International Association of Geodesy symposia :
  symposium no. 115)
    Includes bibliographical references.
    ISBN 3-540-60872-9
    1. Global Positioning System--Congresses.   I. Beutler, G.
  II. Series: International Association of Geodesy symposia :
  symposium 115.
  GA109.5.G68   1996
  526'.1'028--dc20
```

For information regarding symposia volumes 101 and onward contact:
Springer-Verlag GmbH & Co. KG
Heidelberger Platz 3
D-14197 Berlin, Germany

For earlier volumes contact:
Bureau Central de l'Association Internationale de Géodésie
2, Aveneue Pasteur, F-94160 Saint-Mande, France

ISBN 3-540-60872-9 Springer-Verlag Berlin Heidelberg New York

Typesetting: Camera ready by editor/authors
SPIN: 10505866   32/3136 - 543210 - Printed on acid-free paper

Preface

These proceedings include most of the papers presented at the IAG Symposium *GPS Trends in Precise Terrestrial, Airborne, and Spaceborne Applications* held in July 1995 during the XXI-th IUGG General Assembly in Boulder, Colorado. The symposium was jointly organized by the IAG and the International Union of Surveys and Mapping (IUSM).

The symposium was divided into four sessions, namely

(1) *The International GPS Service for Geodynamics (IGS)
    and other Permanent Networks,*
(2) *Spaceborne Applications of the GPS,*
(3) *Kinematic Applications of the GPS,* and
(4) *The GPS and its Relations to Geophysics.*

The main purpose was to give an overview of the state of the art in 1995 of the applications of the GPS to geodynamics, geodesy, surveying, and navigation.

The call for papers generated a flood of originally more than 70 abstracts; quite a few could be redirected to other symposia, but still 56 papers found their way into these proceedings. We thus conclude that the volume gives a rather complete overview of *GPS Trends in Precise Terrestrial, Airborne, and Spaceborne Applications* in the year 1995.

Gerald L. Mader, the convenor of the 1991 IAG Symposium on *Permanent Satellite Tracking Networks for Geodesy and Geodynamics*, held at the XX-th IUGG General Assembly in Vienna, wrote in the preface for the proceedings of that symposium: "The Global Positioning System (GPS) is becoming, and promises to remain for some time, one of the most important geodetic measurement systems. The contributions to date of GPS geodesy are truly revolutionary, encompassing such diverse applications as measurements of crustal deformation, precise positioning of mobile platforms and monitoring of ionospheric conditions. When one considers the accuracy obtainable with GPS and the relatively low cost for acquiring this technology, the full impact of GPS over the next decade is indeed very difficult to estimate."

It is our impression that four years later it is still difficult to estimate the full impact of the GPS in all mentioned areas. Meanwhile the GPS became fully operational; at times there were even more than the 24 planned satellites (due to some long-lasting Block-I satellites). We have also seen that *Anti-Spoofing (AS)* had little effect on the high accuracy interferometric applications of the GPS if modern receiver technology is available. These developments led to a dramatic growth of the GPS user community. What else happened during the last four years?

We have seen the remarkable exploitation of the GPS for the purposes of regional and global geodynamics. Today, the *International GPS Service for Geodynamics (IGS)* routinely produces orbits of sub-decimeter accuracy complemented with IGS pole coordinates accurate to about 0.1-0.2 milliarcseconds. Who would have predicted this performance and this accuracy level in 1991? The IGS contributions to the establishment of the ITRF, the IERS Terrestrial Reference Frame, are getting more and more important, too. The number of GPS-derived reference sites is steadily growing, the accuracy of the estimated coordinates, and, with the time basis increasing, the accuracy of the estimated station velocities are comparable today with the results of the other space geodetic methods. This development is well documented in the first chapter of the proceedings. It also becomes apparent from browsing through this first chapter that very powerful other permanent networks, like the German DGPS service or the dense Japanese permanent GPS networks, are being built up. This development will undoubtedly continue in future and it will stimulate in turn most reliable hardware and very powerful and easy to use software tools. We would not be amazed if four years from now several regional networks for monitoring crustal movements or deformations with several hundred receivers would be fully operational and turn out results in near real time.

The articles contained in the second chapter *Spaceborne Applications of the GPS* have the focus on the Topex/Poseidon mission, where a space-borne GPS receiver together with SLR (Satellite Laser Ranging) and the French DORIS system were used for orbit determination, and on the GPS-MET experiment, where a spaceborne GPS receiver was used for atmosphere sounding. That only few institutions were able to contribute to the research in this area up till now, does not mean that the area is not relevant, but rather that the topic is still "very young". From the overview article (first article) of chapter 2 we conclude that the potential of spaceborne GPS applications is tremendous, indeed. We thus predict that four years from now there might be many more institutions interested in and contributing to this topic.

Kinematic applications of the GPS always are fascinating, sometimes even amusing to browse through. One may get the impression that every day a new application for the GPS is found. If you do not believe it, just read the overview article (first article) of the third chapter – even farmers seem to get involved in space geodesy nowadays! Apart from that it seems that the progress over the last four years in the areas of receiver technology and processing strategies were more pronounced for kinematic than for static applications. It is also interesting to note that the combination of GPS with other sensors is alive in kinematics: the combination of GPS receivers with Inertial Navigation Systems (INS) proves to be most useful.

There are in principle two motivations to study the *atmosphere* refraction using the GPS: one may model the atmosphere with the main goal of improving the estimates for the other parameters of a GPS adjustment (coordinates, orbits, etc.). This was the main motivation in the past, where the atmosphere parameters were considered as nuisance parameters. One may, on the other hand, be primarily interested in atmosphere physics, or more precisely, in setting up an interface between the "GPS world" and atmosphere physics. When we asked for contributions to the fourth session we had applications of the second kind in mind, like e.g. the estimation of the total precipitable water content with a high time resolution (hours to minutes), or the establishment of ionosphere models, or the assessment of the *stochastic behaviour* of the ionosphere. We believed that this would be a "hot topic", in particular in view of the potential offered by a globally distributed network like that of the IGS. Our opinion was shared by the authors of the overview paper for the fourth chapter and by the article written by atmosphere physicists. Most other contributions dealt with *GPS internal* atmosphere modeling questions, like "how to establish GPS-derived ionosphere models with the goal to use them for the processing of single-band GPS data". Again, we believe that the situation might be quite different four years from now.

In an a posteriori evaluation of our symposium we found it problematic that there was *no* theory and method-related session. On the other hand we obtained quite a few remarkable papers belonging to this area. This is why we created a fifth chapter *theory and methodolgy* for these proceedings. We believe that it would have been a mistake to hide fascinating articles like those related to ambiguity resolution, or topics like the antenna phase center variations, which are known to be crucial in the future, or reports about new software developments with interesting characteristics in the other four chapters.

Modern scientists have many obligations and it is not always easy to observe deadlines. Let us therefore express our sincere thanks to the 56 authors of these proceedings making it possible to send the manuscript *GPS Trends in Precise Terrestrial, Airborne, and Spaceborne Applications* to the publisher about four month after the Boulder symposium.

Gerhard Beutler
Günter Hein
William G. Melbourne
Günter Seeber

Table of Contents

Chapter 1
The International GPS Service for Geodynamics (IGS) and other Permanent Networks

Chapter 2
Spaceborne Applications of the GPS

Chapter 3
Kinematic Applications of the GPS

Chapter 4
The GPS and its Relations to Geophysics

Chapter 5
Theory and Methodology

Chapter 1

The International GPS Service for Geodynamics (IGS) and other Permanent Networks

THE INTERNATIONAL GPS SERVICE FOR GEODYNAMICS (IGS): THE STORY

Gerhard Beutler
Astronomical Institute,University of Berne,
CH-3012 Bern, Switzerland
Ivan I. Mueller
Dept. of Geodetic Science and Surveying,
Ohio State University, Columbus, Ohio 43210 USA
Ruth E. Neilan
IGS Central Bureau,
Jet Propulsion Laboratory, Pasadena, California 91109 USA

ABSTRACT

The primary goal of the *International GPS Service for Geodynamics (IGS)* is to provide highly accurate *orbits, earth rotation parameters, and station coordinates* derived from observations of the Global Positioning System (GPS) to the geodynamics community, which currently demand the highest accuracy. A further goal is the extension of the IERS Terrestrial Reference Frame (ITRF) for the purpose of global accessibility. The data of the permanent IGS network consisting of about 80 stations are transmitted on a daily basis to regional and global data centers from where they are retrieved by the IGS Analysis Centers.

Today the IGS produces a *combined orbit* based on the contributions of seven Analysis Centers in addition to the orbit- and earth rotation- series of the individual IGS Analysis Centers. The IGS orbit is available about two weeks after the observations, its accuracy is of the order of 10 cm. The IGS orbits together with the IGS station coordinates and the GPS measurements in the IGS network give the user community a direct access to the ITRF.

The paper gives an overview of the development of the IGS through 1995. The present structure is discussed, the accuracy and the reliability of the products will be mentioned, current developments and future plans will be presented.

DEVELOPMENT OF THE IGS

According to (Mueller, 1993), the primary motivation in planning the IGS was the recognition in 1989 that the most demanding users of the GPS satellites, the geophysical community, were purchasing receivers in exceedingly large numbers and using them as more or less black boxes, using software packages which they did not completely understand, mainly for relative positioning. The observations as well as the subsequent data analyses were not based on common standards; thus the geodynamic interpretation of the results could not be trusted. The other motivation was the generation of precise ephemerides for the satellites together with by-products such as earth orientation parameters and GPS clock information.

Table 1: Chronicle of Events 1989-1991

Date	Event
Aug 89-Feb 90	IAG General Meeting in Edinburgh. Initial Plans developed by I.I. Mueller, G. Mader, W.G. Melbourne, B. Minster, and R.E. Neilan
16-Mar-90	IAG Executive Committee Meeting in Paris decides to establish a Working Group to explore the feasibility of an IGS under IAG auspices. I.I. Mueller was elected as chairman.
25-Apr-90	The Working Group is redesignated as *The IAG Planning Committee for the IGS* in Paris
02-Sep-90	Planning Committee Meeting in Ottawa Preparation of the Call for Participation (CFP)
01-Feb-91	CFP mailed. Letters of Intent due 1 April 1991
01-Apr-91	CFP Attachments mailed to those whose letters of intent were received
01-May-91	Proposals due
24-Jun-91	Proposals evaluated and accepted in Columbus, Ohio
17-Aug-91	Planning Committee reorganized and renamed as IGS Campaign Oversight Committee (OSC) at the 20th IUGG General Assembly in Vienna
24-Oct-91	First IGS Campaign Oversight Committee Meeting in Greenbelt, MD. Preparation of the 1992 IGS Test Campaign scheduled for 21 June - 23 September 1992 and for a two weeks intensive campaign called Epoch 92

These ideas were first discussed in 1989 at the IAG General Meeting in Edinburgh (Neilan, Melbourne and Mader, 1990) and led soon thereafter to a Working Group (later redesignated as the *IAG Planning Committee for the IGS*) with Ivan I. Mueller, then President of the IAG, as chairman. After several meetings the *Call for Participation* was issued by this group on 1 February 1991. More than 100 scientific organizations and governmental survey institutions announced their participation either as an observatory (part of the IGS network), as an analysis center, or as a data center. The Jet Propulsion Laboratory (JPL) volunteered to serve as the Central Bureau, and The Ohio State University as the Analysis Center Coordinator. At the 20th General Assembly of the IUGG in Vienna, August 1991 the IAG Planning Committee was restructured and renamed as *IGS Campaign Oversight Committee*. This committee started organizing the 1992 events, namely the *1992 IGS Test Campaign* and

Epoch'92. Two IGS Workshops (the first at the Goddard Space Flight Center in October 1991, the second in Columbus, Ohio in March 1992) were necessary to organize the 1992 activities. The essential events of this first phase of the IGS development are summarized in Table 1.

The 1992 IGS Test Campaign, scheduled from 21 June to 23 September 1992, focused on the routine determination of high accuracy orbits and ERPs; it was to serve as the proof of concept for the future IGS. Epoch'92 on the other hand was scheduled as a two-week campaign in the middle of the three-month IGS Campaign for the purpose of serving as a first extension of the relatively sparse IGS Core Network analyzed on a daily basis by the IGS Analysis Centers. More background information about this early phase of IGS may be found in (Mueller, 1993), (Mueller and Beutler, 1992).

Two events prior to the 1992 IGS Test Campaign have to be mentioned: (1) the *communications test*, organized by Peter Morgan, Australia, demonstrated that data transmission using the scientific Internet facility had sufficient capacity for the daily data transfer from the IGS stations to the Regional, Operational and Global Data Centers then to the Analysis Centers. (2) The establishment of the *IGS Mailbox* and the *IGS Report* series based on e-mail proved to be very important as information resources and as a tool to insure a close cooperation between the IGS participants. This e-mail service, initially located at the University of Bern, was transferred to the Central Bureau (JPL) by the 1st of January 1994.

The 1992 IGS Campaign started as scheduled on 21 June 1992. About two weeks later the first results of the IGS Analysis Centers started to flow into the IGS Global Data Centers, which made these results available to the user community. The ERP series were regularly analyzed by the IERS Central Bureau and by the IERS Rapid Service Sub-bureau.

Data collection and transmission as well as data analysis continued on a *best effort basis* after the official end of the 1992 IGS Test Campaign on 23 September, 1992. At the third IGS Campaign Oversight Committee meeting on October 15, 1992 at Goddard Space Flight Center (Table 2) it was decided to formally establish the IGS Pilot Service to bridge the gap between the 1992 IGS Test Campaign and the start of the official service. Since 1 November 1992 the orbits of the individual processing centers were regularly compared by the IGS Analysis Center Coordinator (Goad, 1993). An overview of the 1992 IGS events may be found in the Proceedings of the 1993 IGS Workshop (Brockmann and Beutler, 1993).

Two Workshops, the Analysis Center Workshop in Ottawa (Kouba, 1993) and the Network Operations Workshop in Silver Spring, MD, and the first Governing Board (GB) Meeting (also in Silver Spring) took place in October 1993. One important outcome of IGS meetings in October 1993 was the decision to produce an official IGS orbit. This responsibility was given to the IGS Analysis Center Coordinator, who, according to the Terms of Reference must be an analysis centers' representative.

Table 2: Chronicle of Events 1992-1993

Date	Event
17-Mar-92	2nd IGS O S C Meeting at OSU,Columbus,Ohio
04-May-92	Communication Tests
21-May-92	IGS e-mailbox installed
21-Jun-92	Start of IGS Test Campaign
01-Jul-92	First results, about 2 weeks after begin of campaign
27-Jul-92	Start of Epoch-92 (2 weeks)
23-Sep-92	official end of campaign, NOT of data collection, processing
15-Oct-92	3rd IGS O S C Meeting at GSFC,Greenbelt MD
01-Nov-92	Start of IGS PILOT Service Start of routine orbit comparisons by IGS Analysis Center Coordinator
24-Mar-93	1993 IGS Workshop and 4th IGS O S C Meeting at the University of Bern
27-May-93	5th IGS O S C Meeting, AGU, Baltimore MD
09-Aug-93	IAG-Symposium in Beijing. Approval of the Service by the IAG
12-Oct-93	Analysis Center Workshop, Ottawa
18-Oct-93	Network Operations Workshop and 1st IGS Governing Board Meeting in Silver Spring MD
08-Dec-93	GB Business Meeting in San Francisco

In view of the success of the 1992 IGS Test Campaign and of the IGS Pilot Service the IGS Campaign Oversight Committee at its 4th Meeting in March 1993 in Bern decided to take the necessary steps towards the establishment of the official IGS on 1 January 1994. In particular the Terms of Reference for this new service were written (IGS Colleague Directory, 1994), the organizations active in the 1992 IGS campaigns were asked to confirm their participation in the future service, and last but not least the IAG approval for the establishment of the IGS for 1 January 1994 was requested. It was encouraging that most of the key organizations confirmed their participation in the official service: the Central Bureau stayed at JPL, the three Global Data Centers and all but one Analysis Centers continued contributing to the IGS. In view of this encouraging development it was gratifying that the preliminary IAG approval (to be confirmed at the 21st IUGG General Assembly in Boulder, 1995) was given in August 1993.

A key element of the new Service is the Governing Board (GB) consisting of 15 members (current membership in (IGS Colleague Directory)). Another key element is the interface between the IGS and the IERS both being IAG services with many common interests. In practice the IERS relies on the IGS for all GPS operations, the IGS in turn relies on the IERS for the continuous maintenance of the terrestrial reference frame.

Table 3 contains the essential events since the start of the official IGS on January 1, 1994. It was of greatest importance that the Central Bureau Information System (CBIS) (Liu et al, 1994) and the combined IGS orbit (Beutler, Kouba, Springer, 1995) became available with the start of the new service.

The densification of the ITRF through regional GPS analyses was a key issue in 1994. The guidelines for such a densification were defined at the IGS Workshop in December 1994. The topic will continue to be in the center of IGS activities in 1995 and in the years to come.

Table 3: Chronicle of Events 1994- mid 1995

Date	Event
01-Jan-94	Start of official IGS. Production of Combined IGS Orbit. Central Bureau Information System (CBIS) established
21-Mar-94	Combined Workshop IERS/IGS in Paris (1 week)
25-Mar-94	2nd IGS Governing Board Meeting
30-Nov-94	IGS Workshop Densification of the ITRF through regional GPS Analyses in Pasadena
06-Dec-94	3rd IGS Governing Board Meeting in San Francisco
15-May-95	IGS Workshop on Special Topics and New Directions in Potsdam
06-Jul-95	4th IGS Governing Board Meeting in Boulder

STRUCTURE AND PERFORMANCE OF THE IGS IN 1995

The IGS consists of a *network* of permanent high-accuracy GPS tracking receivers, a set of operational, regional, and global *data centers*, currently seven *Analysis Centers*, the *Analysis Center Coordinator*, the *IGS Central Bureau*, and the IGS Governing Board (IGS Colleague Directory).

The network was considerably growing since the 1992 IGS Test Campaign. Figure 1 shows the current permanent IGS tracking network, where the distinction is made between fully operational and planned or future sites. Even today the network is far from ideal : we still have the problem that the regions of the former USSR, of China, India and of the Southern hemisphere are underrepresented; but the situation is *much better today* than in 1992. It should be mentioned that all operational sites in Figure 1 are connected with high performance data links. Usually the scientific Internet facility is used, sometimes satellite data links had to be established (e.g. from Kitab to Potsdam by the GFZ).

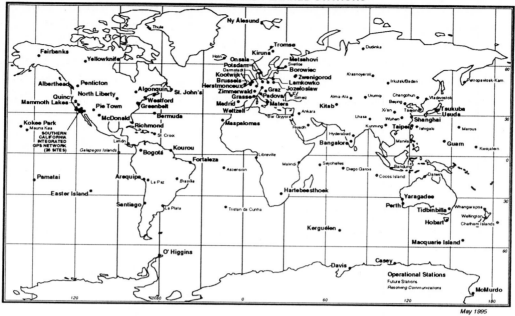

GPS TRACKING NETWORK OF THE INTERNATIONAL GPS SERVICE FOR GEODYNAMICS OPERATIONAL AND PLANNED STATIONS

Figure 1: IGS Permanent Tracking Network in 1995

Table 4: Analysis Centers (A), Global Data Centers (D), Analysis Center Coordinator (C), Central Bureau (B) of the IGS

Abbreviation	Institution	Participating
CODE (A)	at Astr. Institute, Bern, Switzerland joint venture of AIUB, IfAG, IGN, and Swiss Federal Office of Topography	since 21 June 1992
NRCan(A)	Natural Resources, Canada (former EMR)	since September 1992
ESOC (A)	European Space Agency, Germany	since 21 June 1992
GFZ (A)	Geoforschungszentrum, Germany	since 21 June 1992
JPL (A)	Jet Propulsion Laboratory, USA	since 21 June 1992
NOAA (A)	Natl. Oceanic and Atmosph. Adm. USA	since March 1993
SIO (A)	Scripps Inst. of Oceanography, USA	since 21 June 1992
UTX (A)	University of Texas at Austin,USA	21 June to 23 Sep 1992
CDDIS(D)	Goddard Space Flight Center, USA	since 21 June 1992
IGN (D)	Institut Geographique National, France	since August 1992
SIO (D)	Scripps Institution of Oceanography USA	since 21 June 1992
OSU (C)	Ohio State University	21 June 92 - Dec 1993
NRCan(C)	Natural Resources, Canada (former EMR)	since January 1994
JPL (B)	Jet Propulsion Laboratory, USA	since 21 June 1992

8

Table 4 lists the IGS Analysis Centers (A), the Global Data Centers, the current IGS Analysis Coordinating Center (C), and the IGS Central Bureau (B). Table 4 documents the remarkable stability of the essential IGS components. Let us keep in mind that the primary task of the IGS Analysis Center Coordinator (C) is the *production of the combined IGS orbit* based on the contributions of the currently seven IGS Analysis Centers. The Central Bureau coordinates all IGS activities, it is also responsible for the interface with the IERS. The *Central Bureau Information System (CBIS)* introduced in the previous section is accessible through Internet from all over the world.

A thorough discussion of the quality of IGS products as available today would require much more space than is available in this short report. We confine ourselves to a few remarks concerning the orbit quality and refer to other articles in the same volume for a more detailed discussion of the quality of clocks (Mireault et al, 1995) and earth rotation and orientation parameter (McCarthy, 1995). The quality of the annual coordinate solutions stemming from IGS Analysis Centers is e.g. documented in the first IGS Annual Report (Zumberge et al, 1995a).

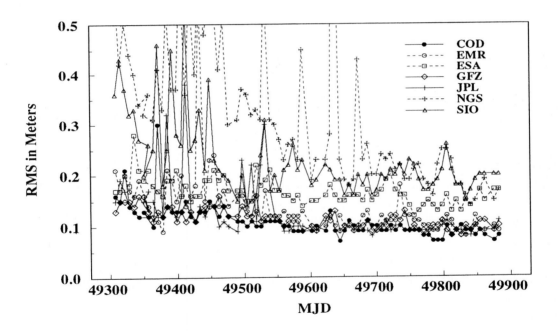

Figure 2: Orbit Quality from IGS Orbit Combination November 1993 - May 1995 (Weighted RMS, Weekly Averages)

Figure 2 gives an impression of the consistency of the orbits stemming from different IGS processing agencies. We can see the rms per satellite coordinate of the fit of individual solutions with respect to the IGS Combined Orbit as it is computed for every day by the IGS Analysis Center Coordinator (Table 4, weekly averages are given in Figure 2). Clearly the best four contributions show an rms-consistency per satellite coordinate of about 10 cm. We also see a clear trend of all processing centers to the same level of accuracy. This fact makes the orbit combination relatively unproblematic, and the detection of occasional outliers relatively easy. In consequence the IGS combined orbit is not only a very accurate, but also a very *robust and reliable* product (the principles of IGS orbit combination are given in (Beutler, Kouba, Springer, 1993)).

This high level of consistency is also reflected by the other IGS products. We believe that today the precision of the individual IGS polar motion series are of the order of 0.2 - 0.5 mas for diurnal estimates (these are the accuracies attributed to the individual IGS series by the IERS), that the length of day estimates (daily means) are of the order of 0.03 msec/day), and that the precisions of the IGS coordinates are about one centimeter in annual coordinate solutions (sub-centimeter values in horizontal coordinates, 1-2 cm in height). With increasing length of the IGS time series the velocity estimates for the IGS tracking network are becoming more and more reliable, too.

RECENT, CURRENT AND FUTURE DEVELOPMENTS WITHIN THE IGS

Let us discuss some results of the latest two IGS Workshops, namely the *IGS Workshop Densification of the ITRF through regional GPS Analyses in Pasadena* and the *IGS Workshop on Special Topics and New Directions in Potsdam*.

The first attempt to densify the relatively sparse permanent IGS tracking network in Figure 1 to a network of relatively uniform density (spacings of 1000 - 2000 km are aimed at) was *the 1992 IGS Epoch Campaign* (Table 2). A large number of sites was observed during the two weeks and processed afterwards, some of the networks measured (e.g. the French DORIS network) even showed a rather nice global distribution. Due to several reasons the overall result was *not* satisfactory, however. Therefore it was decided to devote a three-days Workshop to the topic of densifying the IGS network.

Four position papers were presented at the *IGS Workshop Densification of the ITRF through regional GPS Analyses in Pasadena* (Zumberge and Liu, 1995b). The main conclusions may be summarized as follows:

- The densified IGS network should consist of about 200 globally well distributed sites. A maximum spacing should be specified.

- *Wherever possible permanent tracking should be established.*

- The densification should be realized in cooperation with other groups setting up permanent GPS networks (e.g. the Gloss group).

- Regional analyses will be performed by (new) IGS Associate Analysis Centers, called AAC Type-1 in this context. These will make extensive use of the global products (orbits, ITRF coordinates and velocities of sites well established in the global network). These AACs produce free network solutions on a weekly basis. The seven IGS Analysis Centers may play the role of AAC Type-2, too.

- The weekly free network solutions will be combined by special IGS analysis centers, called AAC Type-2 at present. These centers will analyse the results of the AAC Type-1 and prepare a weekly combined solution, giving all results in a consistent reference frame.

- A Pilot Phase involving *initially* only the seven IGS processing centers of the IGS as AAC Type-1 and *three AAC Type-2* will start on 1 September 1995 and will last for one calendar year. The three AAC Type-2 are:

  - Jet Propulsion Laboratory, Pasadena (J. Zumberge)
  - University of Newcastle (G. Blewitt)
  - Scripps Institution of Oceanography (Y. Bock)

  Later in the experiment a selected number of AAC Type-1 will join the Pilot Project.

- The ITRF section of the IERS will accompany the Pilot Project by analyzing the products of the AAC Type-2 on a monthly basis.

The topics at the *IGS Workshop on Special Topics and New Directions in Potsdam* were quite different. The focus was on one hand on processing aspects (fascinating orbit modeling aspects and antenna problems were discussed) on the other hand on possible new products like the *ionosphere models*, the *station-specific precipitable water content values* with a high temporal resolution using the IGS network. Last but not least *commercial aspect* of the IGS activities were considered. For a more complete report we refer to (Beutler, 1995) and of course to the proceedings of this Workshop which are under preparation. Let us add a few remarks concerning the earth's atmosphere.

All IGS Analysis Centers have to model tropospheric refraction. It was shown that the consistency of estimates stemming from different IGS Analysis Centers is relatively high. It should thus be possible to extract on a routine basis the precipitable water content for the entire IGS network with a high temporal resolution (two hours or finer) – *provided high accuracy barometers are deployed in the IGS network*. Even if the IGS products are available "only" about two weeks after the observations, the results are still most valuable *for climatological studies*. Should the IGS decide to produce very rapid products in addition to what is available today, the same estimates might be used for weather prediction, a topic of highest interest to meteorologists (Bevis et al, 1992).

The motivation to produce GPS derived ionosphere models is manyfold: Local or regional models may be used to remove (reduce) biases in single frequency surveys; regional or global models may be used to calibrate e.g. altimeter measurements (ERS-1 is an example) or to calibrate radio signals from space vehicles in the planetary system;

11

last but not least there are pure research projects (ionosphere maps, extraction of geomagnetic indices). At present we may *not* speak of a high degree of consistency of ionosphere models produced by different IGS Analysis Centers. This will be one aspect to focus on in the near future. The other, more important, aspect for the future development is to define an interface between the IGS and the ionosphere research community. What exactly is needed in ionospheric research, what kind of products should be made available? The IGS definitely needs the help of the mentioned research community.

This report reflects (hopefully) the fact that the IGS is a most stimulating and rewarding adventure for all its participants. This GPS *service* makes essential contributions not only to geodesy an to regional and global geodynamics, but also to scientific areas like meteorology and ionosphere physics.

Undoubtedly the progress made since the XXth IUGG General Assembly in Vienna is far beyond any expectations. Only three years after the first plans the IGS, an IAG service in support of geodesy and geodynamics, became fully operational on January 1, 1994. In view of the complexity of this task such a rapid development is an achievement in itself. It was made possible through the experience, the expertise, and the pioneer spirit in the IGS Campaign Oversight Committee and its working groups. The IGS Oversight Committee was dissolved by the end of the 1993. We should acknowledge its important contribution to the creation of the IGS.

The first one and a half years of the official IGS service were extremely successful, too: the official IGS orbit has become the accepted standard for a highly accurate GPS orbit. The Central Bureau Information System (CBIS) developed into the reliable source of information about the IGS for a growing user community.

REFERENCES

Beutler, G. and E. Brockmann (1993). International GPS Service for Geodynamics. *Proceedings of the 1993 IGS Workshop, 369 pages, Druckerei der Universität Bern*, available through IGS Central Bureau.

Beutler, G., J. Kouba, T. Springer (1993). Combining the Orbits of IGS Processing Centers, *Proceedings of the IGS Analysis Center Workshop*, Ottawa.

Beutler, G. (1995). Summary of the Potsdam IGS Workshop. *IGS Message No. 961, 26 May, 1995.*

Bevis, M., S. Businger, T.A. Herring, Ch. Rocken, A. Anthes, R.H. Ware (1992). GPS Meteorology: Remote Sensing of Atmospheric Water Vapor using the Global Positioning System. *Journal of Geophysical Research*, Vol. 97, No D14, pp. 15787–15801.

Goad, C.C. (1993). IGS Orbit Comparisons. *Proceedings of the 1993 IGS Workshop*, pp. 218-225, Druckerei der Universität Bern, available through IGS Central Bureau.

IGS Colleague Directory (1994). Available through the IGS Central Bureau.

Kouba, J. (1993). *Proceedings of the IGS Analysis Center Workshop, October 12-14, 1993, Ottawa, Canada*, 114 pages, Geodetic Survey Division, Surveys, Mapping and Remote Sensing Sector, NR Can, Ottawa, Canada.

Liu, R., W. Gurtner, (1994). Introducing the Central Bureau Information System of the International GPS Service for Geodynamics. *IGS Colleague Directory*, IGS Central Bureau, JPL, Pasadena, December 1994.

McCarthy, D. (1995). Using GPS to Determine Earth Orientation. *GPS Trends in Precise Terrestrial, Airborne, and Spaceborne Applications*, this Volume.

Mireault, I., J. Kouba (1995). Using GPS to Determine Earth Orientation. *GPS Trends in Precise Terrestrial, Airborne, and Spaceborne Applications*, this Volume.

Mueller I.I. (1993). Planning an International Service using the Global Positioning System (GPS) for Geodynamic Applications, *Proc.IAG Symp. No.109 on Permanent Satellite Tracking Networks for Geodesy and Geodynamics*, Springer Verlag.

Mueller, I.I., G. Beutler (1992). The International GPS Service for Geodynamics - Development and Current Structure, *Proceedings of the 6th Symposium on Satellite Positioning*, Ohio State University, Columbus, Ohio.

Neilan, R.E., W. Melbourne, G. Mader (1990) The Development of a Global GPS Tracking System in Support of Space and Ground-based GPS Programs, *Proc. IAG Symposia No. 102: Global Positioning System: An Overview*, Y. Bock and N. Leppard, (eds.), Springer-Verlag.

Zumberge, J.F., R. Liu (1995a). *Densification of the IERS Terrestrial Reference Frame through Regional GPS Network*. Proceedings of the December 1994 IGS Workshop in Pasadena, available through the IGS Central Bureau.

Zumberge, J.F. et al. (eds.) (1995b). *IGS Annual Report for 1994*, in preparation.

IGS COMBINATION OF PRECISE GPS SATELLITE EPHEMERIDES AND CLOCKS

Y. Mireault, J. Kouba and F. Lahaye
Geodetic Survey Division, Geomatics Canada, Natural Resources Canada
Ottawa, Ont., Canada

ABSTRACT

Since January 1, 1994, the Geodetic Survey Division (GSD) of Natural Resources Canada (formerly Energy Mines and Resources, EMR) has been combining and comparing the GPS satellite ephemerides and clock corrections produced by the seven Analysis Centres contributing to the International GPS Service for Geodynamics (IGS). The IGS ephemeris/clock combination is produced weekly and provides information on the internal and external consistency between results of the IGS Analysis Centres. 7-day orbital fits characterize the internal consistency whereas comparisons between properly re-aligned daily satellite orbit solutions provide the external validation; the statistics are summarized in weekly reports which, along with the combined orbits and the corresponding Earth Orientation Parameters (EOP), are available electronically from the IGS Global Data Centres. Improved weighting, editing and handling of multiple reference clock resets produced significantly improved clock combinations. Steady improvements of results by all Analysis Centres and other improvements of combination techniques have resulted in accuracies approaching 10 cm for IGS combined orbits and 1 ns for combined satellite clocks under Anti-Spoofing (AS).

INTRODUCTION

Precise IGS orbits/clocks simplify significantly GPS data reduction by eliminating the need to process large data sets involving long baselines which usually require complex software. Furthermore, the IGS precise orbits ensure positioning results traceable to the International Terrestrial Reference Frame (ITRF).

Currently, seven IGS Analysis Centres contribute solutions to the IGS orbit/clock combination (see Table 1) to produce Rapid IGS orbits/clocks within one or two days after the last submission or within 10-11 days after the last observation and the Final IGS orbits/clocks which are referred to the final IERS Bulletin B EOP. The Ottawa workshop [Kouba, 1993] recommendations have been followed to produce and distribute orbit/clock combinations. The IGS products comprise of three types of files which are

Table 1. IGS Analysis Centres Contributing During 1994-1995.

Centre	Description
cod	Centre for Orbit Determination in Europe (CODE) Bern, Switzerland
emr	Natural Resources Canada (NRCan) (Formerly Energy, Mines and Resources - EMR) Ottawa, Canada
esa	European Space Agency (ESA) European Space Operations Center (ESOC) Darmstadt, Germany
gfz	GeoForschungsZentrum (GFZ) Potsdam, Germany
jpl	Jet Propulsion Laboratory Pasadena, USA
ngs	National Oceanic and Atmospheric Administration (NOAA) National Geodetic Survey, Silver Springs, USA
sio	Scripps Institution of Oceanography La Jolla, USA

produced weekly: seven daily orbit/clock files, one EOP file and one summary file. The following summarizes the combination procedure for both ephemerides and clocks based on the above recommendations:

1. Long arc ephemerides fit and day-to-day consistency for each Centre (7-day arc).
2. Transformation to the same references (orbits/clocks treated separately).
3. Orbit and clock combination as weighted averages (orbits and clocks treated separately).
4. Long arc ephemerides fit and evaluation of the combined IGS orbits (7-day arc).

LONG ARC ORBIT FIT AND EVALUATION

The long arc evaluation was implemented to detect problems that could affect the daily weighted average combination and to assess the consistency of each Analysis Centre solutions over a one week period. Ephemerides for each Centre are analyzed individually and independently from the combination process (weighted average). The evaluation process is based on programs which were developed at the Astronomical Institute of the University of Bern (AIUB) [Beutler et al., 1994] and adapted at NRCan to perform the IGS orbit combination/evaluation.

To automate the process, the original VAX-VMS script files were converted for an HP-UX platform and modified to include/exclude specific Centres, to choose between Centre specific or Bulletin A/B EOP and to delete satellites for specific days/Centres.

In summary, daily precise ephemerides for a single Centre are transformed into the J2000.0 inertial system using the Centre EOP solutions. A seven-day a priori orbit arc is then generated for each satellite. Using the daily J2000.0 ephemerides as pseudo-observations, the a priori weekly orbit arc fit is improved by adjustment of six Keplerian elements and nine radiation pressure parameters per satellite [Beutler et al., 1993 and Beutler et al., 1994]. If problems like satellite maneuvers or momentum dumps arise, the seven day arc of the satellite in question can be divided in two independent arcs, estimating two sets of Keplerian elements and radiation pressure parameters.

ORBIT/CLOCK COMBINATION BY WEIGHTED AVERAGE

The weighted average orbit combination software was developed by T. Springer and G. Beutler at the AIUB [Springer and Beutler, 1993].

Orbit Combination by Weighted Average

The orbit combination is performed using all Analysis Centre submissions for a given day. Each Centre ephemerides are first rotated to use the same orientation reference by applying the difference between its associated x and y pole coordinate solutions and the IERS reference EOP corrected for the ITRF misalignment [IERS Annual Report for 1992, Table II-3, page II-17 and IERS Annual Report for 1993, Table II-3, page II-19]. The most recent IERS Bulletin A pole coordinates are used as the reference for the Rapid orbit combination whereas the Final orbit combination uses the final IERS Bulletin B daily EOP values.

Since the GPS week 803, the Rapid orbit combination has been performed directly in the ITRF without any alignment to the Bulletin A pole coordinate series. This change has not introduced any noticeable discontinuities in the IGS orbits.

Before this modification, the IGS EOP values provided with the Rapid orbits were those of the Bulletin A whereas now, x and y pole values are the weighted average of all Centres x and y pole solutions using the orbit combination weights, supplemented by the UT1-UTC values from Bulletin A. However, the Final orbit combination still uses the Bulletin B EOP as reference and the EOP values associated with the IGS orbits are those provided in the Bulletin B, corrected for the ITRF misalignment.

The realigned ephemerides are weighted and combined to generate the IGS official orbits. The steps to produce the IGS orbits and the associated statistics are:

1. An unweighted mean orbit is first computed and a 7-parameter Helmert transformation is estimated between each Centre's and the mean ephemerides. These transformation parameters are computed using robust L1-norm estimates and are used to transform

each Centre ephemerides. Centre weights are derived from the mean absolute deviation from the mean ephemeris.

2. A weighted average orbit is then computed using the Centre weights and the transformed ephemerides as defined in step 1.

3. Again, a set of 7-parameter Helmert transformation is estimated (L1-norm) between each rotated Centre ephemeris and the weighted average orbit, but this time using satellite weights. These satellite weights are derived from the RMS of the differences from the weighted average orbit as computed from step 2.

4. Finally, the IGS combined orbits are computed as the weighted average (similar to step 2), utilizing the Centre weights from step 1 and the newly transformed Analysis Centre ephemerides using the last transformation parameters estimated in step 3.

Clock Combination by Weighted Average

The satellite clock correction combination is performed in a fashion similar to the orbit combination. The individual Analysis Centre clock corrections are first aligned to the same time reference by determining clock offsets and drifts for each Centre. Clock resets for the Centre reference clock are removed. Currently, GPS time as provided by broadcast clock corrections is used as the reference. Since under Selective Availability (SA) the broadcast clock corrections have an RMS of about 100 ns, direct alignment of each Analysis Centre to the GPS time can cause the Centres clock corrections to be offset by as much as 10 ns. However, the best submitted clock solutions are consistent at the sub-ns level. Two strategies were used to overcome this problem:

a. A specified Analysis Centre is chosen as the reference. Its clock corrections are aligned to GPS time through L1-norm estimation of clock offset and drift using broadcast clock corrections. The other Centres clock corrections are then aligned to the adjusted clock corrections of the reference Centre, again by L1-norm estimation. The Centre weights are computed from the absolute deviation of the adjusted clock corrections with respect to the unweighted mean. In this manner, the best alignment possible is provided both between Analysis Centres (sub-ns) and with respect to the time reference (10 ns in the case of GPS time). This strategy was used between GPS weeks 730 and 741.

b. Each analysis Centre clock corrections are aligned to GPS time by L1-norm estimation of clock offset and drift using only non-SA satellite broadcast clock corrections (usually 2 or 3 satellites). Centre clock weights are determined from the absolute deviation of this initial alignment using the non-SA satellites. This way, the clock alignments to the GPS time are not affected by SA and corresponding weights are used in the clock combination; this assumes that the non-SA satellite data are representative of each Centre clock solution quality. This strategy has been used since GPS week 742.

The transformed clock corrections are then combined as weighted averages over all submitted solutions. Unlike the orbit combination, no satellite specific weights are used in the estimations. The steps to produce the IGS satellite clock corrections and their statistics are:

1. First, a clock offset and drift between each Centre clock solutions and the broadcast clocks using non-SA satellites is derived to align the Centre clocks to GPS time.

2. Centre clock weights are derived from the mean absolute deviations of the mean clocks (non-SA satellites only).

3. A weighted average clock correction for each satellite and epoch is then computed using the Centre clock weights.

4. A new set of alignment parameters (clock offset and drift) between the weighted clock average (step 3) and each Centre is estimated (one set of parameters for all satellites). Every Centre clock solutions are then realigned using these new parameters.

5. Finally, the IGS combined clock corrections are computed as the weighted average (similar to step 3), using the Centre clock weights from step 2 and the Centre clock corrections generated in step 4.

A more detailed description of the equations involved in the orbit/clock combination and the statistics reported in the weekly summary reports is provided in Appendix I of the **"Analysis Coordinator Report"** [IGS Annual Report for 1994 (in preparation)]. Examples of all tables are included in Appendix II of that report.

IMPLEMENTATION AND GENERAL REMARKS

The ephemeris and clock combination should satisfy the following objectives:

- firstly, the IGS combined ephemeris/clock are to be more reliable than the submitted Analysis Centre solutions;
- secondly, the reported statistics reflect information submitted by the Analysis Centres even if their solutions can not be used for the orbit/clock combination. This provides useful feedback to the Analysis Centres.

Occasional difficulties arise when some submitted solutions perturb the combination and thus have to be excluded in accordance with the first principle. The L1-norm estimation scheme was therefore chosen on the basis of its robustness, i.e. its insensitivity to "outlier data". During the initial phase of IGS combinations, it became clear that for certain severe cases (e.g. when a Centre solution for one satellite in comparison with others shows RMS of several meters) the L1-norm estimation may fail. This is due to limited redundancy of the data set provided by the seven Analysis Centres and the first stage unweighted averaging which provides, in some cases, poor initial estimates. Similar problems arise with the clock combination since only four Centres provide clock solutions. Moreover, the assumption that non-SA satellites are always representative of each Centre clock solution quality is sometimes questionable and limited by the satellite clock stability of 1-2 ns.

1994-1995 RESULTS

The following results have been obtained during the first sixteen months of IGS service, i.e. January 2 to April 31, 1995 (GPS weeks 730 to 798).

Figure 1 displays the weekly averages and standard deviations of the X and Y rotations of satellite ephemerides for a typical Analysis Centre after the daily Helmert transformations with respect to the Final IGS orbits (referred to the IERS Bulletin B). It should be mentioned that the X and Y rotation parameters indicate the stability of the Centre y and x pole series, provided that the Centre orbit and EOP solutions are consistent. The standard deviations shown as vertical lines plotted with the mean values are typically below the 0.4 mas level.

Figure 2 shows the satellite ephemerides RMS for the orbit combination and long arc evaluation of a typical Analysis Centre. Three types of RMS are included in this figure: the combination RMS, the weighted combination RMS (WRMS), and the long arc fit RMS. The WRMS is a weighted version of the combination RMS using the precision codes provided by the Analysis Centres in their SP3 files. Bad or marginal satellite solutions will show up in the Centre orbit RMS but not in its weighted RMS if appropriately acknowledged by the Centre in its SP3 files.

Figure 3 shows an example of a Centre clock combination RMS. Centres used in the clock combination are EMR, ESA, GFZ and JPL. The other Centres are excluded because they either report broadcast clocks (COD, NGS starting on GPS week 753) which are only used in clock alignment and clock weight determination, or clock corrections are not reported (SIO, NGS prior to GPS week 753). For completeness, all clock information is compared with the combined solution.

In Figure 3, some clock RMS values were out of the range of the plot. These outliers generally indicate a bad satellite clock solution. In most cases, the bad satellite clock solutions were excluded from the combination but kept in the RMS computations. On GPS week 754, the high clock RMS was caused by the reference clock reset which could not be removed by the combination programs at that time. This has now been corrected.

The effect of permanent AS (GPS week 734) is apparent from the clock RMS (Figure 3). The daily clock RMS before the GPS week 734, despite the high clock RMS during the GPS week 730, shows that the RMS level increased from 1 ns to about 10 ns after AS implementation. AS was switched off from the GPS week 797, day 3 (Apr. 19) at ~ 20:00 UT until GPS week 800, day 3 (May 10) at ~ 20:00 UT but it is hardly noticeable in the clock RMS due to improvements in the Analysis Centre processing strategies.

Examination of these figures shows that a considerable effort was made throughout the past 16 months by all Analysis Centres to improve the quality of orbit and clock solutions. Towards the end of 1994 (GPS week 770), some clock RMS have again reached the 1 ns level and some Centres satellite ephemerides RMS have reached the 10 cm level, despite AS.

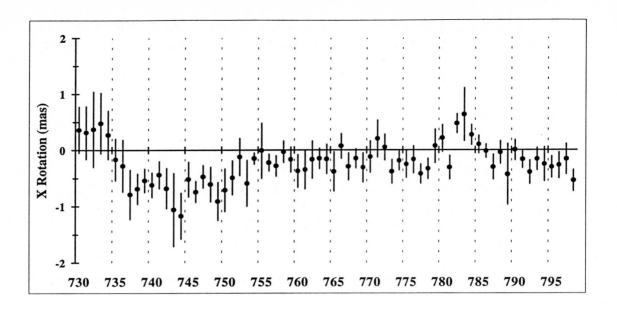

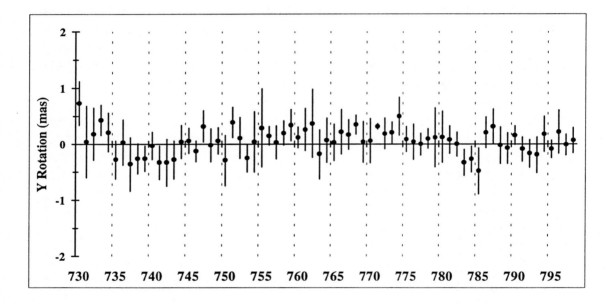

Fig 1. Weekly Mean 7-Parameter Helmert Transformations - X and Y Rotations (EMR). Rotations of X and Y satellite coordinates (which correspond to the y and x pole coordinates, respectively) from the daily Helmert transformations with respect to the Final IGS orbits (referred to the IERS Bulletin B) are shown. The standard deviations shown as vertical lines plotted with the mean values are typically below 0.4 mas.

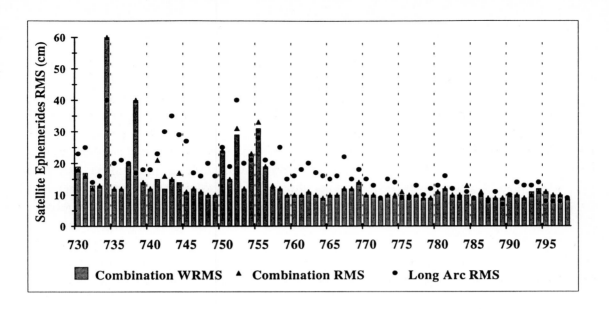

Fig 2. Weekly Mean Satellite Ephemerides RMS (JPL). The combination RMS is based on the estimation of a 7-parameter Helmert transformation for the Centre ephemeris with respect to the Final IGS ephemerides. The WRMS are the combination RMS weighted with the satellite ephemerides quality information provided by each Analysis Centre in the SP3 files. The "long arc RMS" is obtained from the long arc fit evaluation (7-day arc).

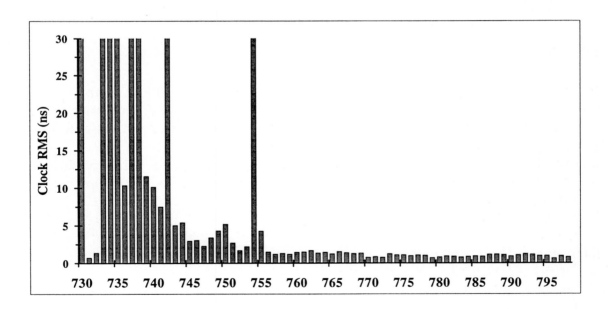

Fig 3. Weekly Mean Clock Corrections RMS (EMR). The clock corrections RMS are based on the estimation of offsets and drifts of the Analysis Centre reference clocks with respect to the Final IGS clock solution.

CONCLUSIONS

Analysis Centre orbit solutions have steadily improved and, towards the end of 1994 (GPS week 770), most contributed orbit solutions show consistency approaching the 10 cm level (coordinate RMS) even under AS conditions. This is confirmed by independent long arc orbit evaluations. The IGS orbit combination attempts to use all submitted Analysis Centres solutions, including days when satellites are being repositioned. The IGS combined orbits are more reliable than the individual Analysis Centre orbits they are based on. Furthermore, the IGS orbits are more consistent in orientation and more precise than the best regional orbits. The satellite clock solution consistency was well below 1 ns during the month of January, 1994 when AS was not invoked. After February 1994, when AS was invoked permanently, the clock solutions deteriorated to the 10 ns level mainly due to biases in pseudo range observations. However, receiver improvements and better solution strategies by the Analysis Centres resulted in the satellite clock solutions returning to the 1 ns level.

ACKNOWLEDGMENTS

The weighted average orbit combination software, adapted for the IGS orbit combination, was provided by Springer and Beutler (1993) and further improved by T. Springer who also kindly provided the UNIX script. The long arc evaluation software was made available by the Astronomical Institute of the University of Bern (AIUB) [Beutler et al., 1993, Beutler et al., 1994] and ported to HP UNIX with the assistance of E. Brockmann of AIUB.

REFERENCES

Beutler, G., E. Brockmann, W. Gurtner, U. Hugentobler, L. Mervart, M. Rothacher and A. Verdun (1994). Extended Orbit modeling Techniques at CODE Processing Center of the International GPS Service for Geodynamics (IGS): Theory and Initial Results, *Manuscripta Geodaetica*, Vol. 19, No. 6, pp. 367-386.

Beutler, G., J. Kouba and T. A. Springer (1993). Combining the Orbits of the IGS Processing Centers, *Proceedings of the IGS Analysis Center Workshop*, Ottawa, Canada, October 12-14, 1993, pp. 20-56.

International Earth Rotation Service (IERS) (1993). IERS Annual Report for 1992, Observatoire de Paris, France.

International Earth Rotation Service (IERS) (1994), IERS Annual Report for 1993, Observatoire de Paris, France.

Kouba, J. (1993) (Edited by). Proceedings of the IGS Analysis Center Workshop, Ottawa, Canada, October 12-14, 1993.

Springer, T. A. and G. Beutler (1993). Towards an Official IGS Orbit by Combining Results of all IGS Processing Analysis Centers, *Proceedings of the 1993 IGS Workshop*, Bern, Switzerland, March 25-26, 1993, pp. 242-249.

Zumberge, J. F., R. Liu and R. Neilan (1995) (Edited by). International GPS Service for Geodynamics, *Annual Report for 1994*, IGS Central Bureau (in preparation).

GLOBAL GPS DATA FLOW FROM STATION TO USER WITHIN THE IGS

Carey E. Noll
Code 920.2, NASA/GSFC
Greenbelt, MD 20771 USA

Werner Gurtner
Astronomical Institute, University of Berne
CH-3012 Berne SWITZERLAND

ABSTRACT

The International GPS Service for Geodynamics (IGS) has been operational since January, 1994. This service was formed to provide GPS data and highly accurate ephemerides in a timely fashion to the global science community to aid in geophysical research. The GPS data flows from the global network of IGS sites through a hierarchy of data centers before they are available to the user at the global and regional data centers. A majority of these data flow from the receiver to global data centers within 24 hours of the end of the observation day. IGS analysis centers retrieve these data daily to produce IGS products (orbits, clocks, Earth rotation parameters, etc.). These products are then forwarded to the global data centers by the analysts for access by the IGS Analysis Coordinator for generation of the rapid and final IGS orbit product and for access by the geodynamics community in general. A discussion of the network data flow, from station to global data center to users, will be presented. Statistics on data quantity, volume, latency, and user access will be given.

FLOW OF IGS DATA AND INFORMATION

The flow of IGS data (including both GPS data and derived products) as well as general information can be divided into several levels (IGS Colleague Directory, 1994) as shown in Figure 1:

- Governing Board
- Central Bureau (including the Central Bureau Information System, CBIS)
- Tracking Stations
- Data Centers (operational, regional, and global)
- Analysis Centers
- Analysis Center Coordinator

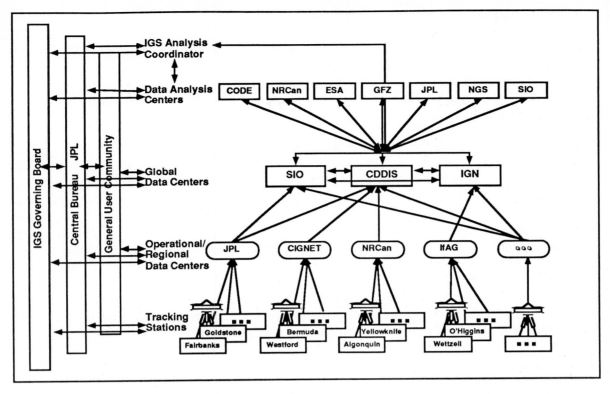

Figure 1. Flow of Data and Information Within the IGS

Governing Board

The IGS Governing Board, consisting of fifteen elected members from the IGS participants, is responsible for the overall management of the IGS and recommending modifications to the organization of the service in order to improve its efficiency, reliability, etc.

Central Bureau

The Central Bureau, located at the Jet Propulsion Laboratory (JPL) in Pasadena, CA, sees to the day-to-day operations and management of the IGS. The Central Bureau facilitates communication within the IGS community through several electronic mail services. The Central Bureau also has created, operates, and maintains the Central Bureau Information System (CBIS) (Liu, et. al., 1995), designed to disseminate information about the IGS and its participants within the community as well as to other interested parties. The CBIS server is accessible over the Internet, via anonymous ftp, and the World Wide Web (WWW). The CBIS contains information about:

- IGS organization and operation
- global network of GPS tracking sites
- general descriptions of GPS receivers and antennas
- access information and data holdings summaries for the IGS data centers
- descriptions of GPS data flow
- up-to-date data and product availability charts

- GPS system status
- IGSMail and IGSReport archives
- software for general use (e.g., UNIX-compatible compress/decompress routines for various platforms)
- IGS combined orbit product archive

Tracking Stations

The global network of GPS tracking stations are equipped with precision, dual-frequency, P-code receivers operating at a thirty-second sampling rate. The IGS currently supports nearly 100 globally distributed stations. These stations are continuously tracking and are accessible through phone lines, network, or satellite connections thus permitting rapid, automated download of data on a daily basis. Any station wishing to participate in the IGS must submit a completed station log to the IGS Central Bureau, detailing the receiver, site location, responsible agencies, and other general information. These station logs are accessible through the CBIS.

The IGS has established three categories of GPS stations (Gurtner and Neilan, 1995): global, regional, and local. Global stations are those whose data are analyzed by at least two IGS analysis centers (located on different continents) and are used for daily estimation of satellite orbits, Earth rotation parameters, and station positions and velocities. The data from the global sites are available at the global data center level. Data from regional stations are analyzed by at least one IGS analysis center for extension of the reference frame. Data from these sites are available at the regional data center level. Local stations are utilized to augment the network of global and regional stations and could be episodically occupied. These stations may be part of dense permanent arrays, such as the Southern California Integrated GPS Network (SCIGN). These data are typically available through a local data center.

GPS data from the global network consists of daily observation and navigation files in Receiver Independent Exchange (RINEX) format. These data are compressed using UNIX compression and forwarded to the appropriate data centers. Approximately 0.6 Mbytes of data are generated by each site on a daily basis; thus, an eighty-station network yields approximately fifty Mbytes per day. These data files are named *ssssddd0.yyt_Z*, where *ssss* is the four-character GPS site name, *ddd* is the three-digit day of year, *yy* is the two-digit year, *t* is the file type, O for observation and N for navigation, and _Z indicates the file is compressed; on UNIX systems, the "_Z" convention is replaced by the second level filename extension, ".Z". A concatenated broadcast ephemeris file, named BRDCddd0.yyN_Z, is created and available at the global data center for users interested in downloading a single file of navigation messages instead of multiple individual, site files.

Data Centers

The IGS has established a hierarchy of data centers to distribute data from the three categories of tracking stations: operational, regional, and global data centers. Operational data centers are responsible for the direct interface to the GPS receiver, connecting to the remote site daily and downloading and archiving the raw receiver data. The quality of these data are validated by checking the number of observations, number of observed satellites, date and time of the first and last record in the file. The data are then translated to RINEX, and compressed using UNIX compression algorithms. Both the observation and navigation files are then transmitted to a regional or global data center within 24 hours following the end of the observation day.

Regional data centers gather data from various operational data centers and maintain an archive for users interested in stations of a particular region. These data centers forward data from global sites to the global data centers within at most 24 hours of receipt. Examples of regional data centers include the Institute fur Angewandte Geodesie (IfAG) for the European region and the Australian Land Information Group (AUSLIG) for the Australian/Southern Pacific region.

The IGS global data centers are tasked to provide an on-line archive of at least 150 days of GPS data in RINEX, including, at a minimum, the data from all global IGS sites. The global data centers are also required to provide an on-line archive of derived products, generated by the seven IGS analysis centers. There are currently three IGS global data centers:

- Crustal Dynamics Data Information System (CDDIS), at NASA's Goddard Space Flight Center in Greenbelt, MD
- Institut Geographic National (IGN) in Paris, France
- Scripps Institution of Oceanography (SIO) in La Jolla, CA

These data centers equalize holdings of global sites and derived products on a daily basis. The three global data centers provide the IGS with a level of redundancy, thus preventing a single point of failure should a data center become unavailable. Users can continue to reliably access data on a daily basis from one of the other two data centers. Furthermore, three centers reduce the network traffic that could occur to a single geographical location.

Analysis Centers

The seven IGS analysis centers retrieve the IGS tracking data from the global data centers on a daily basis and produce daily orbit products and weekly Earth rotation parameters and station position solutions. These solutions, along with summary files detailing data processing techniques, station and satellite statistics, etc., are then submitted to the global data centers within ten days of the end of the observation week.

The format of the daily orbit files generated by the analysis centers is SP3; weekly station position files are in the Software Independent Exchange (SINEX) format. The size of the weekly submission by each analysis center is 1.5 Mbytes; therefore, the total weekly archive of solutions from the seven analysis centers and the combined orbits is approximately thirteen Mbytes. These files are stored in uncompressed ASCII format at the global data centers.

Analysis Center Coordinator

The Analysis Center Coordinator, located at the National Resources Canada (NRCan), retrieves the derived products and produces a combined IGS orbit product based on a weighted average of the seven individual analysis center results. The combined orbit is then made available to the global data centers and the IGS CBIS within two weeks following the end of the observation week.

STATISTICS ON DATA FLOW AND SYSTEM USAGE

The CDDIS has been compiling statistics on system usage of GPS data and products since late 1992. Figure 2 shows that 75 percent of the data from IGS global sites are available at the global data center level within one day and 85 percent are available within two days.

Figure 3 illustrates the monthly volume of data transferred to and from the CDDIS since 1994. As can be seen, the system usage continues to grow. Figure 4 shows the number of host accesses since 1994 as well as the number of distinct hosts accessing the CDDIS per month.

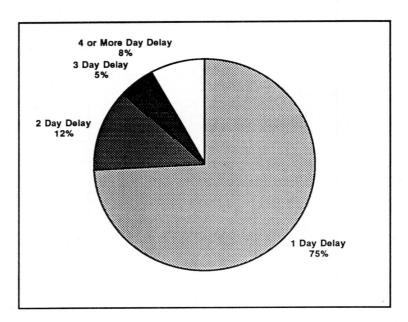

Figure 2. Data Latency Statistics for IGS Global Sites at the CDDIS

REFERENCES

"IGS Terms of Reference", *IGS Colleague Directory* (1994). Available through the IGS Central Bureau.

Liu, R., et. al.. "Introducing the Central Bureau Information System of the International GPS Service for Geodynamics" in *International GPS Service for Geodynamics Resource Information*. January 1995.

Gurtner, W. and R. Neilan. "Network Operations, Standards and Data Flow Issues" in *Proceedings of the IGS Workshop on the Densification of the ITRF through Regional GPS Networks*. March 1995

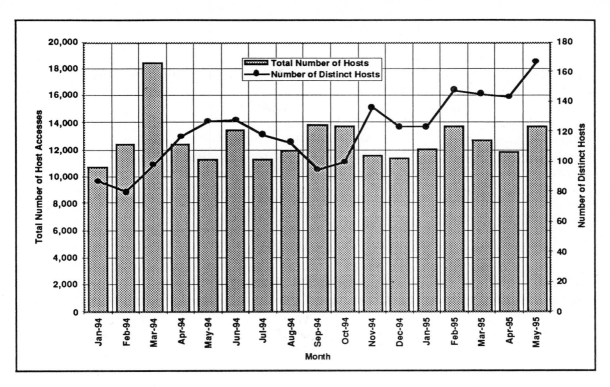

Figure 3. Monthly Host Access Statistics for the CDDIS

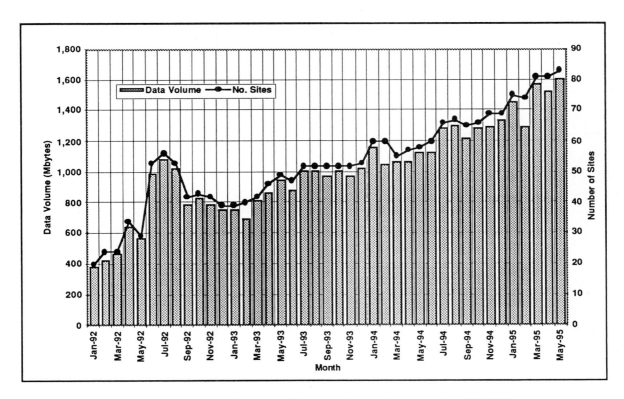

Figure 4. Monthly Data Volume Statistics for the CDDIS

GLOBAL PLATE KINEMATICS ESTIMATED BY GPS DATA OF THE IGS CORE NETWORK

G. Gendt, G. Dick, C. Reigber
GeoForschungsZentrum Potsdam
Telegrafenberg A17 , 14473 Potsdam, Germany

ABSTRACT

The GeoForschungsZentrum Potsdam (GFZ) operates as Analysis Center in the "International GPS Service for Geodynamics (IGS)" and analyses the global GPS data on a daily basis. The high accuracy of the GPS technique allows to determine global tectonic motions even from a small time interval of two and a half year. Baseline rates and station position time series derived from linear fitting of weekly solutions as well as velocity vectors derived from simultaneous adjustment of site position and motion are presented. The determination of site positions and baseline rates gives an accuracy of ±1 - ±5 mm/yr in a global scale. The estimated motions show a good agreement with the geologically derived model NUVEL-1 and with ITRF solution.

INTRODUCTION

The GPS technology has become one of the most important geodetic techniques for regional and global studies of the Earth's kinematics. The development of the "International GPS Service for Geodynamics (IGS)" in the last two years has provided a leading role of GPS in estimation of the present-day global tectonic motions and maintenance of global reference frame.

One of the seven IGS Analysis Centers was implemented at the GeoForschungsZentrum (GFZ) Potsdam. It has participated in the IGS from the very beginning. For the permanent analysis of the global GPS data on a daily basis the automated GPS Analysis Software package EPOS.P.V2 was developed (Gendt et al., 1995). Having now more than two years of data from a permanent network of IGS core stations the tectonic motions on a global scale have been determined with a very high accuracy.

DETERMINATION OF GLOBAL TECTONIC MOTIONS

A global set of station coordinates for 55 IGS core sites was determined (Gendt et al., 1995). Similarity (HELMERT) transformations between various variants (e.g. velocities adjusted and fixed) and comparison with the ITRF93 solution (label SSC(IERS)94C01) show an accuracy of about ±4 mm for horizontal components and ±9 mm for the hight.

In the routine IGS analysis the daily, fiducial free, and unconstrained normal equations for station coordinates are stored into a data base for further analysis. Investigations over long time intervals demand an effective technology for combining of solutions. To reduce the computing times and the amount of files, computation and archiving of weekly normal equations have been performed. By combining daily normal equations into weekly ones, the combination software gives the possibility to produce homogeneous sets of equations based on the same given initial values for station coordinates, eccentricity values and tectonic model. This way it is easy (i) to introduce new initial coordinates, (ii) to use the most recent eccentricities values for the solution, (iii) to change the tectonic model for the coordinate determination. The errors of chosen *a priori* site velocities are negligible for such short time intervals (one week or even one month, if in future data over many years have to be analyzed). The combined normal equations can be extended by parameters for site velocities.

The tectonic motions have been determined in two variants:

1. Baseline rates and station positions from weekly coordinate solutions. The advantage of this method is the control of data quality and eccentricities as well as a good evaluation of solution stability and accuracy. Episodic motions remain visible.

2. Simultaneous adjustment of station coordinates and their velocities from the whole data set. This method gives optimum weighting of the data. Correlations between coordinates and velocities are automatically taken into account. Episodic motions cannot be seen. In this variant the accuracies are too optimistic and have to be scaled according to the first variant.

Time series for station positions of selected sites together with their rates derived from the first variant are given in Fig. 1. The scattering in the horizontal components is between ±4 and ±7 mm, the rates agree with those of NUVEL model.

Some of the baseline rates derived also from weekly coordinate solutions together with determined slopes are shown in Fig. 2. The slope values from the NUVEL model are given for comparison. For baselines of about 1000 km (WETB-MATE) the scattering is ±3 mm, for longer baselines the scattering increases by 1.5 to 2 mm per each 1000 km. The plate tectonic motion can be seen clearly. Fig. 3 gives the baseline rates from KOKB and WETB to their neighboring sites. The accuracies of the shorter baselines in Europe are ±1 mm/yr. In regions where we have a good site distribution (Europe, North America) even for longer baselines accuracies of ±2 mm/yr can be obtained. In these cases the agreement with the velocities from NUVEL or ITRF are in the range of a few mm/yr. For isolated stations, especially on the Southern hemisphere, the accuracy is about ±5 mm/yr. Even here we have a close agreement to other models.

Fig. 4 shows the velocities as resulting from the global simultaneous adjustment of coordinates and velocities. Here again, the agreement with NUVEL and ITRF velocities for Europe and North America is obvious. In the Australian region a small net rotation can be stated, which probably originates from a relatively weak connection to the global network.

On the Southern hemisphere recently a lot of new sites became available. So we can expect to have an accuracy of ±2 to ±3 mm/yr in a global scale within the next few years.

Our results show that data of only two years yield accuracies comparible to those from analysis of SLR and VLBI data over 5 to 10 years. This demonstrates the advantage and great potential of the GPS technique also for global applications. Because we will have a densification of the network in the near future which is unlikely for the SLR and VLBI techniques, the importance of the GPS technique becomes very obvious.

CONCLUSIONS

The results presented here from a relatively short data span of 2,5 years show that GPS will play a major role for realizing and maintaining the global reference frame and its changes in time. The densification of the IGS Network will provide accuracies of ±2 to ±3 mm/yr for all regions of the Earth within the next 2-3 years. Global control networks may be installed by SLR and VLBI to investigate possible systematic effects of the GPS technique and to obtain a high precision in defining the geocenter (SLR).

REFERENCES

Gendt, G., G. Dick, Ch. Reigber (1995). IGS Analysis Center at GFZ Potsdam. 1994 IGS Annual Report, in press.

Gendt, G., G. Dick, Ch. Reigber, W. Sommerfeld, Th. Nischan (1995). Earth Rotation Parameters, Station Coordinates and Velocities from GPS Data. IERS Technical Note, Paris, in press.

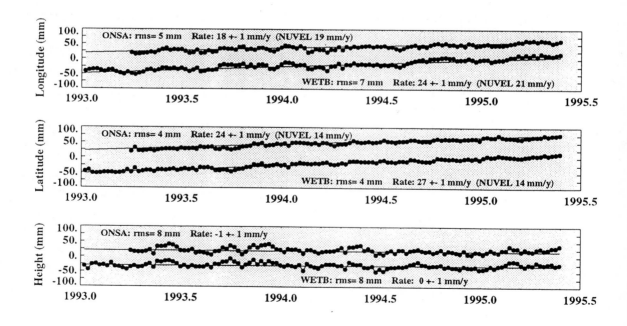

Fig. 1. Time series for ONSA and WETB from weekly coordinate solutions (NUVEL-1 values are given for comparison).

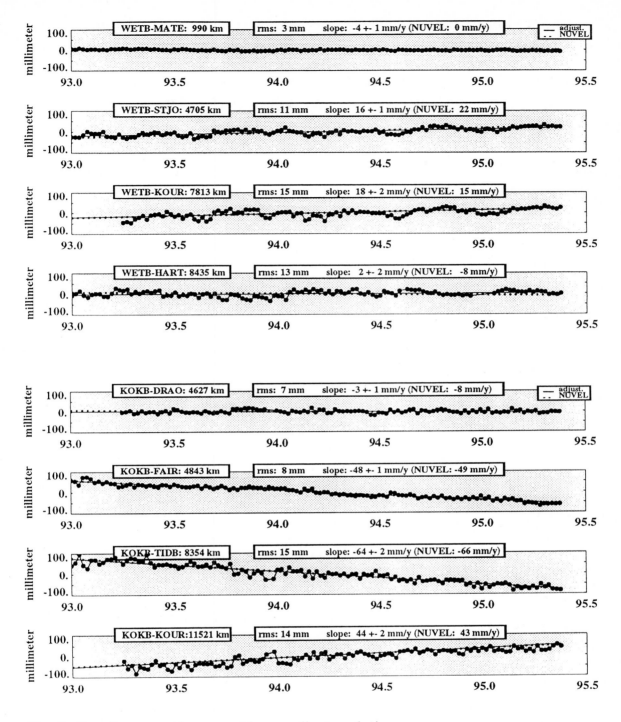

Fig. 2. Baseline rates from weekly coordinate solutions
(NUVEL-1 values are given for comparison).

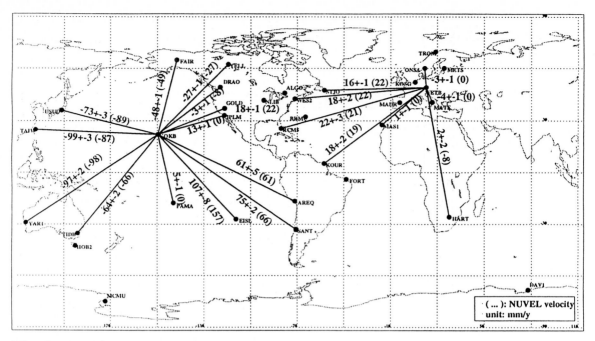

Fig. 3. Baseline rates of KOKB and WETB to neighboring sites
(NUVEL-1 values are given in parantheses).

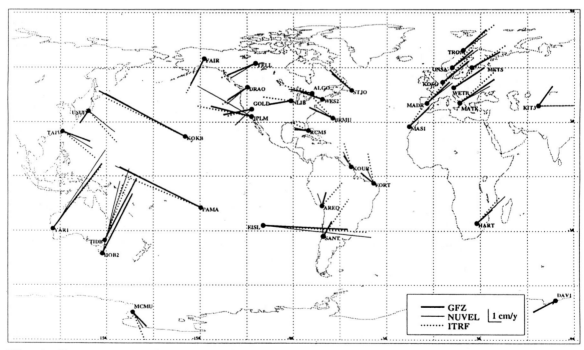

Fig. 4. Site velocities from 2.5 years of IGS data
(NUVEL-1 and ITRF values are given for comparison).

THE IMPACT OF IGS ON THE ANALYSYIS OF REGIONAL GPS-NETWORKS

D. Angermann, G. Baustert, J. Klotz, J. Reinking, S. Y. Zhu
GeoForschungsZentrum Potsdam (GFZ)
Telegrafenberg A17, D-14473 Potsdam
Division "Recent Kinematics and Dynamics of the Earth"

ABSTRACT

The GeoForschungsZentrum Potsdam (GFZ) has established large scale geodynamic GPS-networks in Central Asia and southern South America. The network analysis was performed with the GFZ software EPOS. For the processing of the regional network data IGS products and IGS station data were used. Different analysis strategies have been applied. The high quality of the IGS products and the importance of the global distributed IGS station data for the regional network analysis was pointed out. The repeatability of the network geometry is 1-2 mm horizontal and 5-6 mm for the height.

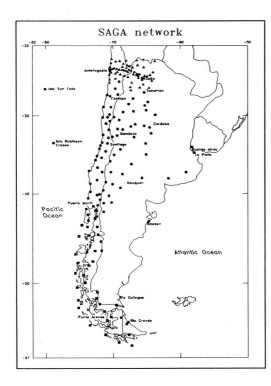

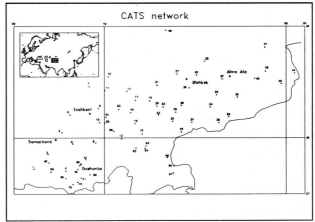

Fig. 1: GFZ / GPS networks in South America and Central Asia

INTRODUCTION

The GFZ Potsdam established and observed in 1992 a 40 site GPS network in the Pamir-Tienshan-Region (CATS=Central Asian Tectonic Science). In 1994 the CATS network was extended to a 67 site network covering Southeast Kasachstan, Kyrgystan, Usbekistan and Tadshikistan. In 1993 and 1994 the SAGA network (SAGA=South American Geodynamic Activities) with 190 sites in Chile and Argentina was established and observed (Reinking et al. 1995). The extension of the SAGA network is 4000 * 2000 km.

The GPS data of the CATS and SAGA network were analysed with the GPS software EPOS (Earth Parameter & Orbit System). First deformation results were derived for the CATS-network (Klotz et al. 1995). The IGS (International GPS Service for Geodynamics) plays an important role for the regional network analysis. Different analysis strategies were applied using IGS products and IGS station data. A description of the IGS is given e.g. in publications of Mueller (1993), Beutler et al. (1994) and Gendt et al. (1995).

DATA ANALYSIS

The GFZ Potsdam has developed the EPOS Software for the processing of GPS networks. EPOS is based on undifferenced phase- and pseudorange measurements and consists of the following main subsystems:

Decoding: The raw data (binary) were decoded and daily RINEX (Receiver Independend Exchange Format) observation files were generated.

Preprocessing: Detection and elimination of cycle slips, outliers and short data spans. Merging of individual station files, data correction and generation of ionospheric free linear combination of phases and pseudoranges in internal EPOS format.

Input-Generation: Arrangement of all model and auxiliary data, setting of control parameters for the network adjustment.

Network Computation: Least squares adjustment of the merged observations in form of daily solutions. Solved for parameters are station coordinates, ambiguities, clock parameters for each observation epoch (the clock of one satellite is fixed), troposheric parameters; optional is the computation of orbit and earth orientation parameter.

Accumulation: The daily solutions were accumulated to the campaign solution.

COMPARISON OF DIFFERENT STRATEGIES

Different strategies for combining the campaign data with the IGS products and IGS station data were applied. The influence of different analysis strategies on the accuracy of the network geometry is shown by the following two examples:

Example 1) The IGS orbits and Earth Orientation Parameters (EOP) were fixed for the network processing. All station coordinates were solved without constraints. The IGS orbits and EOP define the reference system. Three analysis strategies have been applied for the SAGA network:
a) SAGA network stations only (regional solution)
b) SAGA network stations + 4 IGS stations in South America (continental solution)
c) strategy b) + 12 global distributed IGS core stations (global solution)

Table 1: Mean residuals [mm] of the SAGA stations after 3-D-Transformation between different strategies

Comparison	North-South	West-East	Height
a) - b)	1.3	1.5	4.7
a) - c)	1.4	1.8	5.5
b) - c)	0.4	0.8	1.9

A comparison between the regional network solution and the combined solution with the IGS station data shows differences of 1-2 mm horizontal and 5 mm vertical for the network geometry (Table 1). The difference between the continental and the global solution is less than 1 mm horizontal and 2 mm vertical. The three different strategies do not show a difference in the daily repeatability of the regional station coordinates.

Example 2). The SAGA network data were processed simultaneously with the data of 28 global distributed IGS stations. The fixed ITRF (IERS International Reference Frame) coordinates of 13 IGS core stations defined the reference system for the orbit determination. In order to determine the influence of the GPS-orbits on the network geometry three strategies have been applied:
a) standard IGS orbit (GFZ solution)
b) orbit determination with the data of 4 permanent SAGA stations + 28 IGS stations
c) orbit determination with all SAGA stations + 28 IGS stations

Table 2: Mean residuals [mm] of the SAGA stations after 3-D-Transformation between different orbit determination strategies

Comparison	North-South	West-East	Height
a) - b)	0.6	0.5	2.0
a) - c)	0.6	0.6	2.4
b) - c)	0.4	0.4	1.0

The network solutions obtained with fixed IGS orbit and different orbit determination strategies agree very well; the differences of the network geometry are less than 1 mm horizontal and 1-2 mm vertical (Table 2). This result emphasizes the good quality of the IGS orbits.

With respect to this results an efficient and precise method to analyse regional networks with an extension of up to a few thousand kilometers is to process the campaign data simultaneously with the data of selected global distributed IGS stations with fixed IGS orbit and EOP and to solve for all station coordinates without constraints. The reference system of the daily solution is defined by the IGS orbit and EOP. In order to obtain global network coordinates the accumulated network solution is transformed to the ITRF using the included IGS stations.

RESULTS

With the above recommended analysis strategy the following results have been obtained for the accuracy of the regional network geometry and for the global station coordinates.

The quality of the network geometry was estimated by a comparison of the daily solutions and the campaign solution. Fig. 2 shows the results for the CATS94 network.

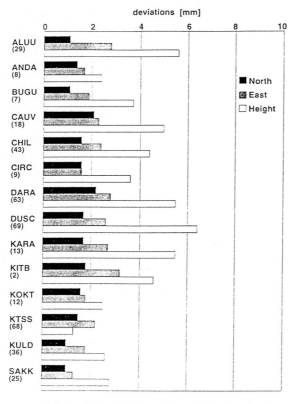

Fig. 2 : Repeatability of the CATS94 network

Table 3 : Residuals of a Helmert–Transformation between the global SAGA94 solution and ITRF93 coordinates

IGS – Stations	Residuals [mm]		
	North–South	West–East	Height
Madrid	− 11.3	− 7.6	28.0
Tromsoe	3.0	− 5.5	− 31.7
Wettzell	− 5.9	− 0.5	− 5.0
Algonquin	1.0	0.0	9.5
Fairbanks	− 4.6	− 9.7	16.7
Santiago	− 0.1	1.7	1.3
Kourou	6.9	− 4.2	− 14.7
Fortaleza	1.1	15.0	− 4.1
Mean	5.5	7.3	17.4

The mean residuals after a Helmert-Transformation between the 15 daily solutions and the campaign solution are 1-2 mm horizontal and 5-6 mm for the height. Same results were obtained for the SAGA network. For a network extension of 2000-4000 km the relative accuracy is about 1ppb ($1*10-9$).

The global solution of the SAGA network simultaneously with 15 global distributed IGS stations was used to estimate the accuracy of global station coordinates. The recommended analysis strategy with fixed IGS orbit and EOP was applied for the global network computation. All station coordinates were solved without constraints. The accumulated network solution was transformed to the ITRF using the included IGS stations. The residuals after a Helmert-Transformation between the computed IGS station coordinates and the ITRF93 coordinates are shown in Table 3. The mean residuals are less than 1 cm horizontal and 1.7 cm for the height for 8 IGS stations in Europe, North- and South America. This results is a realistic estimation of the accuracy of the global network solution since the ITRF93 coordinates are computed by the IERS (International Earth Rotation Service) through a combination of different SLR-, LLR, VLBI- and GPS-solutions (Boucher et al. 1994). It has to be considered that uncertainties of the ITRF93 coordinates are also included in the residuals.

CONCLUSION

The results of the regional network analysis confirm the high quality of the IGS products and the importance of the global IGS station data. The quality of the regional networks with an extension of a few thousand kilometers is 1-2 mm horizontal and 5-6 mm for the height. The accuracy of the global coordinates is better than 1 cm horizontal and 1-2 cm for the height. The good quality of the IGS orbits is emphasized also by the comparison of different orbit determination strategies; the agreement of the network geometry is 1 mm horizontal and 1-2 mm vertical (Table 2).

REFERENCES

Beutler, G., Mueller, I. I., Neilan, R. E., Weber, R. (1994): *IGS - Der Internationale GPS-Dienst für Geodynamic*, ZfV 119.

Boucher, C, Altamini, Z, Duhem, L. (1994): *Results and Analysis of the ITRF93*. IERS Technical Note 18, Central Bureau of IERS, Paris.

Gendt, G., Dick, G., Reigber, C. (1995): *Das IGS Analysezentrum am GFZ Potsdam: Software und Ergebnisse*. ZfV 120.

Klotz, J., Angermann, D., Reinking, J. (1995): *Großräumige GPS-Netze zur Bestimmung der rezenten Kinematik der Erde*. ZfV 120.

Mueller, I. I. (1993): *International GPS Service for Geodynamics, Terms of Reference*. Proceedings of the 1993 IGS Workshop, Bern.

Reinking, J., Angermann, D., Klotz, J. (1995): *Zur Anlage und Beobachtung großräumiger GPS-Netze für geodynamische Untersuchungen*. AVN 6/1995.

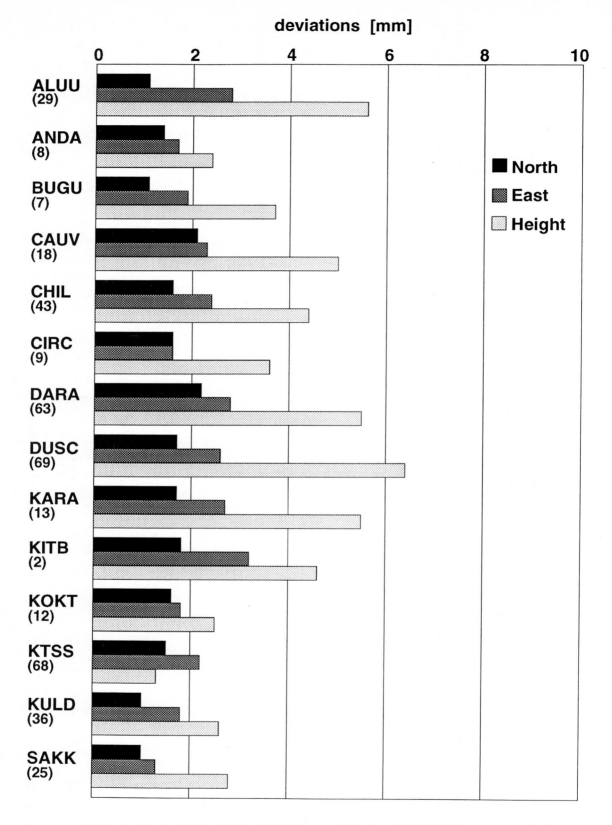

Fig. 2 : Repeatability of the CATS94 network

Table 3 : Residuals of a Helmert–Transformation between the global SAGA94 solution and ITRF93 coordinates

IGS – Stations	Residuals [mm]		
	North–South	West–East	Height
Madrid	− 11.3	− 7.6	28.0
Tromsoe	3.0	− 5.5	− 31.7
Wettzell	− 5.9	− 0.5	− 5.0
Algonquin	1.0	0.0	9.5
Fairbanks	− 4.6	− 9.7	16.7
Santiago	− 0.1	1.7	1.3
Kourou	6.9	− 4.2	− 14.7
Fortaleza	1.1	15.0	− 4.1
Mean	5.5	7.3	17.4

VARIATION IN EOP AND STATION COORDINATE SOLUTIONS FROM THE CANADIAN ACTIVE CONTROL SYSTEM (CACS)

R. Ferland, J. Kouba, P. Tétreault and J. Popelar
Geodetic Survey Division, Geomatics Canada,
Natural Resources Canada (NRCan, formerly EMR), 615 Booth Street,
Ottawa, Canada, K1A OE9

ABSTRACT

Since August 1992, precise GPS satellite ephemerides, Earth Orientation Parameters (EOP) and tracking station coordinates have been generated daily at the Master Active Control Station (MACS) of the Geodetic Surveys Division (GSD) as part of the International GPS Service for Geodynamics (IGS). A modified version of the GIPSY II software developed at the Jet Propulsion Laboratory is used to process data from 6 to 8 CACS sites and up to 20 globally distributed tracking stations. Analysis of daily solutions for the CACS station coordinates shows a centimeter level precision of the mean weekly values and good agreement with the International Earth Rotation Service (IERS) Terrestrial Reference Frame (ITRF). Systematic variations and trends at this level have implications for studies of regional crustal deformations approaching mm or ppb resolution.

INTRODUCTION

The Geodetic Survey Division (GSD) of Geomatics Canada, in collaboration with the Geological Survey of Canada (GSC) has established CACS as an essential component of a modern, fully integrated spatial reference system to support geodetic positioning, navigation and general purpose spatial referencing. CACS, which currently includes 8 stations, contributes to the International GPS Service for Geodynamics (IGS) and facilitates direct integration of the Canadian stations to the ITRF. The CACS network configuration, augmented by about 20 globally distributed IGS stations, provides data for daily determination of precise GPS satellite orbits, station coordinates, clock corrections and Earth Orientation Parameters (EOP). Constrained by 13 globally distributed fiducial VLBI/SLR stations, the CACS products enable positioning with the highest precision for geodetic control networks and studies of crustal dynamics.

CACS PROCESSING

CACS processing at NRCan is based on daily data spans, and solutions for longer time spans are obtained by their proper combinations. The scope of the processing is either global or regional. The global processing uses a worldwide network for precise GPS orbit computations and connect the Canadian core stations to the ITRF. The regional processing uses the precise GPS orbits and data from Canadian stations to densify the ITRF.

Global Daily Solutions

Phase and pseudo-range measurements from a global station network are processed daily in an undifferenced mode using an adapted version of JPL's GIPSY/OASIS II software (Lichten, 1990). An elevation dependent weighting is used to model higher measurement noise at lower elevation angles. Daily solutions use 24 hours of data without overlaps to estimate satellite dependent parameters including orbit initial states, clock offsets and solar radiation pressure corrections. Estimated station dependent parameters include coordinates, tropospheric delay and clock corrections. Other parameters include initial phase biases and daily EOP values. The IGS/IERS processing standards (McCarthy, 1992) are followed.

A priori initial state vectors, tropospheric delay and solar radiation pressure parameters are initialized from the previous day solution. The corresponding variance-covariance information, also taken from the previous day solution, is updated using a random walk process (Kouba et al., 1993). A priori EOP estimates are taken from IERS Bulletin A. UT1-UTC solutions have been estimated since January 1994.

EOP(EMR)95P01 (Daily EOP) - IGS Mean EOP

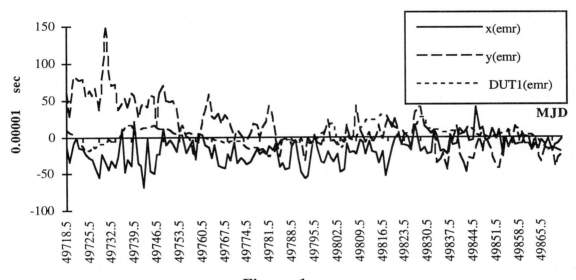

Figure 1

Based on IGS comparisons and combinations (Kouba, 1995), the estimated precision of the daily solutions is about 10-15 cm for ephemerides, 1 cm for unconstrained station coordinates, 0.5-1.0 ns for the satellite and station clocks, 0.2 mas for the pole position and 90 μs for UT1-UTC (Figure 1). The daily ephemerides and EOP solutions are submitted to IGS/IERS with a 3 to 5 days delay. Comparisons of these daily EOP solutions for 1994 (designated as EMR(EOP)94P01 series by IERS) with respect to the IERS Bulletin B show RMS of 0.40 and 0.50 mas for the pole x and y components. Figure 2 shows an annual variation of the y-pole component for the daily values (old) and a much smaller annual period for the yearly combined solutions (new).

1994 EMR(EOP) Pole y (old & new) - IERS Bull B

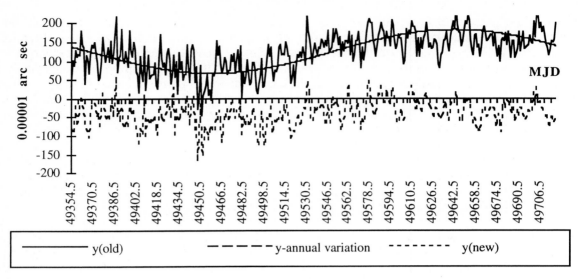

Legend: ——— y(old)   - - - - - y-annual variation   - - - - - - - y(new)

Figure 2

Regional Daily Solutions

CHUR-ALGO (1779k

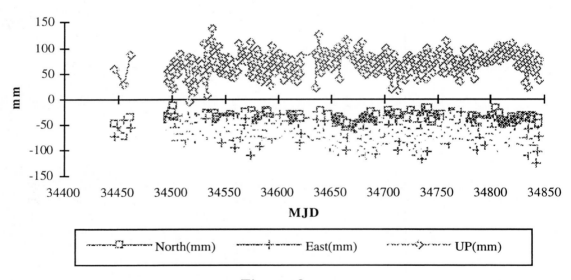

Legend: ----□---- North(mm)   ----+---- East(mm)   ---◇--- UP(mm)

Figure 3

The regional inter station baselines are computed using the CGPS22 software (Kouba and Chen, 1992) processing double differenced phase data. To avoid long baselines, the Canadian regional stations are subdivided into an Eastern and a Western portion with respectively the ALGO and DRAO reference stations. Fixed IGS/CACS orbits provides orientation and scale consistent with ITRF. Figure 3 shows the Algonquin (ALGO)-Churchill (CHUR) baseline results. The station CHUR, is subjected to extreme ionospheric conditions near the North magnetic pole, shows over the relatively short data interval of 13 months, horizontal station velocities of -0.4 and -18.4 mm/y which agree within 3 mm/y with the NUVEL1 model. The derived vertical velocity of 17.4 mm/y is larger by about 10 mm/y than the expected ICE3G model (Tushingham and Peltier, 1991).

Global Weekly Solutions

Each global weekly solution is obtained by combining the global daily solutions as follows: first, the constraints to the 13 ITRF station coordinates are removed; then, the station coordinates and EOP are extracted from the daily solutions and combined; finally, the 13 ITRF station coordinates and variance-covariance constraints are applied to the combined weekly solutions. The weekly combinations improve the daily pole estimates by a factor of approximately 2 when compared to the IGS mean EOP. The combined weekly solutions are submitted to IGS in the SINEX (Solution INdependent EXchange) format to facilitate the IGS ITRF densification project (Blewitt et al., 1995). The RMS for the weekly solutions for weeks 798 to 804 for the Canadian sites (ALGO, NRC1, CHUR, STJO, DRAO, ALBH, WILL and YELL) vary between 2 to 5 mm, with the vertical component being usually larger.

1994 Global Annual Combination Solution

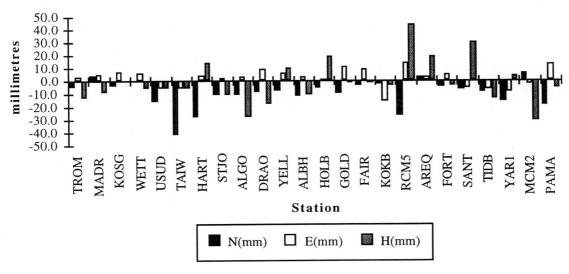

1994 Annual Station Coordinate Solution - ITRF93

Figure 4

The global annual solution is derived from the unconstrained global weekly station and EOP solutions. Long term station movements were modeled by mean linear velocities. To provide a reference for EOP series and station velocities the 13 ITRF station velocities were constrained. The orientation of the annual combination was constrained to ITRF93 coordinates to provide EOP reference. All station coordinates and the remaining station velocities were unconstrained. The differences between the EMR 1994 annual solution and the ITRF93 station coordinates are usually less than 2 cm as shown in Figure 4. The heights have differences of up to 4 cm. Comparisons of the station velocity solutions at the unconstrained stations with ITRF93 show an agreement at the sub cm/y level for the horizontal components and less than 2 cm/y for the vertical components. For 1994, the x and y pole solutions (IERS designation EMR(EOP)95P02) have an RMS of 0.36 and 0.35 mas, respectively.

CONCLUSIONS

CACS produces daily, weekly and annual GPS based results. The precision of the CACS daily solutions is 10-15 cm for the satellite ephemerides, 1 cm for unconstrained coordinates, 0.5-1.0 ns for station and satellite clocks, and about 0.2 mas and 90 μs for EOP as derived from the IGS combinations and comparisons. CACS global weekly combination solutions show improved station coordinate precision of 2-5 mm and pole position precision of 0.1 mas with respect to IGS mean pole. The combination solutions showed significant decrease in the annual periodic EOP variations and RMS reduction from 0.5 mas for the daily solutions to 0.35 mas when compared to the IERS Bulletin B. More research is needed into systematic variations in station and EOP solutions.

REFERENCES

Blewitt, G., Y. Bock and J. Kouba, (1994). Constructing the IGS polyhedron by distributed processing, Proceedings of Densification of the ITRF through regional GPS networks, an IGS workshop held at JPL, Pasadena, Nov. 30-Dec. 2, pp. 21-36.

Kouba, J. and Chen Xin, (1992). Near real time GPS Data Analysis and Quality Control, Proceedings of the Sixth International Geodetic Symposium on Satellite Positioning, held at Ohio State University, March 17-20, pp. 628-638.

Kouba, J., P. Tétreault, R. Ferland and F. Lahaye, (1993). IGS data processing at the EMR Master Control System Centre, Proceedings of the 1993 IGS Workshop, held at Univ. of Bern, Switz., March 25-26, pp. 123-132.

Kouba, J., (1995). Analysis Coordinator Report, IGS Annual Report 1994, (in prep.)

Lichten, S.M., (1990). Estimation and filtering for high-precision GPS positioning applications, Manuscripta Geodetica, Vol. 15, pp. 159-176.

McCarthy, D.D. (ed.), (1992). IERS Standards (1992), IERS Technical Note 13, obsevatoire de Paris.

Tushingham, A. M. and W.R. Peltier, (1991). ICE-3G: A new global model for late Pleistocene deglaciation based on geophysical predictions of post-glacial relative sea-level change, J. Geophys. Res., 96, 4497-4523.

COMMON EXPERIMENT OF THE ANALYSIS CENTRE CODE AND THE INSTITUTE OF GEODESY AND GEODETIC ASTRONOMY OF WARSAW UNIVERSITY OF TECHNOLOGY (IGGA WUT) ON THE COMBINATION OF REGIONAL AND GLOBAL SOLUTIONS

Jerzy B. Rogowski
Mieczysław Piraszewski
Mariusz Figurski
Institute of Geodesy and Geodetic Astronomy
Warsaw University of Technology
Elmar Brockmann
Markus Rothacher
Analysis Centre CODE at the Astronomical Institute of the University of Berne

INTRODUCTION

Since January 1995 the CODE Processing Centre and the Institute of Geodesy and Geodetic Astronomy of Warsaw University of Technology entered into cooperation aimed at testing different processing methods for the Polish network consisting of 3 permanent IGS stations: Józefosław, Borowiec and Lamkówko. The connection to the global reference frame is realised by processing 4 additional European IGS stations: Wettzell, Kootwijk, Metsahovi and Onsala. The results of four different processing strategies are compared with the results achieved by CODE from the global network. The experiment is a practical test of the idea of IGS regional data processing expected for IGS Associated Analysis Centre Type I. Particular attention is paid to the impact of the orbits (CODE vs.IGS) and the influence of the troposphere parameter estimation on the station coordinates. Since end of March 1995 the second receiver a TURBO ROGUE SNR 8000, was included into the permanent network. The results of data processing using the same computation strategies as well as results of the combination with the global solution are shown. The preliminary results of this work are shortly reported in the paper. The experiment of combining global CODE solutions with regional solutions was also tested with the EXTENDED SAGET'94 Network.

CHARACTERISTIC OF THE METHODS AND COMPUTATION STRATEGIES USED IN THE TEST CAMPAIGN

The Bernese software v. 3.5 allows us to obtain normal equation files of a network solution. It is possible to combine these with other normal equation files, for example those of a global solution. Four methods of combination were tested.
The WUT (Warsaw University of Technology) Processing Centre performed an automatic data processing of the Polish Regional Network of the IGS (Józefosław, Borowiec, Lamkówko) using the BERNESE v. 3.5 GPS software, including 4 IGS sites (2 IGS core sites Wettzell and Kootwijk and 2 IGS global sites Metsahovi, Onsala). The map of the network is presented in Fig. 1.
For the 4 common IGS sites, use was made of the following:
- identical naming conventions
- same phase centre corrections
- baselines between IGS Global sites were excluded.

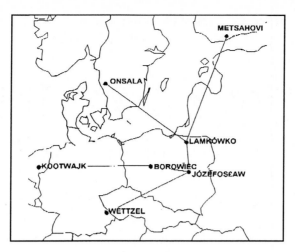

Fig. 1. Map of the network

The combination of the normal equations including troposphere parameters obtained in different centres requires that common processing options are used:
- the same satellite information
- the same number of troposphere parameters, per day and per station (12 per day),
- the same a priori troposphere model (Saastamoinen),
- troposphere for all station estimated,
- use of predominate ambiguities,
- pre-elimination of ambiguity parameters
- use of constraints of 0.0001 m on all Core and Global IGS stations.

Four computation methods were tested.
Method 1 - Coordinates and Troposphere were estimated. CODE orbit solution and CODE ERP were used. Coordinates and Troposphere parameters were saved in NEQ file for combination.
Method 2 - Coordinates and Troposphere were estimated. CODE orbit solution and CODE ERP were used. Only coordinates were saved in NEQ file for combination.
Method 3 - Coordinates and Troposphere were estimated. IGS orbit and IGS ERP solutions were used. Only coordinates were saved in NEQ file.
Method 4 - Coordinates and Troposphere were estimated. Troposphere parameters were estimated only for regional stations. CODE orbit solution, CODE ERP and troposphere parameters estimated by CODE were used for all IGS Core and Global sites. Only coordinates were saved in NEQ file.

RESULTS.

Experiments concerning the Poland Regional IGS Network.

Three Polish IGS permanent stations Józefosław, Borowiec, Lamkówko were chosen for the data processing test campaign. For these stations automatic day to day data processing has been performed using the above mentioned different computation strategies. The results from 100 days in 1995 are presented in Fig. 2, Fig. 3, Fig. 4.
The significant changes in the height of Józefosław station from 1st to 31st day of the test campaign resulted from the installation of a dome during that period on the pillar.
Since these stations are included now in the CODE day to day data processing it is no longer possible to perform combinations of CODE and WUT NEQ files. Mean values of coordinates as well as the rms from the 100 day period are presented in Table 1. The same experiment using observations performed in confirmation

Fig. 2 Comparison of North, East, Up components between methods 1, 3, 4 and CODE solutions for station Józefosław

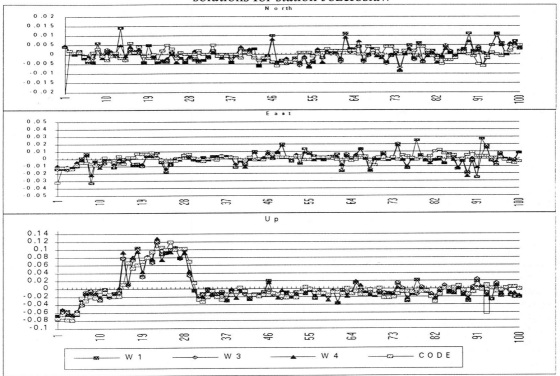

Fig. 3 Comparison of North, East, Up components between methods 1, 3, 4 and CODE solutions for station Borowiec

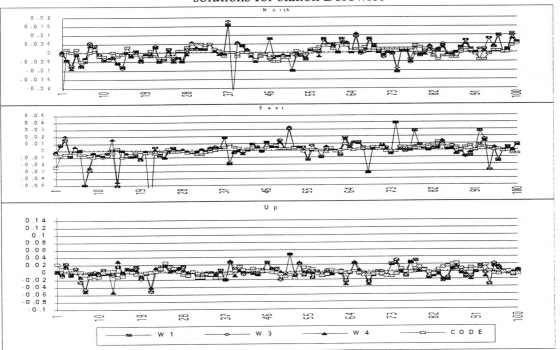

Fig. 4 Comparison of North, East, Up components between methods 1, 3, 4 and CODE
solutions for station Lamkówko

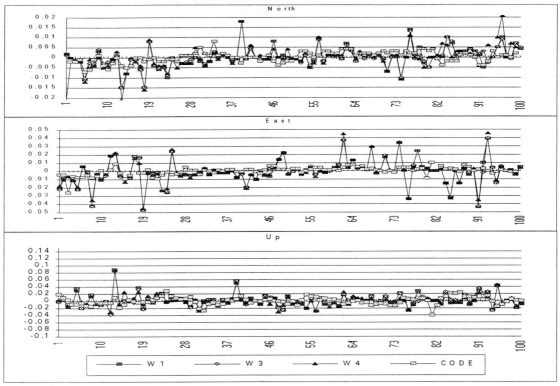

There are no significant differences obtained using the four presented methods. The CODE solutions are very close to each other. However the smallest differences were obtained using the fourth method.

Table 1. Mean value of coordinates and RMS

JOZE	X (m)	RMS	Y (m)	RMS	Z (m)	RMS
W1	3664940.3298	0.0023	1409153.7367	0.0013	5009571.2999	0.0030
W3	3664940.3294	0.0023	1409153.7356	0.0013	5009571.2989	0.0030
W4	3664940.3352	0.0023	1409153.7352	0.0011	5009571.3060	0.0031
CODE	3664940.3406	0.0022	1409153.7396	0.0012	5009571.3078	0.0032
BOR1	X (m)	RMS	Y (m)	RMS	Z (m)	RMS
W1	3738358.6147	0.0012	1148173.5772	0.0017	5021815.6804	0.0012
W3	3738358.6142	0.0011	1148173.5762	0.0017	5021815.6794	0.0011
W4	3738358.6204	0.0011	1148173.5764	0.0016	5021815.6869	0.0012
CODE	3738358.6195	0.0007	1148173.5794	0.0006	5021815.6853	0.0007
LAMA	X (m)	RMS	Y (m)	RMS	Z (m)	RMS
W1	3524523.0834	0.0013	1329693.5187	0.0014	5129846.2597	0.0014
W3	3524523.0835	0.0013	1329693.5182	0.0014	5129846.2593	0.0014
W4	3524523.0859	0.0012	1329693.5180	0.0014	5129846.2638	0.0013
CODE	3524523.0991	0.0008	1329693.5208	0.0006	5129846.2704	0.0009

The combination of EXTENDED SAGET'94 solutions with global IGS solutions.

The experiment of combining regional solutions was tested using the EXTENDED SAGET'94 (J. Śledziński, at all, 1995) Network.
The differences between the solutions obtained from regional data processing and the combination of regional and global IGS solutions are presented in Table 2.

Table 2. Differences in DX , DY , DZ coordinates between the local and global solutions.

STATION	DX [m]	DY [m]	DZ [m]	STATION	DX [m]	DY [m]	DZ [m]
117 JOZE	0.0095	0.0058	0.0034	811 KIRS	-0.0067	-0.0026	-0.0033
151 GRAZ	-0.0268	0.0074	-0.0236	812 LAM0	0.0010	0.0006	0.0054
153 KOSG	0.0001	-0.0003	-0.0001	814 LJUB	-0.0281	-0.0011	-0.0309
155 MATE	-0.0344	-0.0157	-0.0432	815 LVIV	-0.0120	0.0014	-0.0189
157 WETT	-0.0004	-0.0005	0.0004	816 MOPI	-0.0120	0.0035	-0.0132
158 ZIMM	0.0059	0.0051	-0.0016	817 PENC	-0.0104	0.0035	-0.0132
159 ONSA	-0.0047	-0.0112	0.0150	818 POLO	-0.0059	-0.0015	-0.0029
160 METS	-0.0088	0.0110	-0.0022	819 SKPL	-0.0051	0.0031	-0.0110
801 BASO	-0.0281	-0.0012	-0.0309	820 SNIE	-0.0061	-0.0018	-0.0029
802 BOR1	-0.0052	-0.0032	0.0012	821 STHO	-0.0076	0.0032	-0.0124
803 BRSK	-0.0289	-0.0046	-0.0338	823 UZH2	-0.0076	0.0025	-0.0150
804 CSAR	-0.0167	0.0051	-0.0174	831 DARM	0.0003	0.0009	-0.0019
805 DISZ	-0.0170	0.0048	-0.0173	832 KOMO	0.0072	0.0052	0.0046
806 GOPE	-0.0072	-0.0006	-0.0057	833 OKSY	-0.0055	-0.0048	0.0056
807 GRYB	-0.0042	0.0032	-0.0106	835 SIMS	-0.0150	0.0014	-0.0231
808 HOHE	0.0021	0.0024	-0.0018	836 SWKR	0.0023	0.0046	-0.0042
809 HUTB	-0.0097	0.0019	-0.0103	837 WET1	-0.0037	-0.0007	-0.0030

CONCLUSIONS

Generally, we obtained good results while applying the idea of regional IGS networks data processing and its combination with the global IGS solutions using BERNESE v. 3.5 software. However, there are some problems in the automatic data processing displayed as jumps in the results presented above. The authors expect that these distortions will soon be eliminated. The results of the daily computations are accessible 4 days after the orbits become available.

REFERENCES

J. Śledziński, J.B. Rogowski, M. Figurski, L. Kujawa, M. Piraszewski. Final results of EXTENDED SAGET'94 Campaign. Paper presented to the Inetrnational Seminar "GPS in Central Europe", Penc, Hungary 9-11 May 1995.

Using GPS to Determine Earth Orientation

Dennis D. McCarthy
Brian J. Luzum
U.S. Naval Observatory
Washington, D.C. 20392 USA

INTRODUCTION

Analyses of the orbits of the satellites of the Global Positioning System (GPS) by participants in the International GPS Service (IGS) (Mueller and Beutler, 1992) provide daily observations of high-accuracy polar motion. These data are used routinely by the International Earth Rotation Service (IERS) (see, for example, *IERS Annual Report for 1993*). The GPS data have also been analyzed by some centers to produce estimates of UT1-UTC. Currently, the UT1-UTC data are not used because of large systematic errors, but this situation is expected to change in the near future due to the results of ongoing research.

SOURCE OF DATA

Daily estimates of pole positions have been provided by contributors to the IGS. These contributors include the Scripps Institute of Oceanography (SIO), the University of Berne's Center for Orbit Determination in Europe (CODE), Jet Propulsion Laboratory (JPL), the European Space Operations Center (ESOC), National Resources Canada (NRCan), the GeoForschungsZentrum (GFZ), and the National Oceanic and Atmospheric Administration (NOAA). Estimates of UT1-UTC are contributed by the CODE, JPL, and NRCan.

ANALYSIS OF GPS DATA

The time series contributed by each of the institutions mentioned above were analyzed by comparing them with the combination series (McCarthy and Luzum, 1991) of the National Earth Orientation Service (NEOS) which serves as the IERS Sub-bureau for Rapid Service and Predictions. The data used to produce this series are derived from Very Long Baseline Interferometry (VLBI), Satellite Laser Ranging (SLR), Lunar Laser Ranging (LLR), and GPS. Figures 1 and 2 show plots of recent differences in polar motion after the removal of biases. Table 1 shows the statistical analysis of the polar motion and UT1-UTC data.

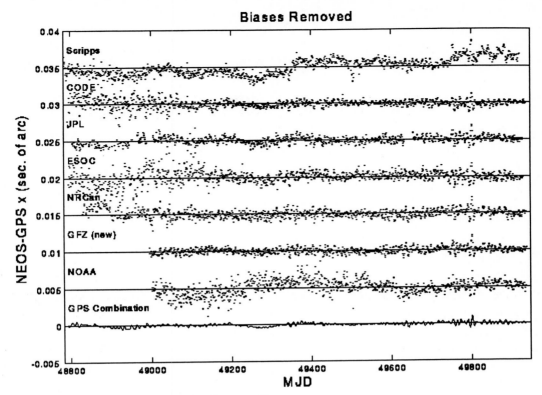

Figure 1. Residuals in x for GPS contributors.

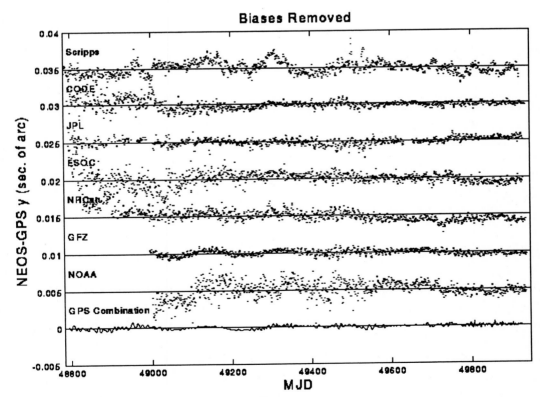

Figure 2. Residuals in y for GPS contributors.

Table 1. Statistics for GPS polar motion data.

Contributor	Data Span	Points	Mean (NEOS-GPS)			Standard Deviation		
			x	y	UT1-UTC	x	y	UT1-UTC
Scripps	48780.5 -49920.5	1138	-0.76	-0.10		1.01	0.82	
CODE	48792.5 -49934.5	1143	0.40	0.14	-2.93	0.55	0.65	4.21
JPL	48794.5 -49927.5	1008	0.11	-0.38		0.47	0.45	
	49648.5 -49927.5	274			-0.82			0.79
ESOC	48794.5 -49927.5	1134	0.15	-0.33		0.89	1.07	
NRCan	48830.5 -49934.5	1059	0.29	0.39		0.53	0.79	
		911			1.87			2.04
GFZ	48997.5 -49936.5	944	-0.02	0.01		0.37	0.35	
NOAA	48988.5 -49934.5	932	-0.17	0.11		0.93	1.05	

USE OF GPS DATA IN IERS BULLETIN A

The NEOS makes use of GPS data contributed to the IERS in producing its weekly IERS Bulletin A. This is done by smoothing the contributed data separately using algorithms similar to that used in the procedure to combine the VLBI, SLR, LLR, and GPS data (McCarthy and Luzum, 1991). Statistical weights are assigned to each of the contributors based on their past agreement with the NEOS combination series. Figures 3 and 4 show the agreement between the smoothed GPS estimates and those derived using data from other techniques for recent times.

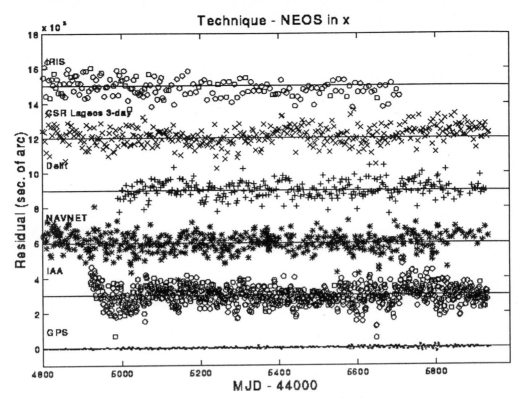

Figure 3. Residuals of contributors in combination solution.

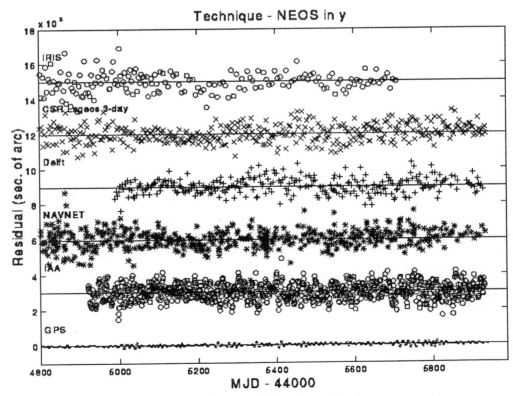

Figure 4. Residuals of contributors in combination solution.

Figure 5 shows the relative contribution of each technique to the three different combined series produced by NEOS. The contribution takes into account the number of data contributeed, their errors, and the speed with which the data are available. The three series produced by NEOS are the "rapid service" derived from near real-time estimates contributed to IERS, predictions based on these data, and a final series derived from the best data contributed by each technique, often more than a few weeks late.

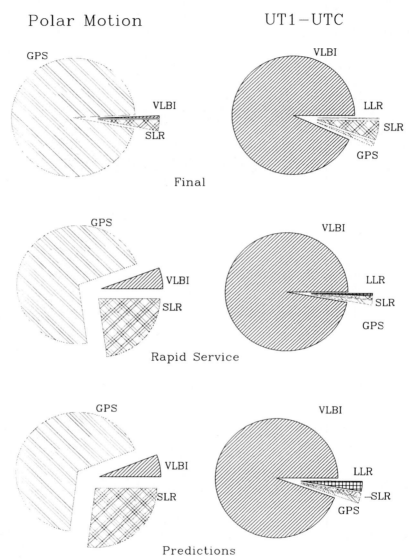

Figure 5. Contribution of various techniques to NEOS products.

ACCURACY

Comparison with the other techniques shows that the combined GPS polar motion series has an accuracy of ±0.15 msec of arc in x and ±0.15 msec of arc in y. Figures 1 and 2 show that serious systematic difference between the contributors remain which must be resolved to obtain further improvement.

UT1-UTC

Estimates of UT1-UTC derived from the analyses of GPS orbits are plagued by the correlation of the motion of the node of the orbital plane with the variation in the rotation angle of the Earth. Plots of UT1-UTC determined from GPS observations compared with NEOS estimates are shown in Figure 6. The precision of these data is ±0.2ms after corrections are made for long-term drifts in the time series. Frequent ties to VLBI observations may be required to obtain useful information regarding the Earth's rotation angle.

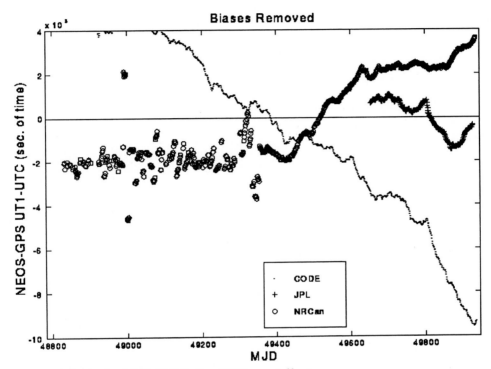

Figure 6. Residuals in UT1-UTC for GPS contributors.

CONCLUSIONS

Polar motion estimates derived from the analysis of GPS orbits form an essential contribution to the work of the IERS Sub-bureau for Rapid Service and Predictions at NEOS. This is particularly true when assessing high-frequency polar motion. Doubts remain about the long-term stability of the polar motion system derived from GPS orbits which can only be addressed by further observation.

GPS UT1-UTC could conceivably make a useful contribution particularly at high frequencies. However, the current estimates suffer from problems in modeling the motion of the orbital planes which have not been overcome. Investigations continue in this area. GPS determinations of both polar motion and UT1-UTC could benefit from a better understanding of the systematic differences between the contributing analysis centers.

REFERENCES

IERS Annual Report for 1993 (1994), Observatoire de Paris, Paris.

McCarthy, D.D. and Luzum, B.J. (1991). Combination of precise observations of the orientation of the Earth, *Bulletin Géodésique*, **65**, 22-27.

Mueller, I.I. and Beutler, G. (1992). The International GPS Service for Geodynamics-Development and Current Structure, *Proceedings of the Sixth International Geodetic Symposium on Satellite Positioning*.

THE PERMANENT GPS STATION AT THE UNIVERSITY OF PADOVA

Alessandro Caporali andAntonio Galgaro
Dipartimento di Geologia, Paleontologia e Geofisica, Universita' di Padova
Via Giotto 1, I-35137 Padova, Italy

ABSTRACT

Since January 1994, in coincidence with the official start of the International GPS Service for Geodynamics (IGS), a fixed GPS station is operational at the Department of Geology, Paleontology and Geophysics of the University of Padova. The equipment is based on a dual frequency receiver TRIMBLE 4000 SSE located on the astronomic observatory on the roof of the University's Main Building, formerly used for astrogeodetic observations. The site is unmanned and the equipment is monitored remotely from our Department or from virtually any other place with a telephone line. Likewise, remote users can access autonomously the local data base and download at any time data in compressed RINEX 2 format. As part of the support of the IGS, daily RINEXing, compression and download via Internet to the Matera Space Geodesy Center for subsequent routing to IfAG and IGS Central Bureau is done automatically with timed batch procedures running in the local PC. The station runs virtually unattended and features a ratio [data yield/operational costs] considerably higher than for other space geodetic techniques. Future enhancements include real time services in support of local DGPS applications and a remotely accessible geodetic/geophysical data base.

INTRODUCTION

At the University of Padova, the Department of Geology, Paleontology and Geophysics contributes to the activities of the Interdipartimental Center for Space Activities "Giuseppe Colombo" (CISAS) in conjunction with the Departments of Astronomy, Physics, Electronic and Mechanical Engineering. Within the CISAS our Department has responsibilities on research and post-graduate teaching on the application of space techniques to Geodesy and Geodynamics. In this rapidly evolving field, one of the most impressive achievements is undoubtely the reliable detection and measurement of plate motion at a continental scale, and the accurate monitoring of changes of the Earth orientation and rotation parameters. These results have been achieved using rather expensive and time consuming techniques such as Very Long Baseline Interferometry (VLBI) and Satellite Laser Ranging (SLR). With the advent of the satellites of the Global Positioning System of the United States, now in a fully operational configuration, it has been demonstrated that data scientifically as useful as VLBI or SLR can be generated with moderate investment and operational costs using GPS receivers. This fact has the potential of triggering a densification of the tracking network, as is already demonstrated by the continuous growth of the number of stations supporting the IGS (Fig. 1). Densification is crucial for certain areas, such as the alpine arc, where a deeper understanding of the collisional processes at the basis of the alpine orogenesis requires more detailed data than those so far used to monitor plate motion on a global scale.

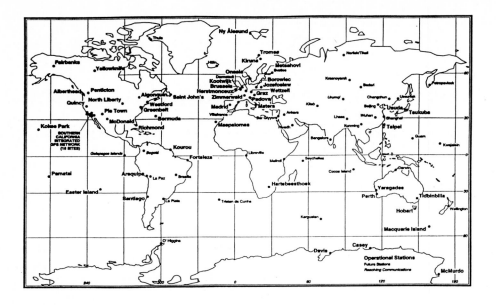

Fig.1. Location of the UPAD station within the IGS tracking network

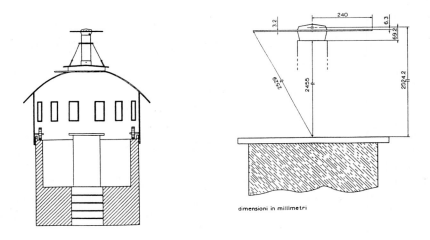

dimensioni in millimetri

Fig.2. Location of the antenna site with DOMES numbers 12750M001

In addition, GPS is expected to become of widespread use in a variety of applications related to applied geology and geophysics, environmental control, positioning and navigation. Consequently, it is natural to address the problem of undertaking initiatives from which pure and applied research, and teaching, can benefit. This paper summarizes how a solution to such problem has been evolving in our academic environment in the last year, and has eventually materialised in the form of a combination of receiving equipment interfaced with communication facilities.

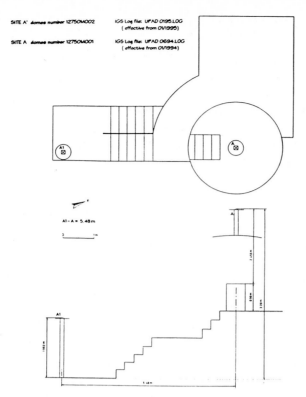

Fig.3.Relative location of the antenna sites with DOMES numbers 12750M001 and

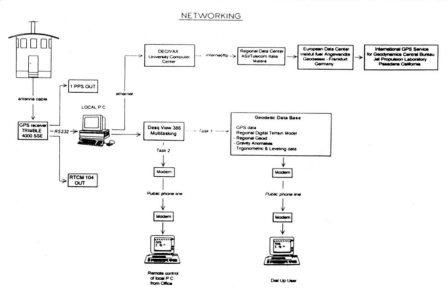

Fig.4. Networking with Internet and public phone line

SITE LOCATION AND EQUIPMENT

As shown in Fig. 2, in 1994 the antenna was mounted on a mast secured by cables on the roof of an astronomic dome formerly used for astrolabe observations.The dome can rotate about a vertical axis by means of a rack and pinion mechanism, but this movement has been disabled to keep the horizontal position of the vertical axis of the antenna free from eccentricity effects.

Part of the dome is covered by a horizontally sliding meridian roof. This feature has been kept, to allow the mast and the antenna to be accessed and serviced. Several tests have demonstrated the repeatibility of such horizontal motion of the sliding roof and we have concluded that the original horizontal position of the antenna can be restored with negligible error. The antenna sees an unobstructed horizon and the housing is on the top of a huge and ancient building downtown the historic center, with no measurable evidence of local movements. In spite of the ground plane the radiation pattern of the antenna proved to be significantly modified by the presence of the dome. For this reason, in December 1994 the antenna was moved to a nearby position, shown in Fig. 3.

The receiving equipment (Table 1) is based on a TRIMBLE 4000 SSE Geodetic Surveyor System, with 9+9 parallel channels, 2.5 Mb of internal memory, providing code and phase data on both L1 and L2 carriers. L2 phase data are generated to full L2 cycles independently of whether Anti Spoofing is on or off. The receiver outputs three types of data: 1 pulse per second (pps) synchronous with GPStime, RCTM SC 104 type 1,2,3 and 9 differential corrections and binary data and navigation files. In the configuration active in 1994, the data and navigation files were logged directly to the hard disk of the local 386 PC using the program LOGST via a serial RS232 C interface. A second serial interface on the PC was connected to a pocket sized modem allowing communications between the PC and the external world, including the VAX of the University's Computer Center which provided a gateway to Internet. This mode of operation in which the PC served as a data logger was abandoned in late 1994 because the computer was kept busy by the data logging 24 hours a day. This implied that it could not be used for other tasks and that the risks of discontinuities in case of power failure were far higher than if the computer had been activated only once during the day. It was decided to abandon the LOGST program and to keep the daily file in the receiver's memory, and download it at the end of the day by means of an undocumented batch operation of the TRIM4000 program, requiring the local PC to be active for just a few minutes. Late in 1994 an Ethernet connection was established at the receiver's location. Communications via modem were then left to remote, "dial up" users and the download of a data file of one day to the VAX dropped from two hours to few minutes (Fig. 4).

Table 1. station logsheet

1. Site Identification of the GPS monument
    Site Name   :UNIVERSITY OF PADOVA, MAIN BUILDING
    4 char ID   :UPAD
    IERS DOMES Number   :12750M002 (since 1995); 12750M001 (during 1994)
2. Site Location
    City   :PADOVA
    Country   :ITALY
    Tectonic Plate   :ADRIATIC-AFRICAN PLATE
3. GPS Receiver
    Type   :TRIMBLE 4000 SSE
    Serial Number   :3247A01926
    Firmware Version :1988-1992 G.S.
    Date   :20/01/1994
4. GPS Antenna
    Type   :TRIMBLE 4000ST L1/L2 GEOD
    Serial Number   :3241A64735
    Date   :20/01/1994
    Vertical Antenna Height :2.455 meters (for DOMES 12750M001); 1.962 meters (for DOMES
               12750M002)
    Antenna Reference Point:BOTTOM OF PREAMPLIFIER

The receiver is scheduled to operate on a daily basis with a sampling rate of 30 sec and an elevation cutoff of 15 degrees above the horizon. The data are stored in daily files with name UPADdoys, where UPAD is the name of the site and stands for University of PADova, doy is the day-of-year and s is the session number, normally zero.

DATA MANAGEMENT

Managing a Daily Session

Data management consists of the tasks: data download , RINEXing and compression of the RINEX files.

Data download from the datalogger is accomplished by the TRIMBLES's TRIM4000 program which unpacks binary ".R00" file types into corresponding .MES, .ION, .DAT and .EPH files. For automatic execution, TRIM4000 is driven in batch mode by a custom binary file which sends to the TRIM4000 program the commands and keystrokes appropriate for data download from the datalogger. The Authors are grateful to dr. Werner Gurtner of the Astronomical Institute of the University of Bern for pointing out this undocumented feature of TRIM4000. Since 1995 the .MES, .ION, .DAT and .EPH files are downloaded directly from the receiver again by a batch procedure automating the execution of the TRIM4000 program. The RINEXing of the data is done with the latest available versions of the TRRINEXO and TRRINEXN programs. Compression of the RINEX files is done with the program COMPR.BAT available through the BBS of the IGS and compatible with the UNIX requirements on the number of line-feeds and carriage returns at the end of each record. The quality control at this time is done with the program QC of UNAVCO which provides quantitative information on the data file (see below). Finally, communications are driven by a "ad hoc" command file which is generated automatically every day by a monitor program.

Results of the 1. quarter of 1995

Typically 92 to 95% of the QC-predicted data are acquired daily. The multipath statistics indicates a daily average of 0.5 m ln L1 and 0.7 m on L2. The statistics of the local oscillator show a mean value of 3.8 msec/hour, typical of a crystal oscillator, and an average time between resets of 15 minutes, implying that the receiver will reset the time counter whenever the receiver time is off relative to GPStime by more than one millisecond, which appears a very reasonable behavior.

AVAILABLE SERVICES AND FUTURE DEVELOPMENTS

At this time most of the effort is devoted to provide data on a regular and timely basis to the IGS. In any case the data are also available to other projects, such as Tyrgeonet, GPS activities in Central Europe or local surveys. Recent data files (typically the last 2-3 weeks) are still available on the local PC. Older files are archived in Matera by the Space Division of Nuova Telespazio on behalf of the Italian Space Agengy and can be recovered there, or on the data base of the IGS.

Because the station has access to the IGS data base, data such as precise ephemeris, ITRF92 station coordinates worldwide, Earth orientation and rotation parameters and other information can be made available to interested local users. In the future, as the use of GPS for survey and navigation becomes more common, other data useful for local geodetic and geophysical applications could include:

◆Digital Terrain Models
◆Gravity and Deflection of the Vertical
◆Geoidal Heights
◆Parameters for coordinate transformation between the conformal Gauss coordinates on the ED79 national datum and UTM coordinates on the global WGS84 datum
◆1 Hz DGPS corrections for precise aerophotogrammetry and navigation
◆lower frequency DGPS corrections for maritime and terrestrial navigation.

As the capability of data dissemination in real time develops, differential corrections for both pseudorange and carrier phase will be broadcast according to the RTCM SC 104 and NMEA 183 standards.

*Acknowledgement.*This research is supported by the Italian Space Agency. We are grateful to dr. Werner Gurtner of the Astronomical Institute of the University of Bern and to Michele Poggi of TELECOM Italia, Matera, for help and advice while setting up the operational procedure

Analysis of Continuous GPS Observations for Geodynamics Purposes
Jan Kostelecký, Pavel Novák, Luděk Skořepa, Jaroslav Šimek[1]

1. Introduction

To estimate the repeatability of baseline determination, the observed data series acquired from five IGS stations have been processed "day-by-day". The individual stages of the experiment were already reported in [1], [3] presenting the results of the analysis of observation time series up to 315 days and discussing different processing strategies. This presentation summarizes the results of the analysis of the period of 585 days. In addition to it an attempt has been made to determine on the basis of the long-term observations the ITRF 93 coordinates of the newly accepted IGS station Pecný (GOPE) and to do the first assessment of its drift within the Eurasian plate.

2. Description of the Experiment

The experiment is based on the continuous processing of GPS observations made at five stations - Wettzell, Graz, Pecný, Onsala and Józefoslaw, see Fig. 1.

As for the stations Wettzell, Graz and Pecný the observation series from September 7, 1993 to April 15,1995 were taken for the analysis. The data acquisition from Onsala and Józefoslaw started in September 1994. The observation series at these two stations are short and therefore the results of their analysis are not presented here.

The main goal of the experiment is to estimate the repeatability of the baseline determination and a trend of baseline time evolution. Analyzed were both the coordinate differences and the derived quantities - baseline lengths and height differences. Of the three data sets only the results for baselines and height differences are presented in detail whereas for coordinate differences only the basic trend characteristics are given.

Fig. 1: Distribution of Evaluated IGS Stations.

3. Observation Data

Phase and pseudo-range measurements on both frequencies at 30 second intervals are recorded at all sites. The following types of receivers are being operated at respective stations: Rogue SNR (Graz, Wettzell, Onsala) and Trimble 4000SSE (Pecný, Józefoslaw). The default tracking mode for all channels of Trimble 4000SSE and Rogue receivers is L1 P-code, L2 P-code.

In the epoch September 1993 -- January 1994 AS was switched off, i.e. the data from this period consist of two phase measurements and two pseudo-ranges estimated with the help of P-code. Since January 31, 1994 the pseudo-range measurements on the first frequency have been done using C/A-code and on the second one the cross-correlation technique has been used (AS on).

4. Data Processing

The data from all five stations were processed by the Bernese GPS Software version 3.5 using so-called *ionosphere-free* linear combination (LC) of the L1 and L2 phase measurements (30 second intervals). Ambiguities were solved by statistical rounding off to integer values (first using L5 LC and then L3 LC). The elevation mask was set to 20 degrees. The standard tropospheric model of Saastamoinen was used in processing determining one tropospheric zenith delay for each site and each 4-hour interval. The a priori constraint with the value of 0.100 m for each zenith delay was used. The influence of elevation dependent phase center variations on the baseline components for the Dorne Margolin (Rogue) and Trimble 4000SST antennas has been already known so that a large part of the mixing errors between these antennas can be corrected by applying elevation angle dependent phase corrections during the data processing (the Bernese Software). As an observation sampling intervals 24 and 2 hours were chosen.

5. Results and Their Analysis

From the computed cartesian geocentric coordinate differences the baselines and ellipsoidal height differences between five IGS stations - Wettzell, Graz, Pecný, Onsala and Józefoslaw - were determined. The solution for the complete cluster of

---

[1] Research Institute of Geodesy, Topography and Cartography, Geodetic Observatory Pecný, CZ-251 65 Ondřejov 244.

stations is available only for a short period and therefore in the following only the results of the analysis of Wettzell, Graz and Pecný will be discussed.

The time series of baselines and height differences were analyzed by an unharmonic analysis [2]. The following topics were investigated in detail:

a)   analysis with sampling interval of 24 hours,

b)   analysis with sampling interval of 2 hours (for a selected period),

c)   dependence of baseline lengths on different types of orbits (broadcast vs. precise, precise orbits from EMR - Ottawa, CODE - Bern, GFZ - Potsdam and IGS orbits),

d)   influence of anti-spoofing on the baseline length determination,

e)   influence of different ambiguity solutions (real, integer) on the baseline length determination.

The items c)-e) were discussed in more detail in [3]. Now we shall deal with the 585-days series of baselines and height differences.

Analysis of 24 hour samples. A 585-days data set of the period September 7, 1993 - April 15, 1995 was analyzed by an unharmonic analysis. After filtering periods with statistical significance greater than 30 percent trends of three baseline lengths were determined.

The RMS errors of one 24-hour data sample are about 3 mm for distances and about 12 mm for height differences. The resulting time changes are given in Table 1 and the corresponding spectrograms are depicted in Figs 2a, b, c and 3a, b, c. The analyzed data were corrected for the effect of polar tides but in the case of heights no corrections were applied either for atmospheric or for ocean loading effect. The last two corrections are a subject of our present investigations.

In the case of the time evolution of baselines a vague periodic term was found with an annual and Chandler period and with the height differences a weak semi-annual term may be detected. The amplitudes of these variations are about 1-2 mm. Taking the zero-hypothesis with 3σ, only the baseline Pecný-Graz indicates a possible secular change. One should be very cautious in interpretation of height changes respecting an influence of tropospheric refraction and the fact that the above mentioned atmosphere and ocean loading corrections were not applied.

The time changes derived from 10-days intervals of the same period of each year are given in Table 2. This may be viewed as a kind of simulation of regularly repeated geodynamical campaigns. Whereas in the case of baselines we can see some statistical agreement with long-term observations, in the case of heights it is obvious that the interpretation of time changes from repeated campaigns of a few days duration is delicate. The time changes of coordinate differences are given in Table 3.

In Table 4 the time variations of the baseline Wettzell-Graz coming from different SLR-based solutions are compared with our GPS-based solution.

Analysis of 2-hour samples. A data set from the time interval of 24 days (April 21 - May 19, 1994) was analyzed by the above mentioned method. The RMS error of one 2-hour sample lies within the limits 7-10 mm. An interesting result was obtained from a separate analysis of "day" and "night" data sets. The a posteriori RMS errors of one baseline determination from the "day" and "night" data sets are 10 and 5 mm respectively. The Fisher's statistical test (testing statistical equivalence of the data sets on the basis of a posteriori RMS errors) indicates that when comparing the "day" and "night" data sets the critical limit was exceeded. In our case we get for $m_I = 10$ and $m_{II} = 5$

$$\frac{m_I^2}{m_{II}^2} > F_{n_1, n_2, \alpha} \tag{1}$$

where the degree of freedom $n1 = 118$, $n2 = 286$ and the criterion of the risk $\alpha = 1$.

The Fisher's test clearly shows that the "night" data gives statistically significantly better results than the "day" one.

6.   Determination of Coordinates of the Station Pecný (GOPE) in ITRF 93

The ITRF 93 coordinates of the station Pecný (GOPE) for the epoch 1995.08 were determined from GPS observations of the station GOPE and of three IGS core stations - Onsala, Wettzell and Graz - in the period January 1, 1995 - February 4, 1995, fixing the coordinates of the three core stations and averaging 34 one-day solutions. The final result is

$$
\begin{array}{lll}
X(1995.08)= & 3979316.2755 & 0.0033 \\
Y(1995.08)= & 1050312.3184 & 0.0015 \\
Z(1995.08)= & 4857066.9970 & 0.0021
\end{array}
\tag{2}
$$

7.   Assessment of the Drift of the Station Pecný (GOPE) Referred to the Eurasian Barycenter

Using the time changes of coordinate differences derived from continuous observations at Pecný, Wettzell and Graz and the shifts of Wettzell and Graz published in [4] an attempt was made to estimate the shift of the station GOPE with respect to the barycenter of the Eurasian plate. In Table 5 the time changes of coordinates are given for all three stations. The values in the first column were obtained as least squares estimates, those in the second column for Wettzell and Graz were taken from [4]. For the station Wettzell both results coincide quite well but for Graz the coincidence is somewhat worse. Anyway, we can infer from these results that relative shifts within the Eurasian plate do not exceed a few millimeters per annum.

On the basis of the values from Table 5 the following shifts of GOPE with respect to ITRF 93, epoch 1995.1 were derived after [4]:

$$dX/dt = \quad -0.0234 \text{ m/y} \quad (-0.0218 \text{ m/y})$$
$$dY/dt = \quad 0.0216 \text{ m/y} \quad (\ 0.0205 \text{ m/y}) \quad\quad (3)$$
$$dZ/dt = \quad 0.0160 \text{ m/y} \quad (\ 0.0130 \text{ m/y})$$

The results in parentheses correspond to the velocities referred to ITRF 93 neglecting the shifts with respect to ETRF 89.

If we take coordinates of GOPE, given in Eq.(1), and the shifts from Eq.(2), make a transformation into ETRF 89 and compare the results with the officially adopted solution of the postcampaign EUREF-CS-H-91, see[5], we get the following differences for the cases of accounting and neglecting the shifts within the Eurasian plate respectively:

$$dX = \quad -3 \text{ mm} \quad\quad dY = 15 \text{ mm} \quad\quad dZ = \ 8 \text{ mm}$$
$$dX = \quad 7 \text{ mm} \quad\quad dY = \ 8 \text{ mm} \quad\quad dZ = -8 \text{ mm}$$

8. Conclusion

The results of the experiment show that

- only negligible drifts and periodic changes have been found with investigated medium-range baselines,
- GPS is an efficient and reliable tool for regional geodynamics investigations,
- repeated regional geodynamics GPS-campaigns can provide worthwhile results as to the indication of position changes,
- reliable interpretation of the height changes is only possible after the effects of tropospheric refraction, atmospheric and ocean tide loading are eliminated.

9. Acknowledgment.

This work has been partly supported by the Grant Agency of the Czech Republic through the Grant No. 103/94/0407 "The Use of the Global Positioning System for Solving High Precise Works in Geodesy." This support is gratefully acknowledged.

10. References

[1] Kostelecký, J. - Novák, P. - Šimek, J.: On the Long Term Behaviour of Longer Baselines Determined by GPS. Initial Stage of Investigations. Reports on Geodesy, No. 2 (10), 1994, Warsaw University of Technology Inst. of Geodesy and Geod. Astronomy, pp. 181 - 186.

[2] Kostelecký, J. - Karský, G.: Analysis of the 1970 - 1983 Circumzenithal Measurements. Bull. Astron. Inst. Czech. 38, 1987, p. 16.

[3] Kostelecký, J., Novák, P., Skořepa, L.: Preliminary analysis of continuous GPS observations Pres. at the 2nd CERGOP Working Conference, Penc - Budapest, Nov. 4-5, 1994;

[4] Boucher, C., Altamimi, Z.: Specifications for reference frame fixing in the analysis of a EUREF GPS campaign. EUREF TWG Circular Letter from March 28, 1995.

[5] Seeger, H., Schlüter, W., Talich, M., Kenyeres, A., Arslan, E., Neumaier, P., Habrich, H.: Results of the EUREF-CS/H'91 GPS campaign, In: Report on the Symposium of the IAG Subcommis. EUREF held in Warsaw 8-11 June 1994. Veröffent. der Bayer. Komm. für die Intern. Erdmess. der BAW, Heft Nr 54, München 1994, p. 87.

[6] Zerbini, S.: Crustal motions from short-arc analysis of LAGEOS data. Centr. of Space Geodesy to Geodynamics: Crustal Dynamics, Geodynamics 23, 1993, p. 371.

[7] Reigber, Ch., Förste, Ch., Ellmer, W., Massmann, F.-H., Schwintzer, P., Müller, H.: Die Lösung (DGFII) 91L02, DGFII Intern. Bericht, München, 1991.

[8] Smith, D. E., Kolenkiewicz, R., Dunn, P. J., Robbins, J. W, Torrence, M. H., Klosko, S. M., Williamson, R. G., Pavlis, E. C., Douglas, N. B., Fricke, S. K.: Tectonic motion and deformation from satellite laser ranging to Lageos, J. Geoph. Res., 95, B13, 1990, p. 22013.

[9] Gendt, G., Montag, H., Dick, G.: Plate kinematics in a global and European scale by LAGEOS laser ranging data from 1983 to 1990. Centr. of Space Geodesy to Geodynamics: Crustal Dynamics, Geodynamics 23, 1993, p. 311.

Solution: May 1995

Table 1: Time Evolution of Baselines & Height Differences - complete analysis
Data from Interval 49237 - 49822 MJD [7.9.1993 - 15. 4. 1995]

[1] - [2]	ds/dt [mm/y]	dh/dt [mm/y] ([2] minus [1])
Pecný - Graz	+2.1 ± 0.3	+3.2 ± 1.2
Pecný - Wettzell	+0.1 ± 0.3	+9.2 ± 1.1
Wettzell - Graz	+0.3 ± 0.3	-6.0 ± 1.0

Table 2: Time Evolution of Baselines & Height Differences - epoch analysis
Data from Intervals 059-070 1994 & 059-070 1995

[1] - [2]	ds/dt [mm/y]	dh/dt [mm/y] ([2] minus [1])
Pecný - Graz	+0.0 ± 1.0	+12.0 ± 3.7
Pecný - Wettzell	-0.2 ± 0.8	+21.9 ± 3.1
Wettzell - Graz	-0.5 ± 0.8	-9.9 ± 3.8

Table 3: Time Evolution of Coordinate Differences,
Data from Interval 49237 - 49822 MJD [7.9.1993 - 15. 4. 1995]

[2] - [1]	Coord.	dDiff./dt[mm/y] ([2] minus [1])	RMS of Diff. [mm]
Pecný - Graz	X	3.6	0.8
	Y	1.9	0.3
	Z	2.2	0.9
Pecný - Wettzell	X	-1.5	1.3
	Y	1.7	0.5
	Z	2.9	1.6
Wettzell - Graz	X	-1.1	1.1
	Y	-0.7	0.3
	Z	-3.8	1.2

Table 4: Time Evolution of the Baseline Wettzell - Graz Derived from SLR and GPS Techniques
Baseline Wettzell - Graz

Technique	ds/dt [mm/y]	Source
SLR (LAGEOS) short arc	-2 ± 5	Zerbini, 1993 [6]
SLR (DGEII) 91/92	-4	Reigber et al., 1991 [7]
SLR (LAGEOS)	0	Smith et al., 1990 [8]
SLR (LAGEOS - 5 day arc)	+1	
SLR (LAGEOS - 30 day arc)	+4	Gendt et al., 1993 [9]
SLR (LAGEOS) semiannual solution	0 ± 3	
GPS (585 days)	+0.3 ± 0.3	Kostelecký et al., 1995

Table 5: Velocities with Respect to the Barycenter of the Eurasian Plate [mm/y]
(1) Kostelecký et al., 1995;   (2) Boucher and Altamimi [4]

	dX/dt		dY/dt		dZ/dt	
	(1)	(2)	(1)	(2)	(1)	(2)
Pecný	-1.6		+1.1		+2.6	
RMS	1.7		0.9		0.9	
Graz	-4.0	-2.9	-0.6	+2.3	+0.6	+0.8
RMS	1.5	1.1	0.8	0.9	0.8	1.1
Wettzell	-3.8	-4.0	-1.3	-1.6	-1.8	-1.7
RMS	0.8	0.4	0.6	0.3	0.5	0.4

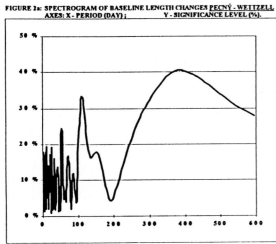

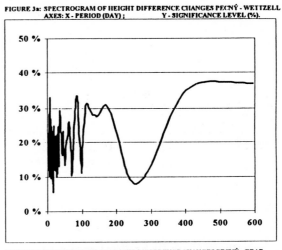

FIGURE 2b: SPECTROGRAM OF BASELINE LENGTH CHANGES <u>PECNÝ - GRAZ</u>
AXES: X - PERIOD (DAY) ;   Y - SIGNIFICANCE LEVEL (%).

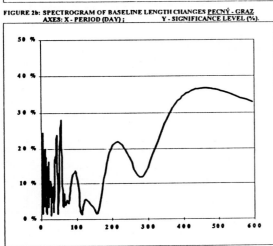

FIGURE 3b: SPECTROGRAM OF HEIGHT DIFFERENCE CHANGES <u>PECNÝ - GRAZ</u>
AXES: X - PERIOD (DAY) ;   Y - SIGNIFICANCE LEVEL (%).

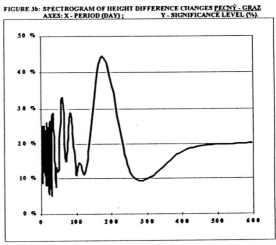

FIGURE 2c: SPECTROGRAM OF BASELINE LENGTH CHANGES <u>WETTZELL - GRAZ</u>
AXES: X - PERIOD (DAY) ;   Y - SIGNIFICANCE LEVEL (%).

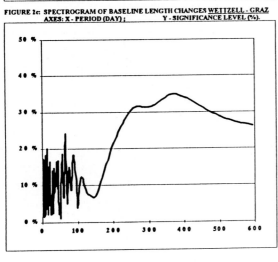

FIGURE 3c: SPECTROGRAM OF HEIGHT DIFFERENCE CHANGES <u>WETTZELL -GRAZ</u>
AXES: X - PERIOD (DAY) ;   Y - SIGNIFICANCE LEVEL (%).

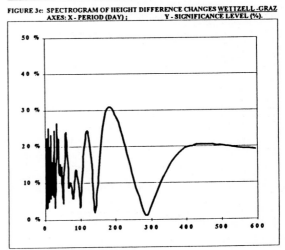

THE WESTERN CANADA DEFORMATION ARRAY: AN UPDATE ON GPS SOLUTIONS AND ERROR ANALYSIS

Xin Chen and Richard B. Langley
Dept. of Geodesy and Geomatics Engineering,
University of New Brunswick, Fredericton, N.B. E3B 5A3, Canada
Herb Dragert
Pacific Geoscience Centre, Geological Survey of Canada,
P.O. Box 6000, Sidney, B.C. V8L 4B2, Canada

INTRODUCTION

The Western Canada Deformation Array (WCDA) is a network of continuous GPS trackers established in the fall of 1992 to monitor crustal deformation across the northern Cascadia subduction zone located in southwestern British Columbia. As of April 1995, the WCDA comprises five tracking sites. The names, site ID's, and the starting dates of operation for the five sites are: Dominion Radio Astrophysical Observatory (Penticton) - DRAO (02/91), Albert Head (Victoria) - ALBH (05/92), Holberg - HOLB (07/92), Williams Lake - WILL (10/93), Ucluelet - UCLU (05/94). The geometric configuration of the WCDA and the basic tectonic setting of the region are illustrated in Fig. 1. Although they both form part of the oceanic side of the Cascadia subduction zone, the Explorer Plate and the Juan de Fuca Plate interact differently with the overlying North American Plate margin. Conventional terrestrial geodetic surveys have been carried out in the past on the southern part of Vancouver Island and have revealed an uplift rate at the outer coast of Vancouver Island of a few mm/yr and a shortening rate of 0.1 ustrain/yr (Dragert et al., 1994).

The Turbo-Rogue receivers (SNR-8000) at the WCDA tracking sites continuously collect data at a 30-sec sampling interval; these data include dual-frequency carrier phases (L1, L2) and pseudoranges (either by P-code or by cross-correlation). The data are automatically downloaded every day shortly after 00:00 UT from the tracking sites to the central facility at the Pacific Geoscience Centre. The software package that has been adopted for the WCDA daily data reduction is CGPS22. It uses the double differencing technique and carries out a batch sequential least squares (LS) adjustment. More details about CGPS22 can be found in Chen (1994).

To date, more than 810 days of data have been reduced using a specific estimation strategy and the precise orbits generated by Natural Resources Canada - NRCan (formerly Energy, Mines and Resources, Canada). In the following sections, the estimation strategy is first described, then the results and error analysis are presented, and finally some conclusions are drawn.

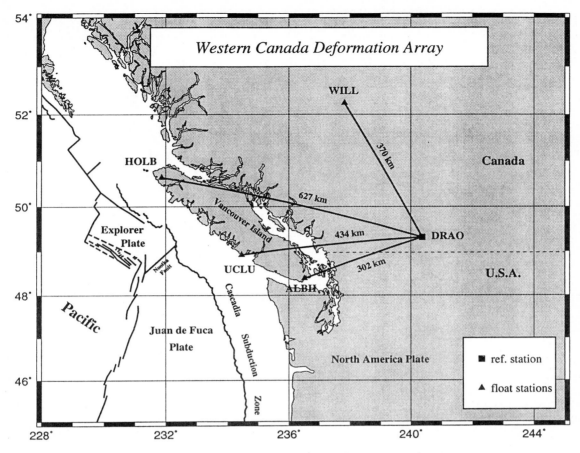

Fig. 1. Western Canada Deformation Array.

ESTIMATION STRATEGY

The estimation strategy used for the WCDA daily data reduction may be briefly summarized as follows:

- DRAO is fixed as the reference station and all other stations are adjusted without any constraints.
- Nominal station coordinates are ITRF standard or well determined from more than two weeks' GPS solutions.
- Satellite orbits are NRCan precise orbits in SP3 format which are held fixed.
- All available satellites above an elevation cutoff angle of 15° are used.
- Analyses are carried out with 120-sec samples for 24 hours of data.
- The solution type is LC phase (ionosphere-free linear combination of L1 and L2) with ambiguities not held fixed.
- Parameters which are estimated are station coordinates, initial phase ambiguities, and local tropospheric scale factors.
- Batch length is 15 minutes.
- The tropospheric parameters are stochastically modeled as coloured noises with a correlation time of 10 hr and a steady state sigma of 5% for all stations.

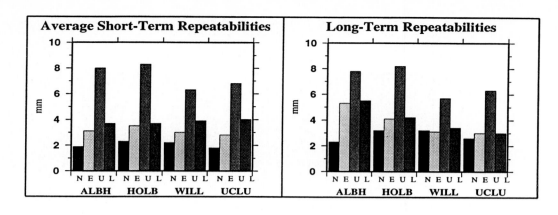

Fig. 2. Precisions of the WCDA Solutions.

- Displacements due to body tides, ocean loading, and pole tide are taken into account.

A detailed explanation of the estimation strategy can be found in Chen (1994).

RESULTS AND ERROR ANALYSIS

To date, more than 810 days of data, spanning 31 months from September 1, 1992 to April 5, 1995, have been reduced. It was found that changes in antenna setup caused significant offsets in the station coordinate solutions. Some of the offsets in the upward (U) components have turned out to be elevation cutoff angle dependent. It is suspected that phase centre variations combined with local multipathing environments are responsible for these offsets; further study is needed to confirm the true cause.

The offsets in the solutions have been estimated using the Least Squares Spectral Analysis (LSSA) technique (Wells et al., 1985) and eliminated from the solution sets. The precisions of the solutions were then evaluated in terms of the short-term and long-term repeatabilities. The short-term repeatability is calculated on a weekly basis as the r.m.s. scatter of the solutions, and the long-term repeatability is the r.m.s. scatter about a best fitting line over the entire solution set. Fig. 2 shows the achieved repeatabilities in the north (N), east (E), upward (U), and the baseline length (L) components of the four WCDA baselines. Due to some non-linear variations caused by unaccounted-for systematic errors, the long-term repeatabilities in most horizontal components and baseline lengths appear larger than the short-term repeatabilities. The long-term repeatabilities in the U components are somewhat smaller than the short-term ones. This indicates that our stochastic model for the tropospheric parameters is in general quite successful, since the vertical component is most affected by tropospheric mismodelling.

Significant variations with approximately annual period have been found in the E and L components of the ALBH-DRAO baseline beginning in late 1993. The amplitude of the annual variation estimated from LSSA is 7.4±0.5 mm and the spectral

Table 1. Estimated linear rates from LSSA (mm/yr).

Baseline	N	E	U	L
ALBH-DRAO	1.8±0.2	8.9±0.4	1.8±0.8	-9.2±0.4
ALBH-DRAO (30°)	N/A	4.6±0.4	N/A	-4.7±0.4
HOLB-DRAO	0.2±0.1	-0.7±0.2	3.3±0.4	0.9±0.2
WILL-DRAO	-0.4±0.3	-2.3±0.4	12.5±1.2	1.7±0.3
UCLU-DRAO	6.6±0.5	4.5±0.6	-12.8±1.3	-5.7±0.6

value, which is a measure of the spectral power density expressed in a range of 0 to 100%, is greater than 40%. The estimated period is 367 days for the E component and 376 days for the L component. Tests have been conducted for ALBH-DRAO data by using different elevation cutoff angles in the analyses. With a 30° cutoff, the estimated amplitude of the annual variation is only 1.8±0.2 mm and the spectral value is reduced to 9%. Since daily-solution residuals indicate strong multipathing effects at ALBH, it is likely that these strong annual variations are an artifact of seasonal modulations of multipathing, although seasonal tilting of the pier cannot be completely ruled out. It should be added that the N components of WILL-DRAO, HOLB-DRAO and UCLU-DRAO also show clear annual variations with amplitudes ranging from 2.1 to 3.4 mm and having spectral values greater than 30%. However, the similarity of the phases of these annual signals from station to station, and their insensitivity to changes in the cutoff angles make local multipathing effects an unlikely cause. It is probable that these signals are artifacts of orbital/EOP errors (Ferland et al., 1995).

In order to eliminate the bias in the regression estimates caused by the anomalous annual variations, we estimated the linear rates by LSSA which can simultaneously estimate and remove specified frequency constituents. These linear trends, together with their formal errors, are listed in Table 1. The rates for the E and L components of the ALBH-DRAO baseline with a 30° cutoff are also listed in order to illustrate the magnitude of the possible bias due to multipathing.

Both ALBH and UCLU show significant eastward to northeastward displacements with respect to DRAO; this is consistent with previous geodetic results. For the E and L components of the ALBH-DRAO baseline, the spread in the trends derived from the nominal 15° cutoff solutions and the 30° cutoff solutions indicate that biases might still be present in these estimates. The true rate may lie somewhere between these extremes. HOLB does not show the same trends in long-term motion with respect to DRAO as the sites on southern Vancouver Island. Although the tabulated rates are reduced from those based on the first 17 months of data (Dragert and Hyndman, 1995), the results for HOLB continue to be consistent with the fact that the nature of the Cascadia subduction zone changes north of central Vancouver Island.

Estimates of the vertical trends need to be treated with great caution due to their comparatively lower precisions and possible biases introduced by antenna changes. The large differential vertical rates derived for WILL-DRAO and UCLU-DRAO contradict theoretical predictions from margin-deformation and glacial-rebound models. WILL-DRAO is known to have suffered two offsets in the vertical component due to antenna setup changes; the magnitude of these offsets proved to be sensitive to the elevation cutoff angle used in the analyses of the data. Although the UCLU-

DRAO data do not exhibit any clear step-functions, the site logs of UCLU are being re-examined to establish times of possible changes in the multipathing environment.

CONCLUSIONS

Based on the WCDA data analysis performed to date, our findings and conclusions can be summarized as follows:

- 2 to 6 mm horizontal precision and 5 to 9 mm vertical precision in terms of long-term repeatability have been achieved in the WCDA data analysis.
- ALBH-DRAO and UCLU-DRAO have shown significant horizontal shortening rates of at least 5 mm/yr; these rates are generally consistent with estimates from previous deformation data.
- HOLB-DRAO has not shown the same horizontal trend as ALBH-DRAO and UCLU-DRAO; this is consistent with most recent tectonic models proposed for the Explorer Plate/North America Plate interactions.
- The large annual variation in the E and L components of the ALBH-DRAO baseline is most likely caused by annual modulations of multipath effects at ALBH.
- The N components of the HOLB-DRAO, WILL-DRAO, and UCLU-DRAO baselines have shown annual variations with 2 to 3 mm amplitudes and similar phases. The NRCan orbit/EOP errors are suspected as the source of these variations.
- Changes in antenna setup have caused unexpected offsets in the solutions. Some of the offsets in the vertical components are dependent on the elevation-cutoff angle and therefore likely due to antenna phase centre variations combined with other effects.

REFERENCES

Chen, X. (1994). Analyses of Continuous GPS Data from the Western Canada Deformation Array, *Proceedings of ION GPS-94*, Salt Lake City, Utah, Sept. 20-23, 1994, pp. 1339-1348.

Dragert, H., R. D. Hyndman, G. C. Rogers and K. Wang (1994). Current Deformation and the Width of the Seismogenic Zone of the Northern Cascadia Subduction Thrust, *J. Geophys. Res.,* Vol. 99 (B1), pp. 653-668.

Dragert, H. and R.D. Hyndman (1995). Continuous GPS Monitoring of Elastic Strain in the Northern Cascadia Subduction Zone, Geophys. Res. Let., Vol. 22 (No. 7), pp. 755-758.

Ferland, R., J. Kouba, P. Tetreault, J. Popelar, and H. Dragert (1995). Variation in EOP and Station Coordinate Solutions for the Canadian Active Control System (CACS), Proceedings of Symposium G1, GPS Trends in Precise Terrestrial, Airborne and Spaceborne Applications, IUGG XXI General Assembly, Boulder, Colorado, Sept. 2-14, 1995.

Wells, D. E., P. Vanicek and S. Pagiatakis (1985). Least Squares Spectral Analysis Revisited, *Technical Report No. 84,* Dept. of Surveying Engineering, Univ. of New Brunswick, Fredericton, N.B.

THE DGPS SERVICE FOR THE FRG
- CONCEPT AND STATUS -

Peter Hankemeier
Freie und Hansestadt Hamburg Baubehörde Vermessungsamt
20302 Hamburg Postfach 300580 Germany

INTRODUCTION

The Federal Republic of Germany is a federation of states and the 16 federal states or *Länder* are responsible for official surveying practice. The duties of the surveying authorities of the individual federal states are defined in each state by legislation. In all statutes a basic duty of the surveying authority is the creation and provision of a uniform reference system in order to be able to satisfy the demands of administration, business and science in a manner which corresponds to contemporary technical standards.

For the purposes of coordination between the various states, implementation of higher-level projects and also the creation and maintenance of uniform surveying principles, the "Arbeitsgemeinschaft der Vermessungsverwaltungen der Länder der Bundesrepublik Deutschland" (AdV) has been set up by the state governments. As a permanent working party it comes under the Conference of State Ministers of the Interior.

The surveying authorities of the European Union recognized and decided early on to create with the help of GPS an adequate and uniform positioning network for Europe. It was decided to introduce the "EUROPEAN TERRESTRIAL REFERENCE SYSTEM 1989" (ETRS 89) as a three-dimensional geocentric reference system. In a joint campaign of the European countries the first phase of this was implemented in 1989 as the EUREF network (the A network). At the instigation of the AdV this network was further extended by a German reference network being set up in 1991 (DREF 91) (B network). At the current time a further level of coverage density is being implemented by virtually all of the German states at state level (C network). These hierarchically structured networks will provide for the first time in Europe a uniform national reference system of high precision and of the classical type (points marking, coordinates) for all applications.

Following the establishment of a working party, "GPS Referenzstationen" (GPS reference stations) and presentation of the first results of work, the AdV decided to combine current DGPS activities and to set up in the territory of the Federal Republic of Germany a uniform service of a similar type called the

"DGPS service of member authorities of the AdV"

as a team project of the AdV. It should help the states to fulfil their own governmental obligations in a more efficient, comprehensive and economic manner than heretofore. Other groups of users are to be introduced to this service so that a uniform reference system for many applications can be provided at high efficiency and in a technically adequate form.

DESCRIPTION OF THE SERVICE AREAS

The concept provides for a permanently operated, full-coverage and multifunctional DGPS service as a basic infrastructural supply which at the same time has high availability. The service should in addition permit an unlimited number of users at the same time where possible. The four service areas of this service are shown in Table 1.

The individual service areas are distinguished by their availability (time), the transmission medium, accuracy, the number of possible simultaneous users and the user interface (Table 2). Each service area will however be served by just one multiple reference station, wherever possible on a regional basis. As distance criteria for the stations which are distributed on a cellular basis the parameters of topography and the transmission are of decisive importance alongside the service areas to be covered. Positive results have been obtained in experiments with distances of between 40 and 70 km.

REAL -TIME POSITIONING SERVICE (EPS)

This sector of the DGPS service is intended for users which have either as yet not carried out positioning activities or have used other and considerably more expensive methods. Users are to be able to determine their positions in real time at low cost. The geodetic reference system thus required will be made available to them via DGPS.

This service can typically be used for the following purposes:

Navigation in traffic (public mass transit systems, taxis)

Safety applications (police, fire department, catastrophe response services)

Fleet management, traffic control systems (public mass transit systems, truck fleets)

Hydrography

Agriculture and forestry

Environmental protection

Data collection for information systems

Table 1 Service areas of the DGPS service

Service	Name
1	EPS (real-time positioning service)
2	HEPS (high-precision real-time positioning service)
3	GPPS (geodetic precision positioning service)
4	GHPS (geodetic high-precision positioning service)

Table 2 Differentiation criteria for the various service areas of the DGPS service

Service	Result	Medium	Accuracy	users	user's port
EPS	Real time	Radio LW,MW,UHF	1 -3 m	∞	RTCM 2.0
HEPS	Real time	2 m band	1 - 5 cm	∞	RTCM 2.1
GPPS	Postprocessing 15 minutes	Telephone	1 cm	n<<∞	RINEX
GHPS	Postprocessing		< 1 cm	∞	RINEX

For the transmission of the correction data (phase-smoothed pseudo-range corrections and pseudo-range changes) procedures are available which are basically distinguished by the transmission method used:

Use of a long-wave transmitter

Use of AMDS (amplitude-modulated data system) in the medium-wave band

Use of RDS (radio data system) in the UHF band of the radio broadcasting authorities

In the final analysis the user himself must choose which of the various transmission media available on the market he wants.

DGPS employing low-frequency transmission

Following the example of time-signal broadcasting, the Institute for Applied Geodesics in cooperation with German Telekom plans to transmit DGPS corrections by employing a low-frequency transmitter. Advantages are ability to cover all Germany with an LF signal with a low interference level. For all GPS satellites which are in view of the permanent GPS station at Mainflingen near Frankfurt am Main a full set of range corrections will be provided within 3 s. The data format used will be RTCM version 2.0. A low-cost LF receiver will send the corrections via an RS232 output directly to a GPS receiver. To check the transmitted DGPS corrections monitor stations will be set up in the area.

Tests have been carried out to demonstrate the accuracy of the DGPS system. The coordinates of first-order network stations all over Germany, which were determined by GPS to the highest precision during the DREF campaign in 1991, have been compared with coordinates derived from observations in DGPS mode also employing low-cost GPS receivers and the transmitted range corrections. A slight range dependancy of the precision was observed but nevertheless the overall results are better than 3 m.

Provision of correction data via AMDS in the medium-wave band

Transmission of RTCM correction data (pseudo-range corrections) in the AMDS channel of the Deutschlandfunk station (756 kHz) should make positioning possible from approximately 1 to 3 m in real time.

AMDS is the data channel in the AM frequency band (medium wave, long wave) which can be used by various services in a similar way to RDS in the UHF band.

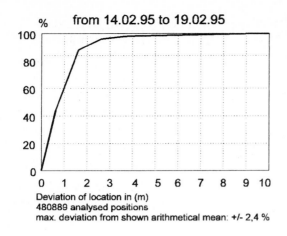

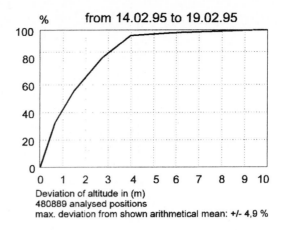

Deviation of location in (m)
480889 analysed positions
max. deviation from shown arithmetical mean: +/- 2,4 %

Deviation of altitude in (m)
480889 analysed positions
max. deviation from shown arithmetical mean: +/- 4,9 %

Diagrams 3 and 4 positioning data of a monitor station

A GPS reference station was set up at the Cremlingen medium-wave transmitter (near Braunschweig) of Deutschlandfunk. The computer generates the RTCM V2.0 correction data from the data coming from the GPS receiver and from the position of the GPS antenna previously calculated in DREF, the German coverage level of the European reference system EUREF.

The RTCM V2.0 correction data are sent by the computer out via a serial interface to a coder; they are coded, integrated into the AMDS structure and then broadcast on the medium-wave transmitter. The corrected data are thus available to interested users, provided they have the corresponding AMDS decoder.

At the present time the entire AMDS canal width of approx. 137 bps is available, which means that an RTCM V2.0 updating rate of approx. 5 seconds is possible for a maximum of 8 received satellites.

Previous studies have shown that the absolute positional accuracy demanded of 1 - 3 m is normally achieved. If however the RTCM correction data are not received clean and the GPS receiver cannot utilize one or more data blocks, this will result in larger deviations from the target position. In such a case the correction data last received will be used, which have themselves already "aged". To prevent excessive ageing, efforts should be made to achieve the highest updating rate possible.

Diagrams 3 and 4 show the positioning data for a monitor station 80 km away.

Provision of correction data via RDS in the UHF band

The existing UHF transmitter network is particularly suited to transmission of DGPS correction data as UHF reception is possible in more than 95% of the German territory. The state surveying authority in conjunction with the German public broadcasting stations (ARD) has developed the RASANT technique (radio-aided satellite navigation technique) for the EPS service. Here RDS is used as the medium of transmission for the data in the pan-European standardization. RDS is supported by virtually all UHF transmitters and can be received with normal radios.

As DGPS correction data, pseudo-range corrections are transmitted in the RTCM V2.0 format. The transmission capacity of RDS is 1200 bps. This means that its usability for non-radio applications is limited. On the other hand, however, accuracy of positioning declines as the correction data become older. For this reason in the RASANT method the RTCM data format is compressed for the duration of transmission and then restored back to the standardized form. With the aid of this interim compression it is possible to reduce the complete data block from 680 bits for 9 satellites, for example, down to 306 bits. In this way in one RDS group per second (34 bits payload capacity) the entire data block can be transmitted within 9 seconds. In addition, with RASANT compression the RTCM data are restructured with the aim of transmitting one satellite's correction data complete within a single group. This makes each individual RDS group independently utilizable by itself.

As receiving device the user will need in addition to an RDS receiver a device for decoding and decompressing the RASANT RDS signal. Prototypes of such user equipment exist from various manufacturers. Tests have shown that with differential GPS with EPS RASANT positions can basically be determined with a precision of better than 3 m.

In Germany GPS correction data are already being broadcast via RDS by a number of radio stations for test purposes. Efforts are being made by the AdV and the ARD to extend this to cover the entire country. At the same time it is intended to publish the RASANT format and to have it standardized by the Comité Europeén de Normalisation Électronique (CENELEC). At the request of the European Broadcasting Union (EBU) an application to this effect is being prepared by the state surveying authority in order that the EPS RASANT method can be used in a standardized manner throughout Europe.

HIGH-PRECISION REAL-TIME POSITIONING SERVICE (HEPS)

With a precision being aimed at of less than a few centimetres, this service can be used in the following areas:

Surveying management and cadastral survey management

Land redistribution, land appraisal

Acquisition for information systems (ATKIS)

Aerial surveying

Hydrography, maritime surveying

Engineering surveying

Utility companies, power-line owners

Agriculture and forestry

Aviation

Differences between RTCM 2.1 and RTCM_AdV	
RTCM 2.1	RTCM_AdV
1 Pseudo-range corrections 2 Delta pseudo-range corrections 3 Reference station coordinates 16 Special messages 20 RTK pseudo-range corrections 21 RTK carrier-phase corrections	1 Pseudo-range corrections 2 Delta pseudo-range corrections 3 Reference station coordinates 16 Special messages 59 Proprietary messages 20 RTK pseudo-range corrections 21 RTK carrier phase corrections
Table 5 Comparison of the RTCM formats	- Record types fully RTCM 2.0/2.1 compatible - Recorded and compressed RTK corrections - Total message length reduced to 50% - Open for future extensions (RTCM LAN/WAN)

These are tasks which before now were performed using very different technologies. DGPS cannot replace these conventional technologies entirely but will be able to perform many of the tasks much more economically.

For this service the correction data of the RTCM V2.0 have been supplemented with carrier phase measurements, compressed in its own special RTCM_AdV format and provided to the user via independently operated transmitters in the 2 m band. For the user the compressed RTCM_AdV format is converted into the RTCM 2.1 format. In Table **5** which follows the differences between the RTCM 2.1 and RTCM_AdV formats are presented.

This independent format has been formulated in response to the following general requirements:

1. Update Rate

Present-day receivers have < 1 sec measurement rate

For high accuracy, corrections should not be extrapolated

➲ Reference station should transmit one complete correction set per second

Format message length	Update rate with		
	2400 bps	4800 bps	9600 bps
RTCM 2.1 > 4800 bits	< 0.5 Hz	< 1 Hz	< 2 Hz
RTCM_AdV < 2400 bits	> 1 Hz	> 2 Hz	> 4 Hz

2. Age of Data

Corrections must be completely available at mobile station to compute PDGPS solution

➲ Use a message length as short as possible in order to minimize the age of the data of the reference signals

Format message length	Age of data		
	2400 bps	4800 bps	9600 bps
RTCM 2.1 > 4800 bits	> 2 sec	> 1 sec	> 0.5 sec
RTCM_AdV < 2400 bits	< 1 sec	< 0.5 sec	< 0.25 sec

The RTCM V2.1 format is used as an interface for the user. In the transmission link the independent compressed RTCM_AdV format is used internally with the HEPS method. This makes it possible to manage with the low transmission speed of 2400 bps. The user will thus be provided with a complete data block (12 satellites) each second.

GEODETIC PRECISION POSITIONING SERVICE (GPPS)

Unlike with the services described so far, the precision strived for here should be provided not in real time but 'near on-line'. This procedure is based on the fact that the completed implementation of the DGPS service has resulted in a relatively close-meshed network of reference stations.

The procedure exploits the fact that at all reference stations the data from all visible satellites are received and held in a standardized format (RINEX). For the 'near on-line' positioning a computing site collects the observation data from multiple reference stations (cell) and carries out preprocessing. The user communicates with the computing site via mobile radio telephone and thus obtains the data for a particular time period. This procedure is currently being worked on as part of a project which is not as yet complete. The project is mainly concerned with distance investigations taking the various fault influences into account. The results flow into a 'quick positioning' software program which is currently being developed and tested by the state surveying authority.

Typical users of the GPPS service are:

Real estate surveys of the surveying and cadastral surveying administrations

Engineering surveying

Fundamental surveying

This user potential represents the core of surveying practice.

GEODETIC HIGH PRECISION POSITIONING SERVICE (GHPS)

This service area represents with the DGPS reference stations a high-precision active reference network such as is required, for example, for geodynamic applications and for monitoring tasks (coastal protection, sea level monitoring). The millimetre accuracy to be provided with this service can only be achieved during post-processing using precise IGS ephemerides.

User applications of this service will include:

Special duties within fundamental surveying

Reference systems of the state surveying authority

Scientific and geodynamic studies

Monitoring tasks (coastal protection, sea level monitoring)

THE REFERENCE-STATIONS: SPECIFICATION AND STANDARDIZATION

With a nationwide service it is not possible to dispense with drawing up technical specifications for the permanently functioning multiple reference stations required. The specifications described here indicate the minimum requirements for a permanently operating station within a homogeneous DGPS service

Reference point. The antenna must be set up with such a degree of stability (forced centering with safeguards) that local changes of position are excluded. Locking marks shall be checked regularly to check stability. External influences (interference from external signals, reflections) must be excluded as far as possible. The location of the antenna must be included in the German reference network (DREF 91).

GPS receivers. Geodetic high-grade dual-frequency receivers must be used which have the following characteristics:

L1 pseudo-distance measurements, uncertainty < 0.2 metres

L2 carrier phase measurements, uncertainty < 0.1 cm

L1 pseudo-distance measurements (from encoded P code), uncertainty < 0.2 metres

L2 carrier phase measurements with full wavelength, uncertainty < 0.1 cm

9 or better 12 channels so as to be able to observe all visible satellites simultaneously

Station software. The reference station software used must include the following features: Checking the functional reliability of the GPS receiver

Data registration

Data quality check

Data preparation of the correction values

Data conversion

Supplying the data in the defined formats to various transmission media

Data archiving

Provision of the internal trans-mission formats (RTCM_AdV, RTCM_RAS)

Output formats for the user: RTCM 2.0, RTCM 2.1 and RINEX

Station computer. The processor must have the power to satisfy high requirements, since all processes run time-critically. For this reason the operating system must support multi-tasking. On all of the stations which have been set up so far the OS/2 Warp operating system has been used.

Current level of expansion of the reference station network

There are currently around 35 reference stations operating in Germany of which only a few are at a multifunctional stage of expansion.

Figure: Survey of reference station currently implemented

A brief descriton of the Hamburg station will now be given as an example of a multi-functional reference station.

Work has been going on for two years setting up this surveying authority reference station and today it satisfies the major demands which the most varied user groups make on a permanently operating multifunctional DGPS station in a conurbation region and at its current level of implementation covers the service areas for the DGPS service.

With the hardware and software currently installed it is possible, with the inclusion of the self-operated radio station in the 2 m band at the Heinrich Hertz Tower (200 m above sea level) to supply an unrestricted number of users with DGPS correction data for position improvement purposes within the Hamburg metropolitan area (approx. diameter of 70 km) to a precision of a few cm and in real time.

For static applications the raw data are saved after conversion into the RINEX format and supplied on request via permanent network connections. The remote radio station and its data broadcasting is permanently monitored by a monitoring station. Check data are temporarily saved for quality control purposes. In addition to broadcasting correction data via the radio station, correction data are broadcast in the RTCM V 2.0 format via the NDR's radio transmitters in the UHF / RDS band. This station has performed convincingly and is being used increasingly for the most varied applications.

CONCLUSIONS

Efforts are being made in the EPS-RASANT area to extend the current test regions to cover all UHF transmitter chains in Germany.

The HEPS and GPPS service areas are to be built up in accordance with demand but will ultimately cover the complete country. The concept for the HEPS area assumes that 5 frequencies in the 2 m band are enough to provide complete coverage. The member administrations of the AdV are currently engaged in setting up all of the organisational and administrative conditions so as to be able to set up a DGPS service for Germany which will satisfy user requirements and be up-to-date technically.

REFERENCES

BICHTEMANN, G and HANKEMEIER, P (1993). DGPS-Dienst in Norddeutschland (SPN 1/93)

DITTRICH, E. KÜHMSTEDT, LECHNER, W. (1994): Experiments with Real-Time Differential GPS Using a Low-Frequency Transmitter in Mainflingen (Germany) - Results and Conclusions. Proceedings of the 3rd International Conference on Land Vehicle Navigation. Dresden.

FRÖHLICH (1994).The High-Precision Permanent Positioning Service (HPPS) in North Germany. Tests and First Results. Proceedings of the Third International Conference on Differential Satellite Navigation Systems. London

HANKEMEIER, P. (1995). Neue Wege zur Bereitstellung geodätischer Bezugssysteme. (SPN4/95)

GPS Dienste der Vermessungsverwaltungen der Bundesrepublik Deutschland (SPN 4/95)

Lindstrot.W. und Plöger W. (1992). Möglichkeiten eines Echtzeit DGPS-Dienstes über Rundfunk. (SPN.S 123)

Interim report of the AdV working party "GPS-Referenzstationen" (unpublished)

STATUS OF THE BRAZILIAN NETWORK FOR CONTINUOUS MONITORING OF GPS (RBMC)

Luiz Paulo Souto Fortes
IBGE/Departamento de Geodésia
Av. Brasil 15671, Parada de Lucas
Rio de Janeiro, RJ, Brazil, 21241-051
fax: 55-21-391 7070; e-mail: fortes@omega.lncc.br

ABSTRACT

The current status of the Brazilian Network for Continuous Monitoring of GPS (RBMC) is presented. This network is going to be formed by seven permanent GPS stations in Brazil, established by IBGE in cooperation with other Brazilian institutions. Besides, two IGS stations (Fortaleza and Brasília) already functioning will be integrated into that structure establishing a set of nine stations in total. This structure is available to the user community that needs GPS L1 & L2 carrier phase and code reference data for post-processing positioning applications as well as for real time applications. It is planned to contribute to IGS network releasing the data of the seven remaining stations. The RBMC stations belong also to the SIRGAS network.

INTRODUCTION

IBGE is carrying out the Brazilian network for Continuous Monitoring of GPS project since 1991, when the first propositions were presented by Fortes and Godoy (1991). Later, the project experienced the modifications described by Fortes (1993), which come up with the present configuration. In this configuration, the network will be formed by seven permanent GPS stations, besides the IGS ones which are already running.

RBMC CONFIGURATION

The stations of the RBMC's final configuration are shown in Fig. 1. It should be emphasized that new stations can be added, in order to satisfy an application which needs a greater density in a specific region, as the Amazon one. For this purpose, extra financial resources would be necessary.

The stations were already selected and materialized, located always at host institutions of the government. These institutions and the corresponding stations coordinates are listed on Table 1. It should be pointed out the high level of collaboration obtained from each institution, which definitively contributed to the present status of the network.

Each station was materialized using a centering forced device, specially designed by IBGE for the RBMC. The standards and specifications published by IBGE (1992) were taken into account in the station selection, as: no obstructions 10 degrees above the horizon; no multipath surfaces close to the station; no electromagnetic sources in the range of the GPS signals (1.2 to 1.6 GHz). Besides of these conditions, the following requirements were satisfied: a very stable structure of the building where the centering forced device could be set up; a restricted public access area; availability of an office close to the station with continuous power supply and telephone facility to be used for data communication and remote control of the station; finally, permanent security.

The IGS stations (Brasilia and Fortaleza) are already functioning. Brasilia station was installed at IBGE regional office, in the Ecological reserve of Roncador, in cooperation with NASA/JPL. Fortaleza station was installed at Brazilian Space Agency (INPE) office, in Eusébio, by CRAAE (Centro de Radio Astronomia e Aplicações Espaciais), in cooperation with NOAA and IBGE. Turbo Rogue receivers are being used at both stations, which are tracking the GPS signals at 30^S sampling rate. These stations were already integrated into RBMC and IBGE is responsible for releasing the observed data in Brazil.

The stations of Presidente Prudente and Curitiba are running experimentally with old GPS receivers from Polytechnic School of São Paulo University (EPUSP) and Federal University of Parana (UFPR), respectively. These TRIMBLE 4000 receivers will be replaced by the definitive ones according to the project schedule.

The stations will be established using basically the equipment described by Fortes (1993). Meteorological registers utilization will be postponed. A current discussion at the IGS community argues the necessity of using those data for tropospheric refraction correction (Noll, 1995).

Table 1. RBMC stations.

Station	Latitude	Longitude	Host institution
Bom Jesus da Lapa	-13.25	-43.42	Agência da Capitania dos Portos
Brasília	-15.95	-47.88	IBGE/DIEAC
Cuiabá	-15.55	-56.07	INPE
Curitiba	-25.45	-49.23	UFPR
Fortaleza	-03.88	-38.43	INPE
Imperatriz	-05.50	-47.47	CEFET/UNEDI
Manaus	-03.12	-60.06	DSG/4ª DL
Presidente Prudente	-22.12	-51.41	UNESP
Viçosa	-20.75	-42.90	UFV

TIME SCHEDULE

Table 2 shows the time schedule of the RBMC project.

Table 2. Time schedule of the RBMC project.

Step	Period
Station specification	July → September/95
Equipment purchase	August → September/95
Building preparation	September → November/95
Equipment installation	October → December/95
Pilot phase	October/95 → March/96
Operational phase	starting in April/96

PERSPECTIVES

The RBMC structure was designed for users that need carrier phase and pseudorange observations in L1 and L2, at reference stations, for relative positioning. Besides fulfilling the post-processing requirements, each station may be used as a reference station for real time differential GPS (DGPS) applications, since a communication link be installed.

It is intended to release the data using Internet facilities in near future.

As all RBMC stations belong to the SIRGAS reference network (SIRGAS, 1994a and 1994b), every new GPS positioning in Brazil will be automatically referred to the new SIRGAS reference system

REFERENCES

Fortes, L.P.S. and Godoy, R.A.Z (1991). Rede Brasileira de Monitoramento Contínuo do Sistema de Posicionamento Global - GPS. *Coletânea de Trabalhos Técnicos do XV Congresso Brasileiro de Cartografia, vol. 3, pp 677-682*, São Paulo.

Fortes, L.P.S. (1993). Rede Brasileira de Monitoramento Contínuo do Sistema GPS (RBMC) - Estágio Atual. *Expanded abstracts of the 3rd International Congress of the Brazilian Geophysical Society, vol. 1, pp 637-642*, Rio de Janeiro.

IBGE (1992). Especificações e Normas Gerais para Levantamentos GPS (Preliminares). Rio de Janeiro.

NOLL, C. (1995). Meteorological Data Available from the CDDIS. *IGS Electronic Mail Message Number 0931*.

SIRGAS (1994a). Newsletter #1. IBGE, Rio de Janeiro.

SIRGAS (1994b). Newsletter #2. IBGE, Rio de Janeiro.

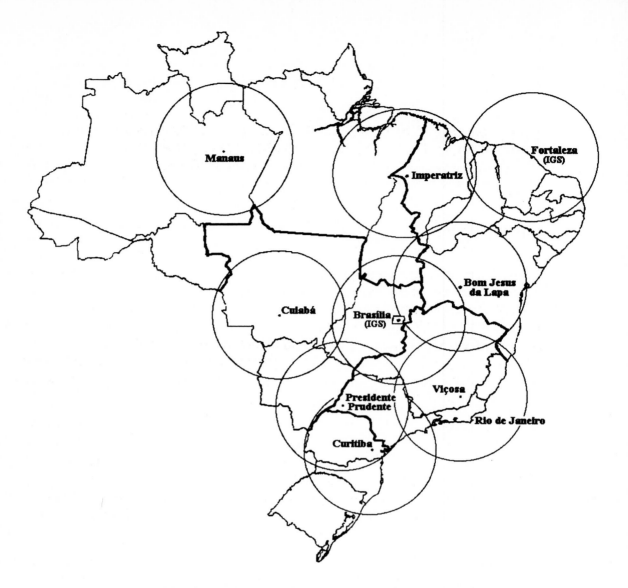

Fig.1. RBMC stations (each circle has a 500 Km radius).

WEEKLY-ARC APPROACH TO IMPROVE THE RESOLUTION OF THE LONG-TERM CRUSTAL MOVEMENTS IN JAPANESE GPS FIXED-POINT NETWORK

Seiichi Shimada
Nat. Res. Inst. for Earth Sc. and Disaster Prevention (NIED)
3-1 Tennodai, Tsukuba, Ibaraki 305 JAPAN

INTRODUCTION

For the studies of the geodynamics, International GPS Service for Geodynamics (IGS) has been producing the precise ephemerides since 1992. However the distribution of the IGS primary fiducial sites is very sparse in and around the Eastern Asia, it is afraid that the precision of the ephemeris is degraded over this area than the North American and Western European area. As a matter of fact, the repeatability of the observations of the Japanese GPS fixed-point network remains 1-2 cm level for the period of 1993 and 1994 (Shimada, 1994). In the paper we examine the Japanese local tracking data incorporated with the IGS global network solutions and the possibility of the improvement of the repeatability by the longer arc approach.

NIED GPS FIXED-POINT NETWORK

NIED has been introduced the GPS fixed-point network in the Kanto-Tokai district, Central Japan, to observe the crustal movements since 1988, and exchanged the receiver to P-code capable Ashtech receiver in fifteen sites in 1993. Fig.1 shows the distributions of the NIED GPS fixed-point network sites with the local IGS sites, Usuda of JPL and Tsukuba of GSI. In the network 30 second sampling continuous tracking has been carried out (Shimada et al., 1989, Shimada and Bock, 1992). In the analysis we adopt the IGS precise ephemeris and fix the orbit parameters, and obtained the baseline repeatability of the NIED network sites with 1-2 cm level for the horizontal component (Shimada, 1994), several times worse than the result obtained in California and Western Europe. IGS precise ephemerides are determined with 13 primary fiducial sites tightly constrained. From the view point of the Eastern Asia, the IGS primary fiducial sites are sparsely distributed in and around the region. Fig.2 shows the distributions of the IGS primary and secondary fiducial sites centering the Central Japan. We suspect the precision of the IGS precise ephemeris over the Eastern Asia is degraded, and examine the

Japanese local tracking data, incorporated with the IGS global tracking data, comparing the shorter and longer arc solutions.

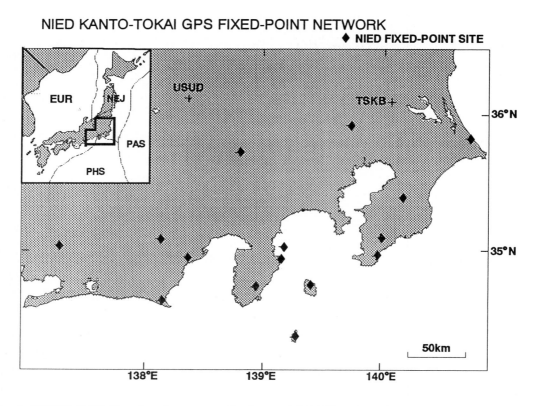

Fig. 1. NIED GPS fixed-point network sites with IGS regional sites(+).

TEN DAYS EXPERIMENT FOR JANUARY 1994 DATA

We examine the global and Japanese local tracking data for ten days in January 10 and 19, 1994, and compare the daily solutions with the ten-day-long solutions of the local site coordinates both incorporated with global solutions.   For the global data, we use all IGS primary fiducial sites and TAIW site in Taiwan, and for the local data, two IGS sites (Usuda and Tsukuba) and 16 NIED network sites in Central Japan.

We incorporate the local data solutions with the global analysis, tightly (1 mm for horizontal and 2 mm for vertical to the ITRF92 coordinates by Boucher et al. (1993)) constraining the coordinates of the IGS primary fiducial sites, obtaining the orbital parameters and the coordinates of all sites.   Fig.3 compares the repeatability of the local site coordinates of the daily solutions with the internal uncertainty of the coordinates of the ten-day-long solutions, indicating the improvement by the ten-day-long solution even though accounting the statistical improvement, square root ten, suggesting the orbital improvement for the longer solution.

Fig.2. IGS fiducial sites centering the Central Japan. ■ denote the primary fiducial, and ● the secondary fiducial sites, respectively.

SIXTEEN WEEKS EXPERIMENT FOR JANUARY AND APRIL 1995 DATA

Therefore we examine the longer data to apply the long arc approach. We use the NIED local and global data for sixteen weeks during January 1 and April 22, 1995, and adopt seven day arc for the long arc, and compare the daily solutions with the weekly solutions of the local site coordinates both incorporated with global solutions. For the global solutions in this experiment, we use that calculated by the Scripps Institution of Oceanography. For the local data, we use seven IGS sites (Kokee Park, Fairbanks, Taiwan, Shanghai, Guam, Usuda and Tsukuba) and 17 NIED network sites in the Central Japan. In the analysis we tightly constrain the coordinates of the IGS primary fiducial sites as same as the former experiment, although the adopted coordinates are ITRF93 by Boucher et al. (1994).

Fig.4 shows the baseline repeatability of the NIED network for the period. Weekly solutions indicate smaller repeatability than the daily solutions for the most of the baseline, although not smaller than the statistic improvement, square root seven. Especially N-S component shows essentially no improvement for the longer period analysis, although the daily solutions prove small scatters. One of the interpretation of the unimprovement of the repeatability is that the other factors than orbit parameters determine the accuracy of the solutions, and ephemeris is well-determined by the solution incorporated with global solutions. As a matter of fact, 1.7 mm + 5 ppb, 3.7 + 9 ppb, and 11.1 mm - 1 ppb repeatability for the N-S, E-W, and U-D components of the daily solutions is absolutely well determined solutions in the state of art. The adopted reference frame, ITRF93, may resolve the inconsistency contaminated in the ITRF92 co-

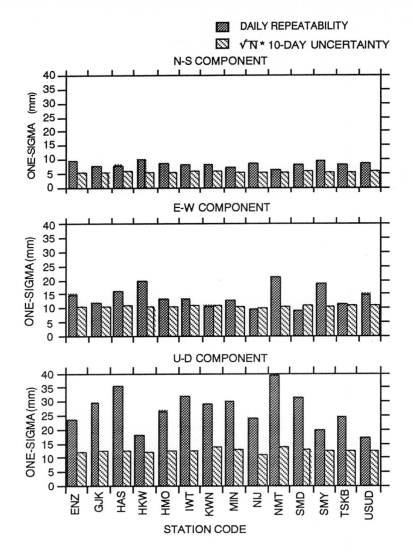

Fig. 3. Daily repeatability and the ten-day solution uncertainty times $\sqrt{N}$.

ordinate system adopted in the former experiment.

We will re-evaluate the analyzing process to farther improve the weekly solutions, including the analyzing procedure and parameters in the Kalman filtering technique we adopted in the local solutions incorporated with the global analysis, developed by MIT and remaining many option procedures in the program.

Acknowledgment.   The author thanks M.Kikuchi, the Association for the Development of Earthquake Prediction, for the routine NIED site maintenance and data management.

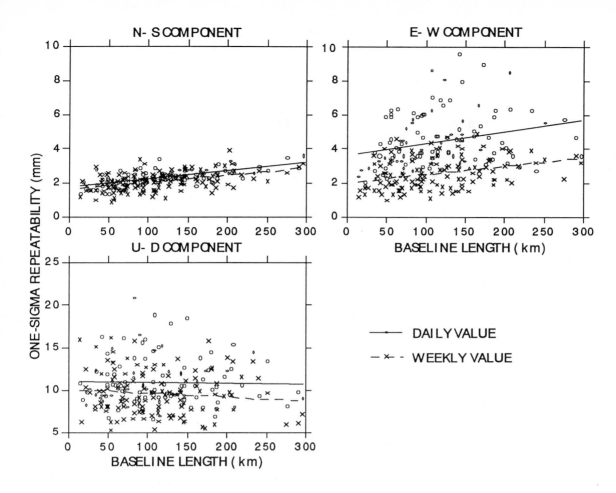

Fig.4. Baseline repeatability of NIED network for 16 weeks during January 1 and April 22, 1995.

REFERENCES

Boucher, C., Altamimi, Z. and Duhem, L. (1993).   ITRF92 and its associated velocity field, IERS Technical Note 15, Observatoire de Paris, Paris.

Boucher, C., Altamimi, Z. and Duhem, L. (1994). Results and analysis of the ITRF93, IERS Technical Note 18, Observatoire de Paris, Paris.

Shimada, S. (1994). Recent crustal movements observed by the NIED Kanto-Tokai GPS fixed-point network—methodology and results, in Proc. 9th Joint Meet. UJNR Panel Earthquake Prediction Tech., Geographical Survey Institute, Japan, 483-505.

Shimada, S., Sekiguchi, S., Eguchi, T., Okada, Y. and Fujinawa, Y. (1989). Preliminary results of the observation by fixed-point GPS simultaneous baseline determination network in Kanto-Tokai district, *J. Geod. Soc. Japan*, **35**, 85-95.

Shimada, S. and Bock, Y. (1992). Crustal deformation measurements in central Japan determined by a Global Positioning System fixed-point network, *J.Geophys. Res.*, **97**, 12,437-12,455.

UNIFICATION OF REGIONAL VERTICAL DATUMS USING GPS

Ming Pan and Lars E. Sjöberg
Department of Geodesy and Photogrammetry
Royal Institute of Technology , S-100 44 Stockholm

INTRODUCTION

The Baltic Sea Level (BSL) GPS project, IAG SSG 5.147, and its long-term scientific goals were described in details by Kakkuri et al. (1990). One of the goals is to unify the vertical datums of the countries surrounding the Baltic Sea. Two BSL GPS campaigns were performed in order to achieve the goals: a two week campaign in October 1990 (the first campaign) and a one week campaign in June 1993 (the second campaign).

Preliminary results of the first campaign were given e.g. by Pan and Sjöberg (1993), and the final results of the first campaign were published by the Finnish Geodetic Institute (Kakkuri, 1994). The results were calculated by four processing groups (Finland, Germany, Poland, and Sweden) using the same software (Bernese ver. 3.3) and the same data. These results allowed the comparison of individual units. But the results were not good as expected (ibid see in detail). In such a case a Second Campaign is necessary and important. Final results of the Second Campaign are discussed in this paper. See also Kakkuri (1995).

THE OBSERVATIONS

All the countries around the Baltic Sea (Denmark, Estonia, Finland, Germany, Latvia, Lithuania, Poland, Russia, and Sweden) participated in the Second Campaign at 35 tide gauges and 10 VLBI and SLR fiducial stations (see Fig.1) with baseline lengths ranging from about 100 km to 2000 km. The Second Campaign was run for five days, namely 7-12 of June, 1993 and measurements were recorded at 30s intervals for 12.5 hours from 15:00 to 3:30 UT per day at each site.

Ashtech or Trimble receivers occupied the tide gauges and some of the fiducial stations. Three kinds of receivers (Ashtech, Trimble and Rogue) were used instantaneously at Metsähovi, Onsala and Wettzell stations. All receivers recorded carrier phase measurements on both L1 and L2 frequencies, observing all GPS satellites in view. P-code measurements were taken on both frequencies by the Trimble and Rouge receivers, except for the first two days (as possibly explained below), whereas different kinds of Ashtech receivers were used, some could provide P-code measurements on both frequencies, some P-code measurements only on L2 frequency, and some were codeless. At the first two days (GPS days: 158 and 159) anti-spoofing was on,

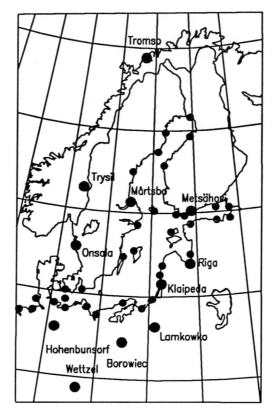

Figure 1. Fiducial and 35 tide gauge stations of the second BSL GPS campaign

which caused some problems, because at that time A/S had been tested during weekends, not weekdays, and those P-code Ashtech receivers did not switch to codeless mode automatically, which caused 11 tide gauges to get no observation data (but only 3.5 hour data of the 11 gauges can be used, since 0-3:30 UT of the second day was in GPS day 290 without A/S).

At the time of the Second Campaign, the GPS constellation consisted of 23 operational Block I, II, and III satellites (SV1, SV2, SV3, SV7, SV12, SV13, SV14 SV15, SV16, SV17, SV18, SV19, SV20, SV21, SV22, SV23, SV24, SV25, SV26, SV27, SV28, SV29, SV31), all of which were observed. However, the data from satellite SV 7 was excluded from the data processing due to information from the International GPS Geodynamics Service (IGS). Surface measurements of temperature, pressure and relative humidity were performed once per hour during the whole campaign.

DATA PROCESSING

The Bernese version 3.4 software(Rothacher et al., 1993) was used for data processing of the Second Campaign.

Strategy

The one session (12.5 hours) solutions were computed using double difference phase observations based on the ionosphere-free linear combination of the L1 and L2 carrier phase measurements. Then weighted averages of all five session solutions were estimated as last results. Due to limited capacity this did not allow a solution of the whole campaign simultaneously. Precise ephemerides were provided by the IGS. Three VLBI stations (Onsala, Wettzell, and Metsähovi) were kept fixed to their International Terrestrial Reference Frame 1993 (ITRF93) coordinates (epoch: 1993 01 01) reduced to observation time (1993 06 07) taking into account the ITRF93 velocity model, while the other fiducial stations (Hohenbünstorf, Riga, Borowiec, Mårtsbo, and Lamkowko) were estimated together with the tide gauges in order to compare with the IGS results.

The Trimble data of the tide gauges and the fiducial stations are processed separately from the Ashtech data to avoid the errors from the elevation-dependent phase centre variations between different antenna types. For a typical session solution the following parameters were estimated: a) Coordinates of the tide gauges and some fiducial stations b) Carrier phase ambiguities c) 4 troposphere zenith delays per session and site.

RESULTS

The coordinate solutions derived from the five-session weighted means and their RMS errors of the means of three coordinate components of the second BSL campaign were obtained. It should be emphasized that these coordinates are given in the reference frame ITRF93 provided by IGS, and the coordinates of Onsala, Wettzel, and Metsähovi were held fixed. The Stockholm tide gauge was occupied with two receivers of different types, Trimble and Ashtech, which operated simultaneously (Stockholm-1 and Stockholm-2). This means that two solutions in Stockholm tide gauge can be readily compared.

Internal Precision

One way to assess the quality of the overall solution is to estimate repeatability, which shows and internal consistency. The repeatabilties cab be fitted as a curve :

$$\sigma = (a^2 + b^2 L^2)^{1/2} \tag{1}$$

where σ is repeatability, L is the baseline length, and a and b are empirical constants, which may be explained as a constant part and a length-dependent part of the ellipsoidal height errors, respectively. For the second BSL campaign, On the average the compared coordinates agreed to 1-2 cm in the horizontal and to 3 cm in the vertical coordinates. Horizontal precision was 8 mm for the east component ans 3 mm for the north component, with an additional length dependence of 1.0 and 0.6 part in 10^8 for the east and north components, respectively. Vertical measurements were less precise than horizontal measurements, 20 mm + 0.7 in 10^8. The result is about a 3-time improvement compared with the first campaign (Pan and Sjöberg, 1993). Obviously, the distance dependent part of the precision is now more or less eliminated. The difference in precision between the two campaigns can be explained by several improvements in the last campaign, such as IGS precise orbit ephemerides, a new version of software used, data of the different kinds of receivers were processed separately, and an increased number of simultaneously observed satellites and improved satellite geometry strength. Overall, the internal precision seems to be very good and reasonable.

Comparison with Fiducial Stations

Another way to assess the quality is to compare the GPS solutions absolutely with coordinates of VLBI or SLR or permanent GPS stations. The coordinate reference frame of compared stations should be the same, otherwise the coordinates cannot be directly compared. In our situation only Riga coordinates are in the same ITRF93 and the same epoch. The GPS-derived coordinates differ absolutely from the IGS results at a level of -0.6,-0.5, and 0.3 cm in the x-, y-, z-components, and with 0.2 cm in the ellipsoidal height. For the stations Borowiec, Lamkowko, and Borowa Gora their frame of coordinate system is in ITRF91 on epoch 1992.5. A Helmert transformation with 3 (translation) parameters was used to compare different solutions. After only translation all three sites agree with the IGS solution within approximately 1cm in horizontal position and 3 cm in height, which is a little worse than those of direct comparison with Riga site. Since only Mårtsbo site is in ITRF89 and all our solutions are ultimately based on holding the ITRF93 coordinates of Onsala, Metsähovi and Wettzell on epoch 1993.44 fixed, the coordinates of Onsala and Wettzel in ITRF93 are directly taken from IGS, and those in ITRF89 are taken from Gurtner et al. (1992). (Metsähovi site is not the same site in ITRF89 and ITRF93). The comparison shows that our solutions are in good agreement with the solutions of IGS and Gurtner et al. (1992), although there are only two sites.

In general terms, the overall agreement between the solutions of the second BSL campaign and IGS is of the order of 1-2 cm in horizontal position and 3 cm in vertical position on the average, both in direct and indirect comparison. For the direct comparison the differences appear less than one cm in all three components, but we have only one station to compare with.

COMPARISON OF HEIGHTS

All the data of the 38 tide gauges were processed, but up to now merely 26 of those gauges have levelling data connected to the national height systems. In the following, the German, Polish, Swedish and Finnish height systems are compared with each other. However, there are some unique phenomena in Fennoscandian area, in which Sweden uses a non-tidal geoid, but Finland a mean geoid. The non-tidal geoid should be converted to the mean geoid by a formula (in cm) (Ekman, 1989):

$$\Delta H_M - \Delta H_N = 29.6\ \gamma\ (\sin^2\varphi_N - \sin^2\varphi_S) \tag{2}$$

where ΔH_M is the height difference above the mean geoid, ΔH_N is the height difference above the non-tidal geoid, γ (0.8) is the elasticity factor of the Earth, and φ_N and φ_S are the latitudes of the northern and southern station, respectively. After considering the postglacial rebound in Fennoscandia and non-tidal geoid and mean geoid, which are used in the definition of the Swedish and Finnish height systems (RH70 and N60), respectively. A 3-parameter (a bias and two tilts) fitting is used to absorb long-wavelength errors remaining in the gravimetric geoid. As a result the difference between the origins of the two height systems was estimated to 15±6 cm. The differences between the German HN76 and Polish H_{1960} height systems and German HN76 and German DHHN85 were estimated to -18±5 cm and 22±5cm, respectively.

Finally, the sea surface topography along the Swedish and Finnish coastlines have been determined both from levelling and from GPS. See Pan and Sjöberg (1995).

CONCLUSION

The second Baltic Sea Level GPS Project was run for one week in June 1993. This study concerns the computation and analysis of the second campaign including 35 tide gauge sites and 5 fiducial stations. The estimated GPS coordinates of the fiducial sites were compared with those provided by the International Geodynamics GPS Service. On the average the compared coordinates agreed to 1-2 cm in the horizontal and to 3 cm in the vertical coordinates. Horizontal precision was 8 mm for the east component and 3 mm for the north component, with an additional length dependence of 1.0 and 0.6 part in 10^8 for the east and north components, respectively. Vertical measurements were less precise than horizontal measurements, 20 mm + 0.7 in 10^8. The result is about a 3-time improvement compared with the first campaign.

We consider the postglacial rebound in Fennoscandia and non-tidal geoid and mean geoid, which are used in the definition of the Swedish and Finnish height systems (RH70 and N60), respectively. A 3-parameter (a bias and two tilts) fitting is used to absorb long-wavelength errors remaining in the gravimetric geoid. As a result the difference between the origins of the two height systems was estimated to 15±6 cm. The differences between the German HN76 and Polish H_{1960} height systems and German HN76 and German DHHN85 were estimated to -18±5 cm and 22±5cm, respectively.

Acknowledgment The financial support from the Swedish Natural Science Foundation (NFR) is gratefully acknowledged.

References

Ekman, M. (1989): Impacts of Geodynamic Phenomena on Systems for Height and Gravity. Bulletin Géodèsique, 63, 281-296.

Gurtner, W., S. Fankhauser, W. Wende, H. Friedhoff, H. Habrich, and S. Botton (1992): EUREF-89 GPS Campaign- Results of the Processing by the " Berne Group"-, International Association of Geodesy, Section 1-Positioning, Subcommission for the European Reference Frame (EUREF) Publication No.1, München 1992.

Kakkuri, J. (1994): On Unification of the Vertical Datums of the Countries on the Baltic Sea, Final results of the Baltic Sea Level 1990 GPS Campaign, 11-15.

Kakkuri, J. (1995) (ed): Final Results of the Baltic Sea Level 1993 GPS Campaign, Reports of the Finnish Geodetic Institute, No. 95:2, Helsinki.

Kakkuri, J., L.E. Sjöberg and J. Zielinski (1990): Baltic Sea Level Project, Proceedings of Second International Symposium on Precise Positioning with GPS, 3-7 September 1990, Ottawa, 572-580.

Pan, M. and L. E. Sjöberg (1993) : Baltic Sea Level Project with GPS, Bulletin Géodèsique, 67,51-59.

Pan, M. And L.E. Sjöberg (1995): Final Results of the 2nd Baltic Sea level GPS Campaign by Swedish Group, Reports of the Finnish Geodetic Institute, No. 95:2, 57-74, Helsinki.

Rothacher, M., G. Beutler, W. Gurtner, E. Brockmann, and L. Mervart (1993): Bernese GPS Software Version 3.4.

GPS DERIVED DISPLACEMENTS IN THE AZORES TRIPLE JUNCTION REGION

Luisa Bastos, José Osório
Astronomical Observatory, University of Porto
Monte da Virgem, 4430 V. N. Gaia, PORTUGAL
Gunter Hein
Institute fur Erdemessung und Navigation, University FAF Munich
Werner-Heisenberg-Weg 39, Neubiberg, GERMANY
Herbert Landau
terraSat GmbH, Munich
Haringstrs. 19, D-85635 Hohen.-Siegerst., GERMANY

ABSTRACT

The Global Positioning System (GPS) has been used to establish a high precision network in the Azores Archipelago for detection of relative displacements between the different islands.

In this region the Azores-Gibraltar plate boundary meets the Mid Atlantic Ridge (MAR), which crosses the Archipelago, forming a triple junction where three main plates converge: the African, the North American and the Eurasian.

From 1988 to 1994 three main GPS campaigns were carried out on the Azores Archipelago. In this paper the results from those campaigns are compared in an attempt to derive conclusions concerning relative point displacements and directions of deformations. Estimates of local drift between the Eurasian and North American plates are presented.

To assess the reliability of those conclusions our results were compared with those obtained by geological and geophysical methods, showing a good agreement.

Monitoring of the Azores-Gibraltar plate boundary is within the goals of WEGENER; strategies for future work in this direction are also presented.

INTRODUCTION

The Azores Archipelago is formed by nine volcanic islands and the Formigas islets. The islands are aligned in a WNW-ESE direction, between latitude 36° to 40° North and longitude 25° to 32° West, crossing the Mid-Atlantic Ridge (MAR) obliquely. Submarine volcanoes also emerge in this area where three main tectonic plates converge: the North American, the African and the Eurasian plates, see Figure 1.The MAR shows here an inflexion, with several transform faults cutting the ridge. There is evidence that the first island appeared already in the Miocene (Luis et al., 1994).

This western limit of the Eurasian plate is not a simple plate boundary. Geological and geophysical evidence indicate the existence of an Azores microplate. This microplate has a triangular form bounded to the west by the MAR and to the south by the East Azores Fracture Zone (EAFZ). To the north the limit is the Terceira Rift, a stucture formed by the islands of Graciosa, Terceira, S. Miguel and the Formigas islets. Some authors support the idea that this boundary is a leaky transform fault passing south of S. Jorge island (Madeira et al., 1990).

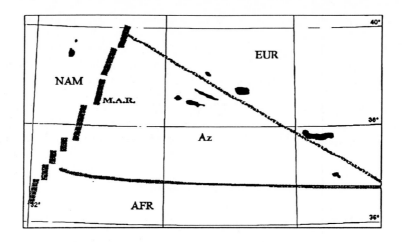

Fig. 1 - Plate boundaries at the Azores triple junction. (NAM-North American plate; AFR -African plate; EUR- Eurasian plate; Az - Azores microplate).

The Azores microplate is cutted by several transform faults whose location is controversial due to the complexity of the system and the lack of a consistent time series of precise geodetic measurements. Tectonic and geophysical data interpretation indicate spreading rates of a few centimeters per year, which are well within the actual capabilities of the GPS technique.

In the scope of the TANGO project several GPS campaigns were organized in Azores. Due to the accuracy achieved we have now the possibility to evaluate annual variabilities for some of the measured baselines. This results are presented here.

AZORES NETWORK AND FIELD CAMPAIGNS

The Azores GPS network is a subnet of the TANGO(TransAtlantic Network for Geodynamics and Oceanography) network, Figure 2. It covers the whole Archipelago (Landau et al., 1989).

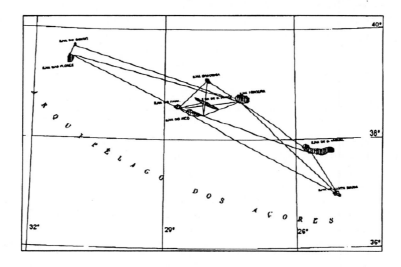

Fig.2 - The Azores GPS network.

Since 1988 three main field campaigns were carried out always establishing at least one station per island (Bastos et al., 1992b):
- TANGO 1 from November 25 to December 5, 1988. TI 4100 dual frequency receivers were used. Station occupation ranged from 2 to 3 nights with observation periods of 3 to 12 hours.
- TANGO 2 from September 30 to October 9, 1991. Dual frequency ASHTECH LD-XII receivers were used. Station occupation ranged from 3 to 5 nights with observation periods between 4 to 8 hours.
- SUPERTANGO from 10 to 16 of October 1994. Dual frequency TRIMBLE SSE receivers were used. Station occupation lasted 5 days with observation periods of 14 hours.

Observations of all campaigns were made between 7 p.m. and 6 a.m. local time to avoid major ionospheric disturbances.

REDUCTION PROCEDURE

Campaigns of 1988 and 1991 were processed with the TOPAS software (VMS version) which uses a multistation multisession approach (Bastos et al., 1992a).

The 1994 campaign and also a recomputation of the one from 1991 was reduced with GEOTRACER, a commercial software working under MS-DOS. Single baseline single session solutions are used in a network adjustment to produce the final solution. Results for 1991 from the two programs were compared and agreed at the 0.3 centimetre level.

NGS and CODE precise ephemeris were used in the different campaigns. The station in S. Miguel island was used as reference as it was tied to a MVLBI station established there in 1992 (Bastos et al., 1992b). ITRF92 (epoch 1988) coordinates of this station were used as reference. All campaigns were recomputed using these coordinates as reference.

RESULTS

Analysis of the results from the GPS campaigns of 1988 and 1991 has been previously reported and is at the several 0.1ppm level (Bastos et al.,1990, 1992, 1994).

Results for the network adjustment of the 1994 campaign (SUPERTANGO) show an improvement of one order of magnitude, accuracy is at the one centimetre level. This is also confirmed by the results obtained in a campaign made in 1993 in some of the Azores stations (Bastos et al., 1994).

In the following, Figure 3, annual baseline variations are shown from comparison between the 1994 campaign and the others. Only the most significant results are shown. Problems in the processing of data from some of the stations in the 1991 campaign, already reported in previous works (Bastos et al., 1994), explain sthe lack of some of the baselines.

From analysis of the figures it is evident the direction of the displacement between Flores (in the NAM plate) and all the other islands (EUR plate). There is a diference in the value of this displacement as computed from comparison of the 1994/1988 or the 1994/1991 campaigns. There is also a disagreement concerning the direction of the variation of the baseline Flores/S.Miguel when considering TANGO 2, but this can be due to the above mentioned problems.

Concerning the other islands there seems to be also some evidence that S. Miguel is moving towards the islands of the central group. In the central group the results are not conclusive but some of the displacements can be related with the so called S. Jorge leaky transform (Madeira et al., 1990). The islands of Graciosa and Terceira are located north of it.

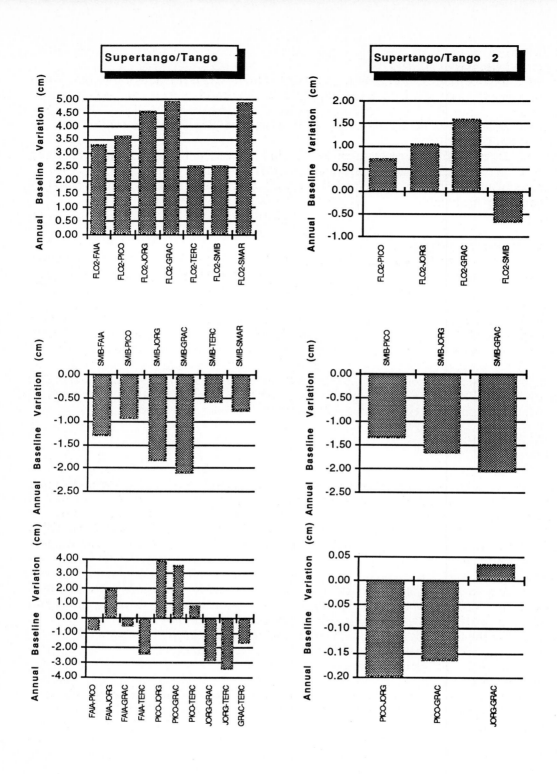

Figure 3 - Annual Baseline variations for the diferent baselines in Azores (FLO2-Flores, FAIA-Faial, JORG-S. Jorge, TERC-Terceira, PICO-Pico, SMIB-S. Miguel, SMAR-Sta. Maria).

CONCLUSIONS

This work focus only on the variation of geometric quantities, namely baseline lengths. When analysing the results from the different campaigns we had to keep in mind some specific problems:

The data was collected within a limited area with a network configuration imposed by the location of the islands;

Expected displacements between consecutive epochs are small compared to the uncertainty of the computed positions in the first campaigns;

Forced changes in station location between some of the campaigns, make sparse time series. The resulting annual spreading rates should therefore be looked with care.

Flores is in a stable plate (NAM). Distance variations between the other islands and Flores give an estimate of the local spreading rate between the North American and Eurasian plates.

The mean values obtained here are of the order of 3.5cm/year, considering TANGO 1, and of 1.5cm/year, considering TANGO 2. The value predicted from magnetic anomalies studies is 2.5cm/year (Luis et al., 1994). The the value given from the NUVEL1 model is within the same range. Our values are slightly different but this can be due to the worst precision of the first campaigns.

Between the other islands the movements are complex and cannot be explained only with the presently available GPS measurements. Faults crossing some of the islands induce other station displacements. There is no agreement concerning directions and intensity of movements of these faults.

Future results from the TANGO project will therefore be a fundamental support for the validation of the geodynamic models proposed for this region.

DIRECTIONS FOR FUTURE WORK

Due to the accuracy attained and the size of the displacements we are looking for, it is important to think of new dedicated benchmark monumentation for all the Azores fiducial stations. The fact that some of the islands are crossed by faults demands for densification of the network with inclusion of more stations per island.

Establishment of permanent stations in the archipelago and annual reobservation of the Azores network is an important issue for the densification of the measurements time series. The use of other space techniques could also be a reliable test for the validation of our conclusions.

Positioning analysis will allow for a crustal dynamic interpretation of point displacement at the centimetre level. Although, the complexity of the deformation in this region brings additional difficulties in the design of a global tectonic model. There is an essential need for a cooperation effort with geologists and geophysicists for the interpretation of the results.

Acknowledgement
We would like to thank Mr. António Barbeito for the effort and excellent collaboration in processing the GPS data. This work has been mainly supported by the Junta Nacional de Investigação Científica e Tecnológica.

REFERENCES

BASTOS, L., OSORIO, J., LANDAU, H., HEIN, G. (1994) - Analysis of Baseline Repeatabilities in a GPS Network in the area of the Azores Triple Junction. Published in the *Journal of the Geodetic Society of Japan*, 1994.

BASTOS, L. (1992a).TANGO 2 - First Results, poster presented at the *Sixth International Geodetic Symposium on Satellite Positioning*, Ohio, U.S.A..

BASTOS, L., OSORIO, J., LANDAU, H., HEIN, G. (1992b). High Precision Geodetic Network for Geodynamic studies in the Azores-Gibraltar Area, *Proc. of the Seventh International Symposium on Geodesy and Physics of the Earth"*, IAG Symposium 112, Potsdam, Germany, Springer-Verlag, 1993, p.46.

BASTOS, L., OSORIO, J., LANDAU, H., HEIN, G. (1990). Transatlantic GPS experiment: results of the TANGO Campaign, *Proc. of the Second International Simposium "GPS 90"*, Ottawa, Canada, p.343.

LUIS, J., MIRANDA, J., GALDEANO, A., PATRIAT, P., ROSSIGNOL, J., VICTOR, L. (1994). The Azores Triple Junction evolution since 10 Ma from an aeromagnetic survey of the Mid-Atlantic Ridge, *Earth and Planetary Science Letters*, Nº125, 1994, p.439.

LANDAU, H., HEIN, G., BASTOS, L., OSORIO, J. (1989). TANGO-TransAtlantic GPS Net for Geodynamics and Oceanography, *Proc. of IAG Symposium 101*, General Meeting of IAG, Edinburgh, Scotand, Springer-Verlag, 1990, p.155.

MADEIRA, J., RIBEIRO, A.(1990): Geodynamic Model for the Azores Triple Junction: a contribution from tectonics. *Tectonophysics*, Nº184, p. 405.

Application of GPS Kinematic Method for Detection of Crustal Movements with High Temporal Resolution

Y. Hatanaka, H. Tsuji, Y. Iimura, K.Kobayashi, and H.Morishita
(Geographical Survey Institute, Kitasato 1, Tsukuba, Ibaraki 305, Japan)

INTRODUCTION

Geographical Survey Institute (GSI) established the GPS Regional Array for PrEcise Surveying (the Grapes) and started the operation from October, 1994. Since the start of the operation, three large earthquakes took place in Japan; the 1994 Kurile Islands (Hokkaido-Toho-Oki) earthquake (M8.1), the 1994 Far-Off-Sanriku earthquake (M7.5), and the 1995 Kobe earthquake (M7.2). We applied a post-processing kinematic analysis for the GPS data from these earthquakes to investigate crustal deformations with high temporal resolution. We will discuss the possibility of using a dense GPS array as seismographs for monitoring crustal deformation in a seismo-geodetic band.

METHOD

BERNESE ver.3.4 and ver.3.5 (Rothacher *et al.* 1993) are used for the analysis with the precise ephemerides produced by International GPS service for Geodynamics (IGS) or GSI. In the case of kinematic analysis, removal of cycle slips and resolution of phase ambiguities are essential. For the data during the coseismic strong motion, we have to be careful in editing the cycle slips since the coordinates of the station vary. The Melbourne-Wübbena linear combinations and ionospheric linear combinations are screened to detect the cycle slips. This procedure is identical to the algorithm of Turboedit (Blewitt, 1992). Detected cycle slips are edited manually on doubly differenced phase. The phase ambiguities are resolved for each arch before and after the mainshock.

RESULTS

Fig.1 shows the time series of the baseline components in every 30 seconds for the typical baselines for each of the three earthquakes. Since the vertical components are noisy affected by many error factors such as atmospheric delays, satellite constellations, etc., we will only focus on the horizontal components in the following discussion.

The time series clearly shows a sudden jump at the epoch just after the P-wave arrival. The amount of the jumps agree well with those obtained from static analyses (Tsuji *et al.*, 1995; Hashimoto *et al.*, 1995) except for the Far-Off-Sanriku earthquake. The static

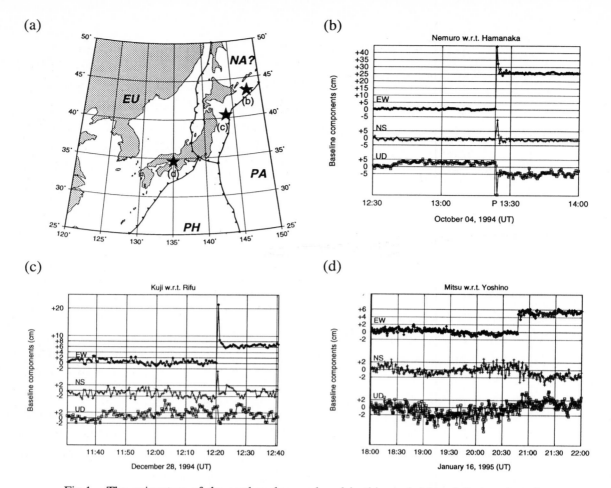

Fig.1   The epicenters of the earthquakes analyzed in this study(a) and time series of a typical baseline for (b) the 1994 Kurile Islands earthquake, (c) the 1994 Far-Off-Sanriku earthquake, and (d) the 1995 Kobe earthquake. The title of each time series shows the stations of the baseline ("w.r.t." means "with respect to"). Deviation of East-West(EW), North-South(NS), and Up-Down(UD) components are plotted.

analysis of the Far-Off-Sanriku earthquake shows 8 cm eastward coseismic displacement of the Kuji station with respect to the Rifu station although the kinematic result shows 6 cm displacement. The difference can be explained by a wavy background noise, which repeats everyday with a 4 minute shift. The noise must be related with satellite constellations. To calibrate this noise, we shift the time series of a previous day by 4 minutes, and subtract it from the time series of the earthquake day (Fig.2). This calibration brings the coseismic displacement of 8 cm, which agrees well with the result of the static analysis.

No precursor can be seen just before the event in every case. The good agreement between the results of static analysis and those of the kinematic analysis also supports that there is no detectable precursory crustal deformation for these cases.

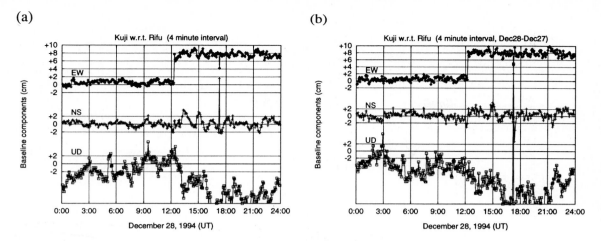

(a)   (b)

Fig.2 The time series of the Kuji station with respect to the Rifu station at the time of
       Far-Off-Sanriku earthquake with 4 minute interval(a). The gradual movement in
       the EW component just after the event disappears after the calibration by
       subtracting the data of the previous day with a 4 minute time shift(b).

The displacement of stations for other stations in the Hokkaido region are obtained
for the Kurile Islands earthquake. Since the fixed station of the network also moves at the
time of the earthquake, we select the Wakkanai station as a fixed point since the coseismic
static displacement is only 3-4 cm (Tsuji *et al.*, 1995). Note that the P-waves are not
arrived at the Wakkanai station at the epoch of 13:23:50 UT when the stations at the east
of Hokkaido area show the initial movement on the time series (Fig.3). The phase
ambiguities are resolved for each independent baseline of the network.

Fig.4 shows the time series of coordinates of several stations which are located on
the northern end of the Hokkaido area. The EW component of the Nemuro station shows

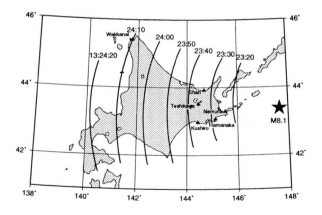

Fig.3 The map of Hokkaido area with the epicenter of the 1994 Kuril Islands earthquake
       (star) and rough contour of the arrival time of P-waves. The contour values are in
       UT.

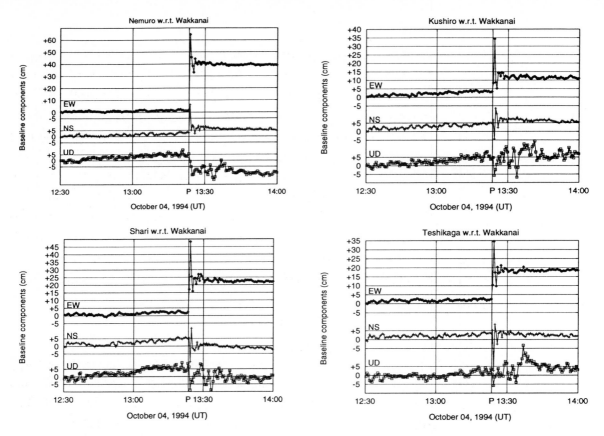

Fig.4 Examples of the time series of relative displacement of stations with respect to the Wakkanai station at the time of the 1994 Kurile Islands earthquake. The mark "P" indicates the arrival time of the P-wave.

pulse-like motion with the peak amplitude of 65 cm followed by a 25 cm backward motion. This is the smallest estimates of the maximum amplitude because of the coarse sampling rate of 30 seconds. The similar motions are also observed in the time series of the Kushiro, Teshikaga, and Shari stations, but the time of the peaks delays by 1 epoch. The epoch of the first movements (13:23:50 UT) of these stations are about 15 seconds after the P-wave arrival. The peak amplitude appears at the next epoch for each time series. This good correlation among the stations indicate that the pulse-like motions, in large part, are not caused by short period seismic waves or the vibration of the antenna pillars. The pulse-like motions are explained qualitatively as follows. In general, the far-field term of the seismic waves is a box-type function of time although the near-field term and intermediate term are step-like functions (Aki and Richards, 1980). The GPS results may represent the mixture of these terms. Quantitative modeling of the seismic waves is needed to confirm it.

DISCUSSION

As shown in the previous section, much information for the coseismic ground motions is included in the kinematic results in spite of low sampling rate of 30 seconds. This suggests a possibility of using GPS arrays as seismographs ("GPS seismograph") for researches of seismic source process if we observe with higher sampling rate. The noise level of the horizontal components is a few cm. At current, the GPS seismograph is capable to detect the relative displacement of more than a few cm for horizontal components with the period longer than (at most) 30 seconds. Ordinary strong-motion seismographs can record the seismic waves with the period shorter than several seconds. On the other hand, geodetic measurements including GPS static measurements are good at resolving the ground motion of one day or longer time scale. The GPS seismograph can fill the gap between the ordinary seismographs and the geodetic measurements. The GPS seismograph may be applied for long period seismology and detection of preseismic deformation and slow/silent earthquakes.

CONCLUSIONS

The GPS kinematic analysis of the Grapes data revealed the coseismic ground motions associated with recent three major earthquakes with high temporal resolution. The short term pre-seismic crustal deformation was not found for these earthquakes. The possibility of the GPS seismograph by using the kinematic method is suggested.

Acknowledgment. We thank to Dr. Y. Yoshida of Japan Meteorological Agency (JMA) for providing the arrival time data of seismic waves of the 1994 Kurile Islands earthquake observed at JMA stations.

REFERENCES

Aki, K. and P. G. Richards (1980): *Quantitative Seismology: Theory and Methods*, W. H. Freeman and Company, San Francisco.

Blewitt, G. (1992): An Automatic Editing Algorithm for GPS data, *Geophys. Res. Lett.*, Vol. 17, 199-202.

Hashimoto, M., T. Sagiya, H. Tsuji, Y. Hatanaka, and T. Tada (1995): Coseismic displacements of the 1995 Kobe earthquake, *J. Phys. Earth*, submitted.

Rothacher, M., G. Beutler, W. Gurtner, E. Brockmann, L. Mervart (1993): Documentation of the Bernese GPS Software Version 3.4, May 1993, Astronomical Institute, University of Berne.

Tsuji, H., Y. Hatanaka, T. Sagiya, and M. Hashimoto (1995): Coseismic crustal deformation from the 1994 Hokkaido-Toho-Oki earthquake monitored by a nationwide continuous GPS array in Japan, *Geophys. Res. Lett.*, in press.

Chapter 2

Spaceborne Applications of the GPS

SPACEBORNE GPS FOR EARTH SCIENCE

Thomas P. Yunck and William G. Melbourne
The Jet Propulsion Laboratory
California Institute of Technology, Pasadena, CA 91109 USA

INTRODUCTION

With the recent completion of the Global Positioning System constellation and the appearance of increasingly affordable spaceborne receivers, GPS is moving rapidly into the world of space flight projects. Indeed, owing to the great utility and convenience of autonomous onboard positioning, timing, and attitude determination, basic navigation receivers are coming to be seen as almost indispensable to future low earth missions. This development has been expected and awaited since the earliest days of GPS. Perhaps more surprising has been the emergence of direct spaceborne GPS science and the blossoming of new science applications for high performance geodetic space receivers.

Applications of spaceborne GPS to Earth science include centimeter-level precise orbit determination (POD) to support ocean altimetry; Earth gravity model improvement and other enhancements to GPS global geodesy; high resolution 2D and 3D ionospheric imaging; and atmospheric limb sounding (radio occultation) to recover precise profiles of atmospheric density, pressure, temperature, and water vapor distribution. Figure 1 offers a simplified summary of the Earth science now emerging from spaceborne GPS.

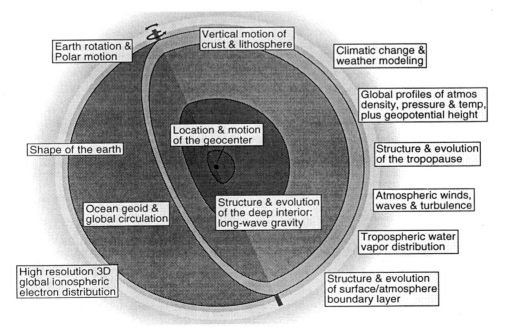

Fig. 1. Some key science applications for a spaceborne array of GPS receivers.

Consistent with these different uses, there has developed in recent years a two-tiered user community for GPS in space: those seeking basic, moderate-performance GPS navigation, timing, and (in some cases) attitude determination, and those pursuing the more demanding science activities requiring the highest performance dual-frequency receivers. As the mission-dependent requirements within each group are diverse, a variety of receiver models for space use has emerged. While that healthy situation is likely to continue, from the standpoint of the scientists it may be hoped that in the future the high end instruments will reach levels of size, cost, and generality of function that will allow them to serve both user classes economically, thus converting the most utilitarian satellites into potentially powerful science instruments.

Conventional single- and dual-frequency GPS receivers have been flown in space for basic navigation and (increasingly) attitude determination on a number of recent missions. These include RADCAL, Christa-SPAS, Orbcomm, and MicroLab I, which carried the Trimble TANS Vector receiver for positioning and attitude determination [*Cohen et al*, 1993; *Lightsey et al*, 1994]; and several Space Shuttle flights, which carried a dual frequency Rockwell-Collins 3M receiver. A variant of the Rockwell receiver, the AST V, was flown on two U.S. military satellites, the Air Force's TAOS (Technology for Autonomous Operational Survivability) and the Advanced Research Projects Agency's experimental DARPASAT [*Cubbedge and Higby*, 1994]. In addition, a 12-channel single-frequency P-code receiver built by Motorola was flown aboard NASA's Extreme Ultraviolet Explorer [*Gold et al*, 1994], and a 5-channel Japanese C/A-code receiver was flown on Japan's OREX mission [*Tomita et al*, 1994].

As GLONASS becomes established as a reliable navigation system we can expect to see considerably more commercial resources devoted to developing the technology for both the ground and space. A high performance spaceborne GPS/GLONASS receiver for navigation and science applications is currently under development by the European Space Agency and may fly within two years [*Silvestrin et al*, 1995].

The utilitarian spaceborne GPS applications represent, in essence, a fulfillment of the GPS vision. They exploit GPS, sometimes in clever ways, for purposes for which it was expressly intended. For the growing class of high-precision spaceborne science users surveyed here, the same cannot be said. GPS was not conceived with such uses in mind (indeed, their feasibility was generally recognized only after GPS deployment was well underway), and has not been altered in any way to accommodate them. Within these diverse scientific enterprises we find many examples in which GPS innovators have, through ingenuity and industry, coaxed a reluctant system to perform unexpected feats, thereby expanding the GPS mission. In the face of the seriously confounding security features known as selective availability and anti-spoofing, they have extracted from GPS levels of performance undreamed of by its architects. The following sections summarize recent highlights in spaceborne GPS science and sketch a picture of its promising future.

PRECISE ORBIT DETERMINATION AND GRAVITY IMPROVEMENT

The first of these unconventional GPS applications to be seriously examined was precise orbit determination (POD) in support of high precision ocean altimetry. A global differential GPS technique for achieving sub-decimeter orbit accuracy on the joint U.S.-French Topex/Poseidon mission was first proposed at the Jet Propulsion Laboratory in 1981. The basic elements of the proposed differential GPS system—a small global ground network, a precision flight receiver, the GPS constellation, and an analysis center—are depicted in Fig. 2. Over the years, a variety of refinements to the proposed orbit estimation technique, evaluated through simulation studies and covariance analysis, revealed the surprisingly rich potential of GPS for few-centimeter tracking of orbiters at low altitudes [*Yunck et al*, 1990; *Wu et al*, 1991].

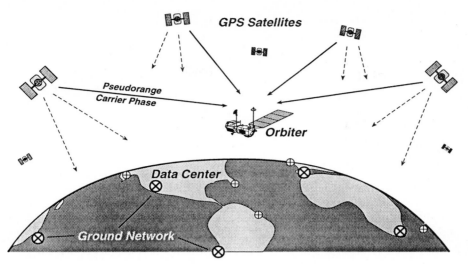

Fig. 2. Key system elements for precise orbit determination with differential GPS.

The Topex/Poseidon ocean altimetry satellite was launched into a 1300 km orbit on an Ariane rocket in August of 1992. It carried an experimental dual-frequency P-code receiver built by Motorola to test these new tracking techniques [*Melbourne et al*, 1994]. The Topex GPS POD demonstration has now surpassed pre-launch expectations of 5-10 cm radial orbit accuracy by about a factor of three.

A number of aspects of this experiment are notable: (1) conventional dynamic differential GPS orbit solutions were essentially equivalent to dynamic solutions obtained with laser and DORIS (Doppler) tracking data, with radial accuracies of 3-4 cm RMS [*Schutz et al*, 1994]; (2) reduced dynamic orbit solutions, in which the unique geometric strength of GPS data is used to minimize sensitivity to force model errors [*Wu et al*, 1991] consistently improved upon dynamic solutions (judged primarily by altimeter crossover agreements) to yield radial orbit accuracies of 2-3 cm RMS [*Yunck et al*, 1994; *Bertiger et al*, 1994; *Hesper et al*, 1994]; (3) University of Texas investigators used GPS data from Topex/Poseidon to improve the Earth gravity model over what had earlier been achieved by tuning with laser and DORIS data, leading to significantly reduced geographically correlated dynamic orbit error [*Bertiger et al*, 1994]; (4) dynamic orbits with a GPS-tuned gravity model surpass those with a laser/Doppler-tuned model, but fall short (by ~1 cm RMS) of GPS reduced dynamic orbits [*Bertiger et al*, 1995]; (5) GPS-based orbits of the highest accuracy are now obtained with a fully automated, unattended processing system; (6) analysis based on Topex results suggests that reduced dynamic orbit accuracies of a few centimeters should be achievable for future missions at altitudes below 500 km [*Melbourne et al*, 1994; *Bertiger et al*, 1994]; (7) recent unpublished results by Ron Muellerschoen at JPL indicate that carefully tuned onboard dynamic filtering could yield real time non-differential orbit accuracies of a few meters under nominal levels of selective availability.

Since the Topex/Poseidon receiver cannot decrypt the Y-codes, the GPS demonstration has been partially in abeyance since anti-spoofing came on nearly full time in January of 1994. Routine processing continues, however, with L1 C/A-code data, yielding radial accuracies in the range of 4-5 cm RMS, itself a somewhat surprising result. In the wake of the Topex success, GPS-based POD has been adopted for several future altimetry missions, including the U.S. Navy's Geosat Follow-On, which will carry a Rockwell MAGR and is slated for a 1996 launch, and the Topex/Poseidon Follow-On, proposed for launch later in the decade.

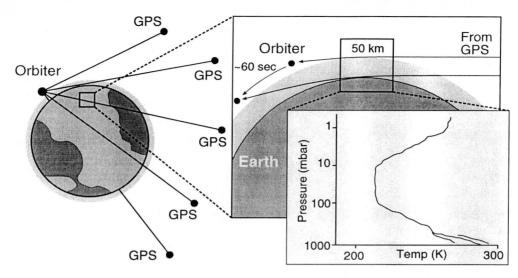

Fig. 3. Illustration of atmospheric temperature profiling by GPS occultation

GPS ATMOSPHERIC OCCULTATION

The probing of planetary atmospheres by radio occultation dates to the early 1960s when Mariners 3 and 4, viewed from Earth, passed behind Mars [*Kliore et al*, 1964 and 1965]. In this technique a radio signal from a spacecraft moving behind a planet is tracked until blockage. As the signal cuts through the planet's refractive atmosphere, its lengthening path delay, revealed by the observed change in phase delay or Doppler shift, can yield a precise profile of the atmospheric density, pressure, temperature or water vapor, and, to some degree, composition and winds. Amplitude variations can expose atmospheric turbulence and wave structure.

While radio occultation has probed many planets and moons throughout the solar system, it has as yet found no useful application to Earth, for two reasons. First, the observation requires both a radio source and a suitable receiver off the planet, outside the atmosphere; seldom have we had such matched pairs in Earth orbit. Second, to be of use in studying Earth's atmosphere, whose nature we know well, such measurements must be continuous, comprehensive, synoptic. We therefore need many transmitters and receivers aloft at once, densely sampling the global atmosphere every few hours. Until the arrival of GPS and low cost microsats, the evident cost of such an enterprise made it impractical within Earth science programs.

In the late 1980s, a group at JPL proposed observing GPS signals from space to make atmospheric soundings by radio occultation, as shown in Fig. 3 [*Yunck and Melbourne*, 1989]. Briefly, the observed Doppler shift in the GPS signal induced by atmospheric bending permits accurate estimation of the atmospheric refractive index. From that one can retrieve, in sequence, profiles of the atmospheric density, pressure, and temperature (or, in the lower troposphere, water vapor) with high accuracy (<1 Kelvin in temperature) and a vertical resolution of a few hundred meters [*Melbourne et al*, 1994b; *Kursinski*, 1994]. Figure 4 shows the predicted accuracy of atmospheric temperature profiles as a function of altitude, based on extensive simulation studies performed at JPL. Notice that in the lower part of the troposphere, the uncertainty in water vapor content, particularly in the tropics, leads to a large error in the recovered temperature. In that region, since it is water vapor that is of greater consequence in weather modeling, it becomes advantageous to adopt nominal temperature lapse rates and instead recover water vapor profiles.

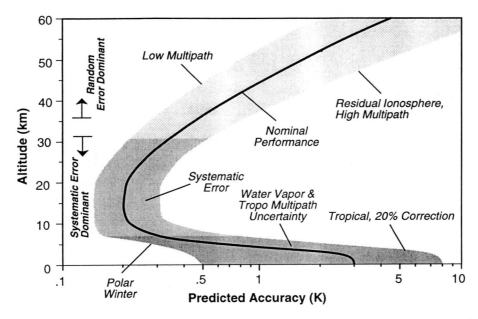

Fig. 4. Estimated GPS-derived atmospheric temperature accuracy vs altitude.

A single satellite can recover more than 500 profiles each day, distributed almost uniformly around the globe; a large constellation would recover many thousands of profiles, which could one day have a profound impact on both long term climatological studies and short term weather modeling. In addition, such an array would enable high resolution 3D tomographic imaging of the ionosphere (see next section) and would serve many geodetic uses (e.g., gravity recovery, geocenter monitoring) as well.

Stimulated by the original JPL proposal, a group led by the University Corporation for Atmospheric Research in Boulder, CO, succeeded in obtaining sponsorship from the U.S. National Science Foundation for a low-cost demonstration experiment called GPS/MET (for meteorology), to fly as an add-on payload to a NASA experiment (an Optical Transient Detector) aboard Orbital Sciences Corporation's MicroLab I satellite. Additional mission support was provided by NOAA (the National Oceanic and Atmospheric Administration) and the FAA (Federal Aviation Administration), and supplemental analysis support was obtained from NASA. To acquire the occultation data, Allen Osborne Associates, manufacturer of the TurboRogue geodetic GPS receiver, developed a ruggedized flight version known as the TurboStar. JPL, a collaborator on the experiment, revamped the receiver software for autonomous operation and occultation scheduling in space. The TurboStar produces 50 Hz dual frequency data samples during occultations using the P-codes when antispoofing is off and less precise alternative methods when it is on.

The MicroLab I was launched successfully aboard a Pegasus rocket in April 1995. While there have been minor problems with the satellite itself, the receiver has performed almost flawlessly from the beginning. Upon power-up, the TurboStar automatically conducts a blind "open sky" search to acquire GPS satellites, uses those to set its internal clock and initialize its orbit solution, computes its Earth-relative position and velocity, schedules hundreds of daily atmospheric occultation passes based on its own computed position, the positions of the GPS satellites, and the known positions of several ground support stations, and feeds a steady stream of data back to the ground. Many hundreds of occultation passes have now been acquired and analyzed. Figure 5 shows a typical temperature profile computed at JPL, along with nearby radiosonde measurements for comparison. For a more comprehensive presentation of results from this experiment see *Feng et al* [1995] and *Hajj et al* [1995] in this volume.

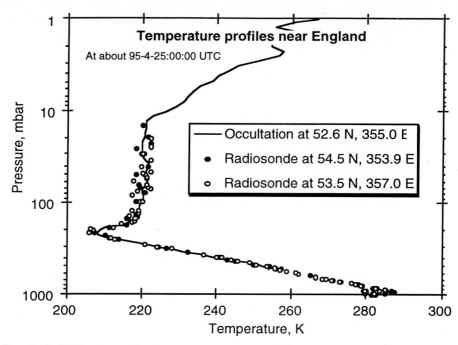

Fig. 5. Typical GPS atmospheric temperature profile compared with two radiosondes.

The best occultation data are acquired with P-code tracking during occasional periods when antispoofing is off, and JPL has been able to negotiate several such periods, each lasting typically a few weeks, with the Department of Defense. Initial profiles recovered by groups at JPL, UCAR, and the University of Arizona are extremely encouraging, in many cases with estimated accuracies of about 1 Kelvin over a wide range of altitudes [e.g., *Hajj et al*, 1995]. This performance is expected to improve steadily as analysis refinements are introduced. Ionospheric studies with the GPS/MET data are just now beginning and as yet no results have been reported.

IONOSPHERIC IMAGING WITH SPACEBORNE GPS

The dual frequency GPS signals offer a direct means of measuring the integrated or total electron content (TEC) along the line of sight from the receiver to the GPS satellites [e.g., *Yunck*, 1993]. Today, ionospheric measurements from the global GPS ground network are used to generate accurate global maps of zenith TEC [*Mannucci et al*, 1993]. Such maps are valuable both for calibration of tracking data from other satellites and for scientific study of the ionosphere. While ground-based zenith TEC maps represent a big advance in our ability to image the ionosphere, they have their limitations. Horizontal resolution is still relatively crude, though that will improve with more ground sites, and information on the vertical electron distribution is entirely absent. Various efforts have been made to recover vertical information from ground based TEC data by means of two-dimensional tomography, but the basic observing geometry severely limits the vertical resolution that can be achieved [*Raymund et al.*, 1994]. That limitation can be readily removed by the introduction of horizontal cuts through the ionosphere afforded by spaceborne receivers. Much like atmospheric occultations, such observations will slice through the ionosphere to provide exquisite vertical resolution; combined data from large numbers of space- and ground-based receivers will enable high resolution two- and three-dimensional snapshot imaging of the global ionosphere [*Hajj et al*, 1994].

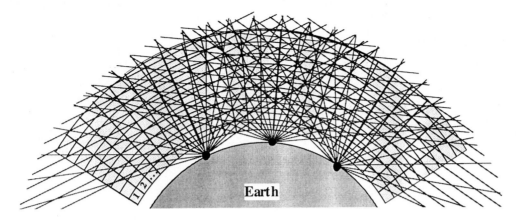

Fig. 6. Ionospheric sampling with combined ground and spaceborne GPS receivers.

Figure 6 illustrates in two dimensions the ionospheric sampling that can be achieved from the ground and space. Simulation studies performed by *Hajj et al* [1994] demonstrate quite dramatically the value of space-based GPS data in ionospheric imaging. With ground data alone, virtually nothing of the vertical electron distribution is revealed; with space data alone, good quality vertical and horizontal images are recovered. Combining space and ground data provides finer detail and overall resolution, however, the improvement over purely space data is rather slight. This is because in addition to providing the vertically slicing cuts needed to recover vertical information, the space links cross one another over a much wider range of angles, supplying much of the information needed for the full image.

Spaceborne GPS imaging will have a profound impact on ionospheric science. The ionosphere is a complex, mutable matrix containing an assortment of transient structures, including troughs, waves, bulges, plumes. The dynamic behavior of the midlatitude trough, for example, is directly related to magnetospheric processes such as substorms caused by solar flares, which in extreme cases can severely disrupt communications and power grids and cause economic losses in the tens of millions of dollars [*Lerner*, 1995]. The ability to image ionospheric structures continuously in three dimensions will help scientists to examine in detail the evolution of "space weather," to trace the formation of geomagnetic storms, and perhaps one day to predict when an observed solar flare will cause disruption on Earth. Today, despite decades of effort, we can only weakly illuminate the ionosphere's complex structure and behavior. An array of orbiting receivers will expose the ionosphere to examination; what is now elusive or hidden will stand revealed, opening a new chapter of ionospheric investigation.

THE FUTURE OF SPACEBORNE GPS SCIENCE

The success of Earth gravity model tuning on Topex/Poseidon has boosted the prospects of various proposed GPS-based missions devoted to further gravity model improvement. Mission concepts proposed by groups at JPL and the Goddard Space Flight Center, among other places, include (1) constellations of GPS-bearing microsats to chart the long wavelength components of the gravity field, up to about degree and order 25, and their time variation; and (2) pairs of low-orbiting microsats flying in formation on which modified GPS receivers would make both conventional "high-low" GPS measurements to observe the long-wavelength gravity components, and more precise "low-low" satellite-satellite range and Doppler measurements (with accuracies of ~10 μm and ~1 μm/s) to observe the shorter wavelength components, up to about degree and order 80. No such mission has yet been approved.

Several planned international microsat missions will carry versions of the TurboStar for atmospheric occultation and gravity modeling. These include the Danish Ørsted and the South African Sunsat missions, set to be launched together on a Delta rocket in 1997, and a possible Brazilian SACI mission in 1998. In addition, a TurboStar is expected make occultation observations from the Wakeshield facility, to be deployed and retrieved by the Space Shuttle this year; another will be placed aboard the Russian MIR space station in 1997 for a timing experiment in cooperation with the Smithsonian Astrophysical Observatory. At least a half-dozen other TurboStar flights are in the discussion stage.

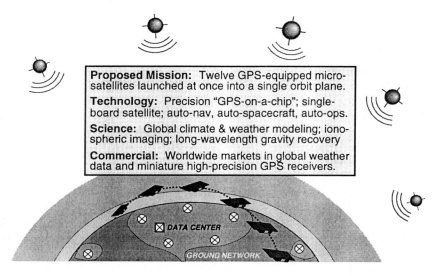

Fig. 7. Concept for a pilot constellation of spaceborne GPS receivers for Earth science.

While the individual occultation missions will serve to advance the GPS technology and validate its capabilities, they will do little for atmospheric science. It is the fervent hope of the growing GPS occultation community that a small pilot constellation of a dozen or so microsats will be sponsored either by government agencies or by commercial interests (eyeing a potential worldwide market in GPS weather products) in the very near future. One such concept being developed at JPL is shown in Fig. 7. This could be the prelude to an array of hundreds of tiny, autonomous satellites continuously monitoring the global atmosphere and ionosphere three-dimensionally, with high resolution in space and time (while also improving the gravity model), within a decade. The results from GPS/MET have made the prospect of such a mission tantalizing, and the prospects for its eventual deployment highly promising.

DISCUSSION

GPS is quickly achieving a routine presence in space for the basic utility functions of real time onboard state determination, precise time and frequency transfer, and moderate precision attitude determination, and is likely to be the method of choice for those tasks for many future Earth satellites, both American and international. At present, real time onboard position accuracies fall in the 50-100 m range, limited by the instantaneous effects of SA dither, which typically introduces 20-30 m errors in measured pseudorange. Experiments on EUVE and Topex/Poseidon (and an earlier simulation study by *Bar-Sever et al* [1990]) show that robust onboard dynamic filtering can smooth real time position error to a few meters, with SA dither at its nominal level. Similarly, onboard time determination, typically accurate to a few tenths of a microsecond today, can be improved to a few tens of nanoseconds through filtering.

The best current GPS-derived attitude accuracies are reported to be a few tenths of a degree, or about 20 arcmin, with antenna baselines of 1 m. A recent study by *Young* [1995] suggests that continued improvements in passive and active multipath suppression combined with real time dynamic filtering can reduce attitude error by more than an order of magnitude, to about 1 arcmin, with performance ultimately limited by the accuracy of the antenna phase center calibration.

Demonstrations of combined GPS/GLONASS in space are still more than a year away. In 1994 the European Space Agency began development of a high performance dual frequency GPS/GLONASS receiver for spaceborne science applications. The receiver will have a minimum of 12 parallel channels, each able to track either GPS or GLONASS signals at high rates, and will be able to track by means of the P-codes, C/A-codes, and several codeless techniques. The first prototypes could be available by the end of 1995 [*Silvestrin et al*, 1995]. While the future of GPS for spaceborne use appears to be secure, it remains to be seen whether GPS/GLONASS will gain a solid foothold in the market.

Spaceborne GPS for Earth science is in the exciting early phase of invention, with promising developments underway in geodesy, climatology, weather modeling, and ionospheric imaging. The advancement of spaceborne GPS science is rapidly becoming an international venture, with small missions in preparation in a number of countries. These science applications invariably require high performance dual-frequency receivers with capabilities well beyond the utility needs of most civilian missions. In the near term there will therefore remain two distinct classes of GPS use in space. It may one day come to pass, however, that as science programs move towards the creation of a large constellation of GPS microsats, the size and cost of high end flight receivers will approach that of utility models, and many more satellites will then be able to contribute to spaceborne GPS science.

Acknowledgment. We thank Bruce Haines, Ed Christensen, Michael Watkins, George Hajj, and Rob Kursinski, all of JPL, for results presented here. A portion of the work described in this review was carried out by the Jet Propulsion Laboratory, California Institute of Technology, under contract with the National Aeronautics and Space Administration.

REFERENCES

Bar-Sever, Y. E., T. P. Yunck and S. C. Wu, GPS-based orbit determination and point positioning under selective availability, *Proc. ION GPS-90*, pp. 255-263, Sep 1990.

Bertiger, W. I., et al, GPS precise tracking of Topex/Poseidon: results and implications, *J. Geophys. Res., 99*, 24449-24464, Dec 1994.

Bertiger, W., T. Yunck, K. Gold, J. Guinn, A. Reicher and M. Watkins, High precision and real timetracking of low earth orbiters with GPS: case studies with Topex/Poseidon and EUVE, 3d Int Workshop on High Precision Navigation, Stuttgart, Apr 1995.

Cohen, C. E., E. G. Lightsey, W. A. Feess and B. W. Parkinson, Space flight tests of attitude determination using GPS, Proc. ION GPS-93, pp. 625-632, Sep 1993.

Cubbedge, S. and T. Higbee, Design, Integration, and test of a GPS receiver on an inertially pointed satellite: A case study, Proc. ION GPS-94, Salt Lake City, UT, pp. 1701-1710, Sep 1994.

Feng, D. S., B. M. Herman, M. L. Exner and B. Schreiner, Spaceborne GPS remote sensing for atmospheric research, IUGG XXI General Assembly, IAG Symposium G2, Boulder, CO, July 1995.

Gold, K., A. Reichert, G. Born, W. Bertiger, S. Wu and T. Yunck, GPS orbit determination in the presence of selective availability for the Extreme Ultraviolet Explorer, Proc. ION GPS-94, Salt Lake City, UT, pp. 1191-1199, Sep 1994.

Hajj, G. et al., Initial results of GPS-LEO occultation measurements of Earth's troposphere and stratosphere obtained with the GPS-MET experiment, IUGG XXI General Assembly, IAG Symposium G2, Boulder, CO, July 1995.

Hajj, G., R. Ibanez-Meier, E. Kursinski and L. Romans, Imaging the ionosphere with the Global Positioning System, *Int. J. Imaging Systems & Tech., 5*, 174-184, 1994.

Kliore, A. J., T. W. Hamilton and D. L. Cain, Determination of some physical properties of the atmosphere of Mars from changes in the Doppler signal of a spacecraft on an Earth occultation trajectory, TR 32-674, Jet Propulsion Laboratory, Pasadena, 1964.

Kliore, A. J., D. L. Cain, G. S. Levy, V. R. Eshelman, G. Fjeldbo and F. D. Drake, Occultation experiment: results of the first direct measurement of Mars' atmosphere and ionosphere, *Science, 149*, 1243-1248, 1965.

Kursinski, R., Monitoring the earth's atmosphere with GPS, *GPS World*, Mar 1994.

Lerner, E. J., Space weather, *Popular Science*, pp. 54-61, Aug 1995.

Lightsey, E., C. Cohen and B. Parkinson, Development of a GPS receiver for reliable real time attitude determination in space, Proc. ION GPS-94, Salt Lake City, UT, pp. 1677-1684, Sep 1994.

Mannucci, A., B. Wilson and C. Edwards, A new method for monitoring the earth's ionospheric total electron content using the GPS global network, Proc GPS-93, Inst of Navigation, 1323-1332, Sept 1993.

Melbourne, W. G., E. S. Davis, T. P. Yunck and B. D. Tapley, The GPS flight experiment on Topex/ Poseidon, *Geophys. Res. Lett., 21*, 2171-2174, 1994.

Melbourne, W. G. et al, *The Application of Spaceborne GPS to Atmospheric Limb Sounding and Global Change Monitoring*, JPL Pub. 94-18, NASA/JPL, April 1994b.

Raymund, T. D., S. J. Franke and K. C. Yeh, Ionospheric tomography: its limitations and reconstruction methods, *J. Atmos. Terr. Phys., 56* (5), 637-650, Apr 1994.

Schutz, B. E. et al, Dynamic orbit determination using GPS measurements from Topex/Poseidon, *Geophys. Res. Lett., 21*, 2179-2182, 1994.

Silvestrin, P. et al, Development of a spaceborne GNSS receiver for atmospheric profiling applications, URSI Conference 1995, Working Group AFG1, Copenhagen, June 1995.

Tomita, H. et al, Flight data analysis of OREX onboard GPS receiver, Proc. ION GPS-94, Salt Lake City, UT, pp. 1211-1220, Sep 1994.

Wu, S., T. Yunck and C. Thornton, Reduced-dynamic technique for precise orbit determination of low earth satellites, *J. Guid., Control & Dynamics, 14*, 24-30, 1991.

Yunck, T. P., Coping with the atmosphere and ionosphere in precise satellite and ground positioning, in *Environmental Effects on Spacecraft Positioning and Trajectories*, A. Vallance-Jones, ed., Geophysical Monograph 73, IUGG Volume 13, pp. 1-16, 1993.

Yunck, T. and W. Melbourne, Geoscience from GPS tracking by Earth satellites, Proc. IUGG/IAG Symp. No. 102, Edinburgh, Scotland, Aug 1989.

Yunck, T. P., S. C. Wu, J. T. Wu and C. L. Thornton, Precise tracking of remote sensing satellites with the Global Positioning System, *IEEE Trans. Geosci. & Rem. Sensing, GE-23*, 450-457, 1990.

Yunck, T. P. et al, First assessment of GPS-based reduced dynamic orbit determination on Topex/Poseidon, *Geophys. Res. Lett., 21*, 541-544, 1994.

TOPEX/Poseidon Precision Orbit Determination With SLR and GPS Anti-Spoofing Data

**Laureano Cangahuala, Ronald Muellerschoen, Dah-Ning Yuan
Edward Christensen, Eric Graat, Joseph Guinn
Jet Propulsion Laboratory, California Institute of Technology,
Pasadena, California 91109 USA**

To take advantage of the quality of the TOPEX/Poseidon sea-level measurements, the radial orbit component must be known to better than a decimeter. Orbits have been produced using Global Positioning System (GPS) and satellite laser ranging (SLR) tracking data. These orbits are produced with small radial position errors (< 5 cm RMS), on a short production schedule (≤ 4 days), with minimal resources. The models and estimation strategies for different data type combinations are outlined. Of special interest are the solutions which contain GPS Anti-Spoofing (AS) data. These orbits are compared to existing precision orbit ephemerides to demonstrate their relative accuracy as an orbit product.

INTRODUCTION

The TOPEX/Poseidon (T/P) spacecraft was launched in August 1992, and is in the final months of its primary mission, with a three-year extended mission ahead of it. The mission objective is to measure sea level (and orbit height) to such an accuracy that small-amplitude, basin-wide sea level changes caused by ocean circulation can be detected. Precision orbit ephemerides (hereafter referred to as POEs), are created once per ten-day cycle, thirty days after the tracking data has been collected, using SLR and DORIS data. In addition, SLR data is used to construct daily fits, called "Medium Precision Orbit Ephemerides," or MOEs [*Cangahuala, et al. 1995*], within 3-5 days after-the-track. The amount of SLR data available for MOE production has decreased throughout the past year due to shrinking funding for the SLR network.

GPS data is collected as quickly (if not quicker) than SLR data, which made it an excellent candidate for MOE production. During non-Anti-Spoofing (AS) tracking, the T/P GPS Demonstration Receiver (GPSDR) uses P-code to obtain GPS pseudorange and carrier phase observables at L1 and L2 frequencies, providing ionosphere-free pseudorange and phase observables. While AS is on, the GPSDR can only track the L1 C/A signal from the GPS constellation and is thus unable to calibrate the observations for the ionospheric delay. A technique was developed to address the lack of information about the ionosphere.

This paper documents the successful implementation of the GPS/SLR "quick-look" orbit determination task. The orbit determination models, data, and filter methodologies for the different data type combinations are described. Results of the initial proof-of-concept are demonstrated, along with an assessment of the GPS/SLR MOEs produced to date.

ORBIT DETERMINATION MODELS

The MOE modeling and parameter estimation scheme is similar to that used for POE orbit determination [*TOPEX/Poseidon POD Team, 1995*]. For POE production, the nonconservative force models account for the spacecraft's attitude history, geometry, and material properties. They are collectively known as the 'Macromodel,' and are tuned with tracking data from cycles 1-48. For MOE production, these forces are not modeled; it has been shown that an appropriate set of empirical acceleration estimates (in this case constant downtrack and once per orbit downtrack and crosstrack estimates) does effectively compensate for this lack of detailed modeling.

DATA DESCRIPTION

The SLR quick-look data is collected from the Crustal Dynamics Data Information System via FTP. Typically, SLR data for a given pass is available within 4 days after-the-track. The SLR data weights used for MOE production are a function of station of origin, and range from 1.0 cm to 200 cm.

The GPS ground data used in this effort comes from a global network of 16 stations, a subset of that collected and reduced at the Jet Propulsion Laboratory as part of an effort with the International GPS Geodynamics Service. The GPSDR data is also processed on-lab. The GPSDR and GPS ground data are both available well under 48 hours after-the-track. The GPSDR phase and pseudorange data are deweighted (10 to 80 cm for carrier phase and 80 to 240 cm for pseudorange measurements) as part of the strategy when the GPS constellation is in Anti-Spoofing mode.

FILTER METHODOLOGY

The filtering is performed with the GPS Inferred Positioning System Orbit Analysis and Simulation Software (GIPSY-OASIS) set developed by the Tracking Systems and Applications Section at the Jet Propulsion Laboratory. Nominally, MOEs based on GPS data are created from a 30 hour data fit. Solutions from adjacent days would thus have an overlap which provides an opportunity to perform a quality check upon the latter solution. The overlap agrees in the radial component to well under ten centimeters RMS.

For solutions involving GPS data, the basic GPS orbit determination strategy usually involves a dynamic fit with the simultaneous adjustment of: the GPS and T/P orbits, station and satellite clock parameters, selected station locations, zenith tropospheric delays, and solar pressure coefficients (scale factors and Y-Bias), and carrier phase biases. In practice, AS data is the norm, so it is more reliable to first solve for the GPS orbits and clocks, then determine the T/P orbit using the previously determined GPS orbits. With the GPS orbits fixed, solving for the T/P orbit (and GPS clocks to account for Selective Availability) with Anti-Spoofing data becomes more reliable. The data editing process does change with the AS status; for example, when AS is on, a higher elevation cutoff angle is used. The origin of this technique [*Muellerschoen et al. 1995*] comes from work performed in the Tracking Systems and Applications Section at JPL [*Blewitt et al. 1988*].

ORBIT DETERMINATION EVALUATION

To demonstrate the proof-of-concept of using GPS Anti-Spoofing data to the T/P project, a battery of solutions using different data type combinations was created over a complete 10 day ground track cycle. To evaluate these solutions, their agreement with the GSFC POE is examined, with the radial and 3-dimensional RSS values of the comparison being the significant quantities. Since the model structure of both the MOE and POE are similar, this comparison is not heavily corrupted by modeling differences. Also, it is necessary to find a figure of merit which is orbit independent. The crossover variances of these orbits is such a measure, since high variances indicate corruption of altimeter data by geographically-correlated orbit error, all else being the same. In addition to the proof-of-concept results, recently created MOEs are compared to the corresponding POE.

Proof-of-Concept Results

Orbits were created for T/P cycle 90, during which the GPS constellation was in Anti-Spoofing mode, and the T/P spacecraft passed from one attitude regime to another (fixed yaw to yaw steering), providing a typical level of spacecraft activity to be encountered during most cycles. The orbits in Figure 1 are differenced against the GSFC DORIS/SLR JGM-3 POE[1], which is considered the most accurate of the set. The comparison of the JGM-2 and JGM-3 POEs demonstrates the magnitude of the orbit solution change brought about by the geodetic model updates. Likewise, going from the JGM-2 (the original MOE) to JGM-3 SLR-only solutions shows some improvement in the agreement, but not as much as the DORIS/SLR solutions. The GPS-only solution has a level of agreement similar to the SLR-only solution; merging the two data types together results in an orbit that approaches the JGM-2 POE agreement with the JGM-3 POE. The altimeter crossover results in Figure 2 tell a similar story.

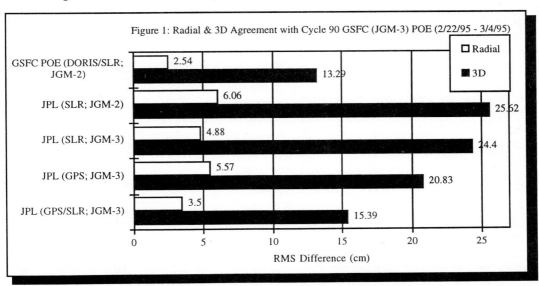

Figure 1: Radial & 3D Agreement with Cycle 90 GSFC (JGM-3) POE (2/22/95 - 3/4/95)

---

[1]In a mid-mission update to the models used for MOE and POE production, the change of gravity field (from JGM-2 to JGM-3) yielded the most dramatic reduction in geographically correlated orbit error. As a result, orbits based on the former and latter model sets are referred to as the "JGM-2" and "JGM-3" orbits, respectively.

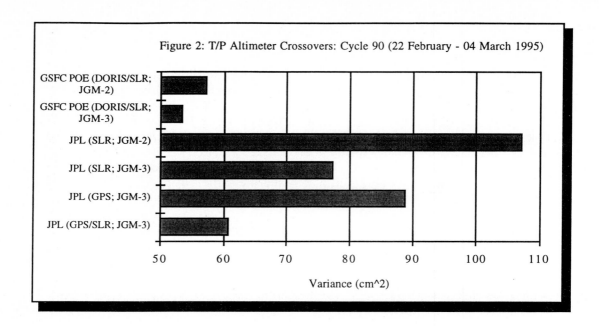

Figure 2: T/P Altimeter Crossovers: Cycle 90 (22 February - 04 March 1995)

Actual MOE Results

GPS/SLR MOE production mode began on 01 June 1995. From late May to late June 1995, MOE production passed through three data type combinations: SLR-only (with JGM-2 models), GPS(AS on)/SLR, and GPS(AS off)/SLR. In Figures 3-4, the radial and 3D RMS agreements between MOEs and POEs are plotted for the daily solution, respectively. The trend amongst the three different solution types is as expected, with the GPS (non-AS)/SLR solution having the best agreement with its corresponding POE. The difference in the agreement between the MOEs with AS GPS data and those with non-AS GPS data can be considered a measure of the orbit degradation brought about by the ionosphere.

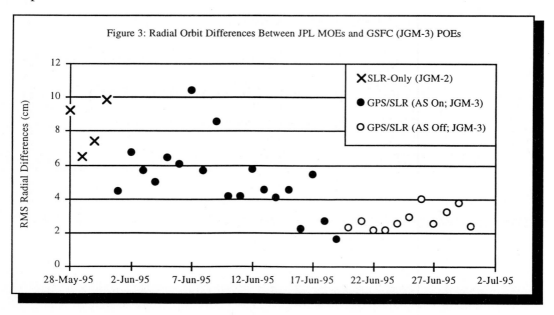

Figure 3: Radial Orbit Differences Between JPL MOEs and GSFC (JGM-3) POEs

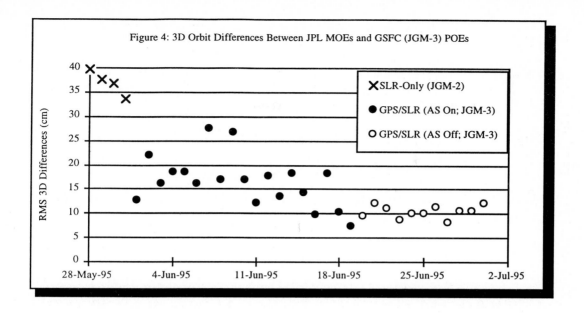

Figure 4: 3D Orbit Differences Between JPL MOEs and GSFC (JGM-3) POEs

SUMMARY

This paper documents the cooperative effort between the Tracking Systems and Applications Section and the Navigation and Flight Mechanics Section to provide precision orbits with minimal resources. Orbit comparisons and crossover statistics show that the accuracy of these "quick-look" orbits approaches that of the POEs supplied to the project during most of the prime mission, even under the degrading effects of Anti-Spoofing.

Acknowledgment. The research described in this paper was carried at the Jet Propulsion Laboratory, California Institute of Technology, under contract with the National Aeronautics and Space Administration.

REFERENCES

Blewitt, G. and Lindqwister, U., "GPS Parameter Estimation with Constrained Carrier Phase Biases," JPL Internal Memorandum 335.4-88-149, 1988.

Cangahuala, L. A., Christensen, E. J., Graat, E. J., Williams, B. G., Wolff, P. J., "TOPEX/Poseidon Precision Orbit Determination: 'Quick-Look' Operations and Orbit Verification," Paper AAS 95-228 presented at AAS/AIAA Spaceflight Mechanics Meeting, Albuquerque, NM, 13-16 February, 1995.

Muellerschoen, R. J., Lichten, S., and Lindqwister, U., "Results of an Automated GPS Tracking System in Support of TOPEX/Poseidon and GPSMET," to be presented at ION Meeting, Palm Springs, CA, September 1995.

TOPEX/Poseidon POD Team, "Precision Orbit Reprocessing for the Improved MGDR," Memorandum, University of Texas at Austin, 28 March 1995.

TOPEX/Poseidon Precision Orbit Determination Using Combined GPS, SLR and DORIS

Joseph Guinn, Ronald Muellerschoen, Laureano Cangahuala,
Dah-Ning Yuan, Bruce Haines, Michael Watkins, Edward Christensen

Jet Propulsion Laboratory, California Institute of Technology, Pasadena, California

TOPEX/Poseidon (T/P) is a joint spaceborne oceanographic mission of U.S. NASA and France CNES design launched August 10, 1992. The satellite has a variety tracking systems for both operational and precision orbit determination. Three precise tracking systems: Satellite Laser Ranging (SLR), Doppler Orbitography and Radiopositioning Integrated by Satellite (DORIS), and Global Positioning System (GPS) provide high quality measurements essential for reconstructing the T/P orbital height with centimeter precision. This paper presents results of simultaneously processing all three data types to exploit the inherent strength of each in a combined solution. SLR and DORIS are routinely combined to provide orbit solutions for the T/P science team. GPS orbit solutions are produced as part of the first demonstration flight of a high quality spaceborne GPS receiver. Coordinate frame and software system differences between the combined SLR/DORIS orbits and the GPS orbits induce orbital height differences of 2 to 3 centimeters. Combining the three data types within a single software system permits removal of software system differences while obtaining coordinate frame calibration information. These calibrations will aid future spaceborne GPS missions that are not complemented with SLR and/or DORIS.

INTRODUCTION

The TOPEX/Poseidon (T/P) satellite carries a high precision Global Positioning System (GPS) receiver as part of a proof-of-concept precision orbit determination experiment. Resulting orbit solutions yield height accuracies below 3 centimeters (1σ) [*Bertiger, et al., 1994*]. At a similar accuracy level are independently determined orbit solutions derived from Satellite Laser Ranging (SLR) and Doppler Orbitography and Radiopositioning Integrated by Satellite (DORIS) observations [*Tapley, et al. 1994*].

Comparisons between these orbit solutions consistently yield height agreements below 3 centimeters but with a bias in the Earth-Fixed Z-component. This "Z-shift", as it will be referred to in this paper, is believed to be associated with a reference frame misalignment between the GPS defined frame and that of the SLR and DORIS systems. The intent of this paper is to combine observations from the three tracking systems in a single solution to obtain a better understanding of the Z-shift and to calibrate the GPS reference frame relative to the SLR and DORIS frames.

SOLUTION STRATEGY

We use the reduced-dynamic [*Wu, et al., 1991*] filtering technique for the combined orbit determination solutions. In addition to estimating all of the GPS space vehicle states simultaneously with the T/P state, the Earth-Fixed geocenter offsets to the GPS station positions are adjusted. These geocenter offsets apply only to the GPS stations and give the translational contribution of the GPS to SLR/DORIS frame tie.

Reference station locations for use in processing the GPS ground observations are derived from a fiducial free adjustment of a global network of about 50 stations for the years 1991 to 1995. These station positions and velocities are closely related to the International Terrestrial Reference Frame (ITRF) of 1993. They are referred to as: JPL95P02 and are the submission from JPL to the International Earth Rotation Service (IERS) ITRF94 solution. SLR and DORIS station locations, also closely tied to ITRF93, are from solutions computed at the University of Texas Center for Space Research (UTCSR). The SLR station coordinates (CSR95L01) are based on LAGEOS and LAGEOS-2 observations between 1976 and 1992. DORIS positions (CSR95D02) are derived from T/P data from 1992-1994.

Observations weights are unchanged from the uncombined solutions. Table 1. gives the weight for each observation type.

Table 1. Observation Weights

Observation Type	Weight
GPS Spaceborne Carrier Phase	0.02 m
GPS Spaceborne Psuedorange	2 m
GPS Ground Carrier Phase	0.01 m
GPS Ground Psuedorange	1 m
SLR Ground	0.01 - 100 m
DORIS Ground	3.2 mm/s

ORBIT COMPARISONS

Five ten day cycles have been processed with virtually no change to the orbital height differences with respect to the GPS only solutions. Figs. 1. and 2. show the height and Z-shift orbit differences for the uncombined solutions. Also, little change is seen in the Z-shift when comparing the combined solutions. This is believed to be the result of the GPS observations dominating the combined solution due to the abundance of observations and overweighting relative to the SLR and DORIS data.

ALTIMETER CROSSOVER COMPARISONS

Altimeter crossover differences for the uncombined solutions are shown in Fig. 3. A noticeable correlation is observed between the T/P beta prime angle (angle related to the Earth/Sun position and the T/P orbit plane) and the altimeter crossovers. This suggest some sort of dynamic mis-modelling in one of the orbit solutions. For the combined solutions, the improvement in the altimeter crossover variances computed is small compared to the GPS only solution. Fig. 4. shows the resulting crossover variances for various combinations of observations during groundtrack repeat cycle 43.

GPS STATION COORDINATE ALIGNMENT

Geocenter estimates for the GPS ground station network produce an average 2 centimeter Z-shift. Fig. 5. shows geocenter estimates from two time periods in 1993 and 1995. The 1995 solutions incorporate improved GPS space vehicle attitude modelling during Earth shadow events. Estimates of the X and Y geocenter offsets compare well with values determined with GPS ground observations only (i.e., no T/P data). However, the Z-component estimates appear to much better determined when including the T/P observations.

CONCLUSIONS

Preliminary results of combining GPS, SLR and DORIS observations show that small improvements in the altimeter crossover variance can be obtained and an average 2 centimeter geocenter offset in the Z-component of the GPS station coordinates is observed. Future study will involve processing more observations and optimizing the relative data weights.

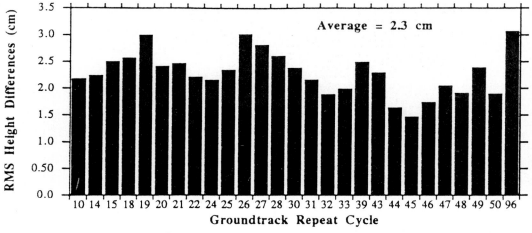

Fig. 1. Orbit Height Comparisons (SLR+DORIS minus GPS)

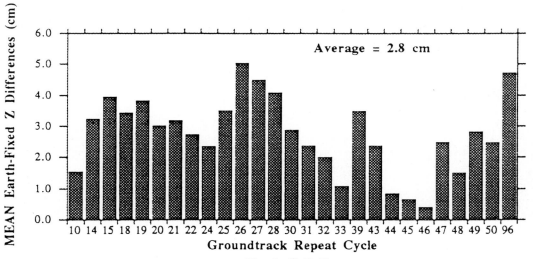

Fig. 2. Z-Shift

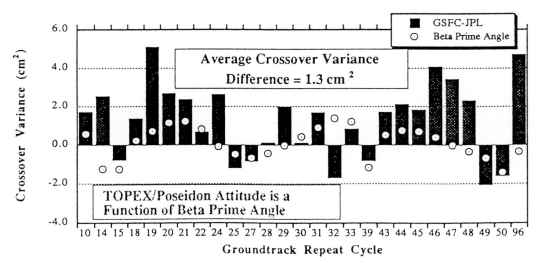

Fig. 3. Altimeter Crossover Comparisons (SLR+DORIS minus GPS)

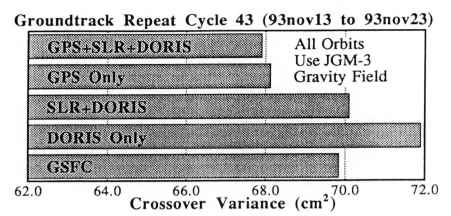

Fig. 4. Altimeter Crossover Results

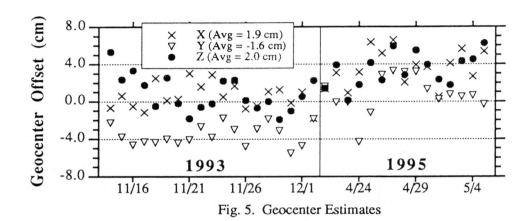

Fig. 5. Geocenter Estimates

ACKNOWLEDGEMENTS

The research described in this paper was carried out by the Jet Propulsion Laboratory, California Institute of Technology, under contract with the National Aeronautics and Space Administration.

REFERENCES

Bertiger, W., et al., GPS precise tracking of TOPEX/POSEIDON: Results and Implications, *J. Geophys. Res.,* 99, 24449-24464, 1994.

Tapley, B., et al., Precision orbit determination for TOPEX/POSEIDON, *J. Geophys. Res.,* 99, 1994.

Wu, S.C., et al., Reduced-dynamic technique for precise orbit determination of low Earth satellites, *J. Guid., Control and Dynamics.,* 14, 24-30, 1991.

Observations of TOPEX/POSEIDON Orbit Errors Due to Gravitational and Tidal Modeling Errors Using the Global Positioning System

B. J. Haines, E. J. Christensen, J. R. Guinn and R. A. Norman
Jet Propulsion Laboratory, California Institute of Technology
Pasadena, California 91109 USA

J. A. Marshall
NASA Goddard Space Flight Center
Greenbelt, Maryland 20771 USA

INTRODUCTION

Satellite altimetry is faced with the challenge of measuring subtle variations in the dynamic topography of the world's oceans with cm-level accuracy. The TOPEX/Poseidon (T/P) mission was designed to resolve these signals by measuring the radial component of the orbit with an accuracy of 13 cm, or better, in a root-mean-square (RMS) sense. Owing to major advances in precision orbit determination, the actual level of performance is estimated to be nearly an order of magnitude better than that [e.g., *Tapley et al.*, 1994a]. This is primarily due to improvements in the gravity model for the Earth, including the tide model, and the effectiveness of the 3 precision tracking systems carried on the spacecraft (see next section). This paper summarizes the results obtained from a comparison between two distinct types of T/P orbits: classical dynamic orbits and GPS-based reduced-dynamic orbits. Surface manifestations of the relative spatial and temporal behavior of these orbits are described in terms of their effect on altimetric observations of dynamic topography.

DYNAMIC AND REDUCED-DYNAMIC ORBITS

The orbit height measurements that appear on the T/P mission geophysical data records are computed using data from a global network of international satellite laser ranging (SLR) stations and French radiometric Doppler (DORIS) beacons [e.g., *Tapley et al.*, 1994a]. Inasmuch as they are computed using a classical dynamical orbit determination technique, their error characteristics are determined in large part by the force models used to integrate the equations of motion. On the other hand, kinematic orbits depend only on the tracking metric to define the trajectory of a satellite and are therefore limited solely by periods of restricted observability and errors associated with the tracking data.

To take advantage of the continuous 3-D data coverage afforded by the GPS Demonstration Receiver (GPS-DR) on board T/P, the strengths of the dynamic and kinematic methodologies have been combined in what is referred to as the reduced-dynamic technique, wherein small, local geometric corrections are made to a previously computed dynamic orbit [*Bertiger et al.*, 1994]. Insofar as reduced-dynamic GPS orbits have a kinematic component, comparisons between GPS and SLR- DORIS dynamic orbits can reveal deficiencies in the dynamic models and errors associated with the tracking systems.

METHODOLOGY

For the current analysis, a time series of radial orbit differences between the NASA Precise Orbit Ephemerides (POE) [*Tapley et al.*, 1994a] and GPS reduced-dynamic orbits for the time span from February 28, 1993 to January 30, 1994 was examined. This time period covers the T/P 10-day repeat cycles 17 through 50 and was chosen because it represents a nearly contiguous span of high quality GPS-DR data. (The GPS-DR tracked in precise dual-frequency mode 86% of the time.) We then averaged the data over ~10-day moving windows centered at 3.3 day intervals (the length of a T/P sub-cycle). This resulted in 99 frames of global radial orbit differences spanning a period of almost one year.

The data in each frame were interpolated onto a uniform global geographic grid (5° X 5°) by employing a least-squares collocation technique [e.g., *Moritz*, 1980] using a Gaussian signal covariance function with a maximum value of 4 cm^2, a decorrelation distance of 6°, and a white-noise covariance of 4 cm^2. Empirical orthogonal functions (EOFs) [e.g., *Priesendorfer*, 1988] were then computed to provide insight into the dominant modes of variability.

RESULTS AND CONCLUSIONS

Depicted in Figures 1 and 2 are geographic distributions of the mean and standard deviation of the radial orbit differences based on the 99 gridded maps. Maps corresponding to both the original NASA POE (based on the JGM-2 gravity model [*Nerem et al.*, 1994]) and the new POE (based on JGM-3 [*Tapley et al.*, 1994b] and improved tide models) are shown. Prominent in Figure 1(a) is a large meridional feature which is due primarily to errors in the JGM-2 gravity model [e.g., *Christensen et al.*, 1994]. This is corroborated by Figure 1(b) which shows that the meridional feature is significantly attenuated when the new JGM-3 POEs are used. Note that the map still has a dominant north-south hemispherical feature which is a manifestation of a slight shift along the Earth spin axis [see also *Marshall et al.*, 1995]. The source of this "Z-shift" is still under investigation.

Superimposed on these stationary features are temporally varying geographically correlated orbit errors (Figure 2). Using EOF analysis to segregate the variability into orthogonal components, we determined that the dominant modes of variability correspond to periodic shifting in the center-of-figure. Shown in Figure 3, for example, are geographic representations of the first 6 modes of variability for the differences of the JGM-2 POE and GPS-based orbits. The first mode—explaining 25 % of the overall variance—is comprised primarily of a long-term variation in the "Z-shift". This is corroborated in Figure 4, which shows a comparison of the Mode 1 amplitude time series and the cycle-by-cycle averages of the body-fixed Z-coordinate differences (JGM-2 POE vs. GPS-based orbit). The second and third modes correspond to shifts along orthogonal axes in the equatorial plane and explain 18 and 11 % of the overall variance respectively.

Modes 5–6 are much less energetic, each explaining 3–6 % of the overall variance. Examination of the spectra for these modes reveals a dominant peak at the ~60 day period (Figure 5). Using Fourier analysis, *Marshall et al.* [1995] have shown that the temporal errors at this frequency are attributable to aliased errors in the non-resonant (i.e., background) components of the principal lunar (M2) and solar (S2) tides which are applied in the POE computations [see also *Bettadpur and Eanes*, 1994].

We repeated the EOF analysis, replacing the JGM-2 POE with the JGM-3 POE. Large-scale hemispherical variations still characterize the first 3 modes. The first two modes share most of the energy—27% and 23% respectively—but neither correspond closely to the Z-shift. Like the JGM-2 comparison, 60-day variations are present in higher modes, though the overall energy at the M2/S2 alias frequency is reduced owing to the application of an improved background tide model [see also *Marshall et al.*, 1995]. A preliminary comparison of the JGM-3 POE with a new set of improved GPS-based orbit shows even less variability, indicating that the solutions based on entirely different data types and orbit determination techniques are converging as the force and measurement models improve.

Also noteworthy are the EOF results which suggest that the most energetic spatio-temporal variabilities associated with the orbit errors are not tide related, rather they have their origin in the definition of the ostensible geocenter. The Z-shift variations are very important, because they can introduce basin-to-basin errors in the ocean topography that directly impact estimates of seasonal steric changes. We note that, at this writing, we have no evidence to suggest that the geocenter problems are due to dynamic model errors associated with the POE; in fact, they likely arise from measurement model errors attributable to either or both of the orbit solutions contributing to the difference. These issues are currently under investigation [e.g., *Guinn et al.*, these proceedings].

Acknowledgment. The research described in this paper was carried out in part at the Jet Propulsion Laboratory, California Institute of Technology, under contract with the National Aeronautics and Space Administration. We are grateful to Ernst Schrama at the Delft University of Technology for providing the least-squares collocation software.

REFERENCES

Bertiger, W. I., *et al.*, GPS precise tracking of TOPEX/Poseidon: Results and implications, *J. Geophys. Res. 99*(C12), 24,449–24,464, 1994.

Bettadpur, S. V., and R. J. Eanes, Geographical representation of radial orbit perturbations due to ocean tides: Implications for satellite orbit determination, *J. Geophys. Res. 99*(C12), 24,883–24,894, 1994.

Christensen, E. J., B. J. Haines, K. C. McColl and R. S. Nerem, Observations of geographically correlated orbit errors for TOPEX/Poseidon using the global positioning system, *Geophys. Res. Ltr. 21*(9), 2175–2178, 1994.

Guinn, J., *et al.*, TOPEX/Poseidon precision orbit determination using combined GPS, SLR and DORIS, these proceedings, 1995.

Marshall, J. A., *et al.*, The temporal and spatial characteristics of TOPEX/Poseidon radial orbit error, *J. Geophys. Res.*, in press, 1995.

Nerem, R. S., *et al.*, Gravity model development for TOPEX/Poseidon: Joint gravity models 1 and 2, *J. Geophys. Res. 99*(C12), 24,421–24,447, 1994.

Moritz, H., *Advanced Physical Geodesy*, Herbert Wichman Verlag, Karlsruhe, Germany, 1980.

Priesendorfer, R. W., *Principal component analysis in meteorology and oceanography*, Elsevier Science Publishers, 1988.

Tapley, B. D., *et al.*, Precision orbit determination for TOPEX/Poseidon, *J. Geophys. Res. 99*(C12), 24,383–24,404, 1994a.

Tapley, B. D., M. M. Watkins, *et al.*, The JGM-3 gravity model, *Annales Geophysicae, 12, Suppl. 1, C192*, 1994b.

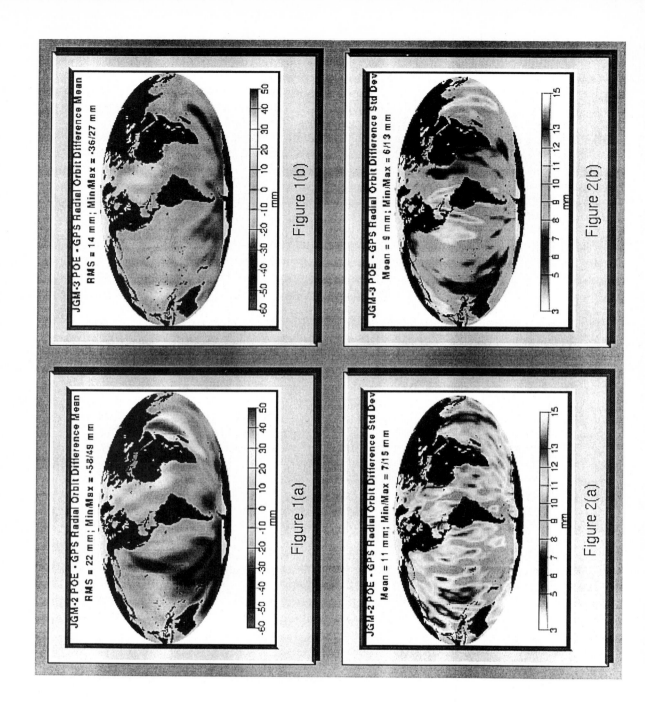

Figure 1(a)

Figure 1(b)

Figure 2(a)

Figure 2(b)

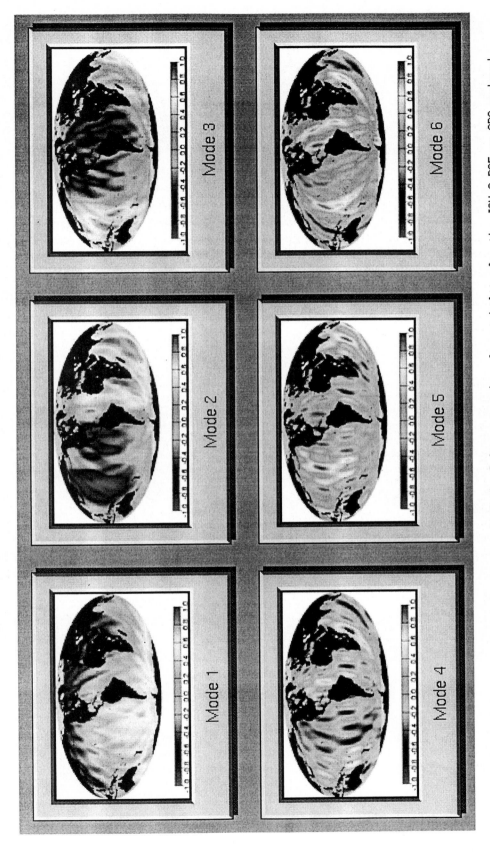

Fig. 3. Spatial representations (eigenvectors) of the 6 largest modes of variability for the JGM-2 POE vs GPS reduced dynamic orbit comparison from empirical orthogonal function analysis. The maps have been normalized so that the largest excursion is +/- 1. Note that the first 3 modes are characterized by large-scale features indicative of shifting in the center of figure (i.e., geocenter). Modes 4 through 6 contain shorter-wavelength features, some of which are attributable to errors in the JGM-2 POE background models of the principal lunar and solar tides. (This is further evidenced in Fig. 5.)

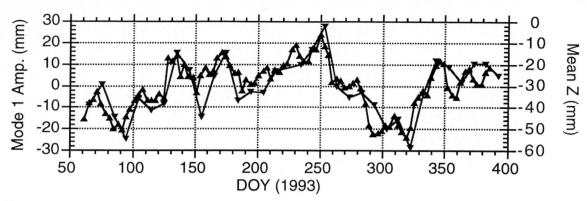

Fig. 4. The blue line shows the amplitude time series of the primary EOF mode for the JGM-2 POE – GPS reduced dynamic orbit differences. The red line shows the cycle-averaged body-fixed Z-offset for the same two orbit solutions. The high correlation corroborates that long-term changes in the Z-shift (compare also Figure 3a) comprise the primary mode of spatio-temporal variability.

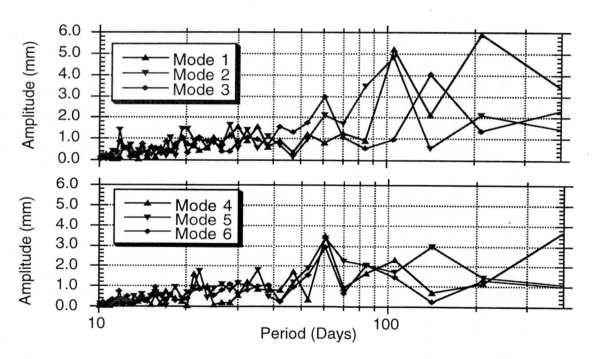

Fig. 5. Spectra of the amplitude time series for the EOF modes 1–3 (top panel) and 4–6 (bottom panel) for the JGM-2 POE – GPS reduced dynamic orbit differences. The first three modes, corresponding to center-of-figure motion, are characterized by long-term variations (> 100 d) while modes 4-6 exhibit 60-day variability associated with the M2/S2 tidal alias.

PRELIMINARY RESULTS FROM THE GPS/MET ATMOSPHERIC REMOTE SENSING EXPERIMENT

D. Feng, B. Herman
Institute of Atmospheric Physics, University of Arizona, Tucson, AZ 85721, USA
M. Exner, W. Schreiner, R. McCloskey, D. Hunt
University Corporation for Atmospheric Research, Boulder, CO 80301, USA

INTRODUCTION

Temperature, pressure, density and water vapor are the most significant variables describing the state of the atmosphere. Therefore they have always been the major goals of direct and indirect atmospheric remote sensings. Radiosonde, satellite infrared, and microwave soundings are the traditional means of collecting temperature, pressure and humidity data today. Radiosonde observations typically yield acceptable accuracy and high vertical resolution, but the geographic coverage is uneven with almost no coverage over oceans. Passive satellite remote sensing techniques yield good global coverage and high horizontal resolution, but the vertical and temporal resolutions are rather coarse. Due to these limitations, these data collection techniques have not been adequate for many weather and global climate applications today.

Since the 1990's, the Global Positioning System (GPS) has deployed dozens of transmitting satellites (24 satellites in operation now) in the sky. These satellites, ~21,000 km above the Earth's surface, emit 2 frequencies at 1.5 Ghz and 1.2 Ghz. The original goal of the GPS was to provide global and all-weather precision positioning and navigation for the military. However, more and more civilian applications and services are finding their places in the GPS frame.

The GPS/MET project is a proof-of-concept experiment to utilize those transmitted signals to determine atmospheric refractivity (and consequently temperature) profiles. It is sponsored by NSF, FAA, NOAA, and NASA. By nature, it is a radio (microwave) occultation experiment. On April 3, 1995, the Microlab-1 satellite carried a shoe-box size high performance GPS microwave receiver into a 750-km low earth orbit (LEO). Occultations occur wherever the radio link between any one of the 24 GPS transmitters and the LEO GPS receiver progressively descends or ascends through the earth's atmosphere. Since April, hundreds of occultations have occurred and large amounts of data have already been collected. Preliminary results indicate that microwave occultation is a powerful and successful technology for determining refractive index and temperature. According to these early results, for dry air, typically the retrieved temperature is accurate to within about ±1°C from 5–7 km up to 35–40 km. Its global coverage, low cost, high accuracy and high vertical resolution make the GPS/MET technique one of the most desirable and promising atmospheric remote sensing methods.

PRINCIPLES OF ATMOSPHERIC RADIO OCCULTATION REMOTE SENSING

Figure 1:
Radio Signal Delay Between GPS and LEO Satellites

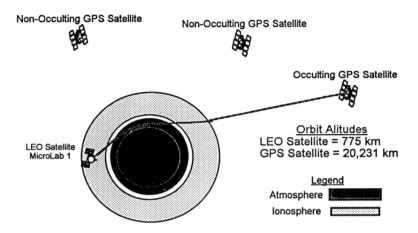

Radio occultation for atmospheric remote sensing is not a new technique. The Center for Radar Astronomy at Stanford University first developed this technique for planetary atmosphere probing some 3 decades ago (Tyler, 1987). During the Pioneer, Mariner and Voyager missions, the Stanford University group along with the Jet Propulsion Laboratory (JPL) group successfully applied this technique to the atmospheres of Mars, Venus, Jupiter, Saturn, Uranus and Neptune, as well as to some planet satellites. This occultation technique, however, had not been applied to the Earth's atmosphere until the GPS/MET project.

The GPS/MET radio occultation experiment basically involves a radio transmitter aboard one of the 24 GPS satellites and the GPS receiver aboard the LEO satellite. The GPS satellite is selected so that the radio link between the GPS and the LEO passes through the Earth's atmosphere. Because the refractive index of the atmosphere is a function of temperature, pressure and water vapor, which all vary with altitude in the atmosphere, the radio path in the atmosphere undergoes both propagation delay and bending. The deeper the radio path in the atmosphere, the larger the delay and bending of the radio beam. The GPS and LEO satellites are both orbiting, and thus the GPS receiver records a constantly varying signal both in frequency (phase) and in amplitude. Figure 1 depicts the GPS/MET radio occultation configuration. The bending of the radio path in this figure has been greatly exaggerated for clarity.

The retrievals of atmospheric refractive index from the GPS/MET data are based upon the so-called Abel equations (Phinney and Anderson, 1968; Fjeldbo et al., 1971). Assuming a spherically symmetric atmosphere, the index of refraction, μ, is related to the bending angle, α, by the inverse Abel equation:

$$\pi \, \ln \mu(r_i) = \int_{a=a_i}^{a=\infty} \frac{\alpha(a)}{\sqrt{a^2 - a_i^2}} \, da = \int_{\alpha=0}^{\alpha=\alpha(a_i)} \ln\left\{ \frac{a(\alpha)}{a_i} + \left[\left(\frac{a(\alpha)}{a_i} \right)^2 - 1 \right]^{\frac{1}{2}} \right\} d\alpha, \qquad (1)$$

where a is the impact distance of the radio path, and r_i is the *ith* perigee radius of the radio path associated with the *ith* impact distance a_i. This inverse Abel equation allows one to calculate the atmospheric index of refraction, μ, from the bending angles α and impact distance a_i, which are in turn deduced from the GPS measurements.

It is convenient to define the refractivity, N, by

$$N = (\mu - 1) \times 10^6 \qquad (2)$$

Thus after introducing the numerical values of the relevant physical constants, N is related to atmospheric density, ρ, by the equation

$$N = 77.6 \times \rho R + 3.73 \times 10^5 \times \frac{P_w}{T^2} \qquad (3)$$

where P_w is the water vapor pressure in mb, T is the atmospheric temperature in Kelvin., and R is the gas constant for dry air. It is clear that eq. (3) can be solved only when $P_w=0$, unless a separate measurement of vapor pressure is available. For the dry air case, we may solve eq. (3) for ρ, and then from the hydrostatic relationship, i.e., $\Delta P = -\rho g \Delta z$, by starting at an altitude high enough so that $P=0$, we may solve for $P(z)$. Combining the hydrostatic equation, the equation of state, and eq. (3), we get an expression for temperature profile $T(z)$,

$$T(z_i) = \frac{1}{R} \times \frac{\int_{z_i}^{\infty} g(z) N(z) \, dz}{N(z_i)} \qquad (4)$$

where $g(z)$ is the gravitational acceleration, and z is the altitude in the atmosphere.

RESULTS FROM THE GPS/MET OCCULTATION EXPERIMENTS

Six of the retrieved temperature profiles are presented here along with nearby radiosonde data and NMC analysis data (if available) for comparisons. The panels in each figure show the day, time, latitude and longitude of the observations. In these panels, CL and RS stand for the CLASS and radiosonde soundings, respectively. These results are all for 'dry' air, i.e., the P_w term in eq. (3) was ignored. This is justified for the retrievals higher than 5-7 km. However, for many applications, such as monitoring of global climate changes and weather forecasting, directly employing the total refractivity N, rather than temperature T, may be more preferable and accurate (Eyre, 1994; Zou et al., 1995); therefore, water vapor would no longer pose any problems.

The first retrieval using the GPS/MET data was for a tropical temperature profile shown in Fig.2. It was obtained under several simplifications, including using 1 Hz data (usually 50 Hz data are used) and a simplified ionospheric correction. Notice the water vapor below ~8 km caused the retrieved T to be shifted to the left (cooler than actual T.)

Fig.3 shows the GPS/MET temperature profile compared to nearby radiosonde data. Again, the cold bias below 6 km due to water vapor is obvious.

Fig.4 shows a sounding down to ~ 1 km. From the cold bias at the bottom, we have estimated a water vapor pressure of 4.6 mb. The corresponding nearby radiosonde observations gave a water vapor pressure of 5.0 mb. Thus two numbers are in fairly good agreement.

Fig.5 was taken over western Lake Superior, with the high-resolution CLASS sounding for comparison. The fine structure of the GPS/MET temperature profile is well exhibited.

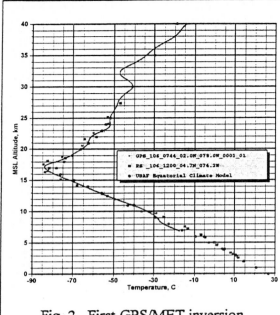

Fig. 2. First GPS/MET inversion.

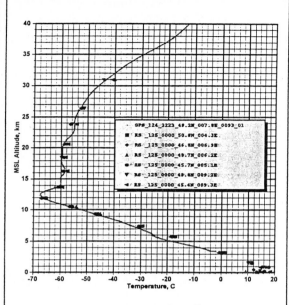

Fig. 3. Day 124 Occ #093.

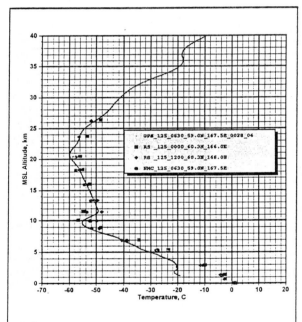

Fig. 4. Day 125 Occ #028.

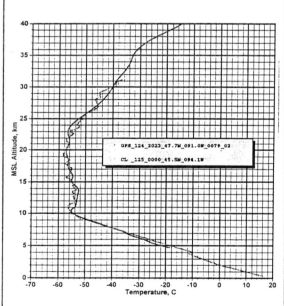

Fig. 5. Day 124 Occ #079.

Figure 6 represents a GPS/MET result north-west of Beijing, China. Ten radiosondes plus the NMC analysis are coplotted. They present the range of variability of radiosonde profiles. The GPS/MET profile seems to run through the middle of these radiosondes.

Notice that all figures show the locations and temperature of the tropopause very well.

SUMMARY

From the preliminary results in the previous section, we conclude that accurate vertical temperature profiles (and refractive index profiles as well) may be obtained using the radio occultation technique from 35–40 km down to ~5 km. The agreements between GPS/MET and nearby radiosondes are usually within 1° or 2°C. The vertical resolution of the GPS/MET is ~1 km or less.

There are two problems that need to be addressed. Above ~40 km, the current GPS/MET results are not very reliable because of the ionosphere. Ionospheric corrections are a subject in need of further investigation. Below 5–7 km, water vapor affects the accuracies of temperature retrievals. Although the retrieved N(z) contains the effects of water vapor, the GPS/MET measurements alone can't independently separate the water vapor effects from the temperature effects. Additional information or observations are necessary for resolving the water vapor. This subject is now under study. Nevertheless. it is clear that the radio occultations from space afford a promising and powerful atmosphere remote sensing technique. Weather forecasting and global climate change research would directly benefit from this technique because of its high data quality, complete global coverage, and low cost. The GPS/MET data can also be used to determine total ionospheric electron content and therefore may contribute to studies of the ionosphere.

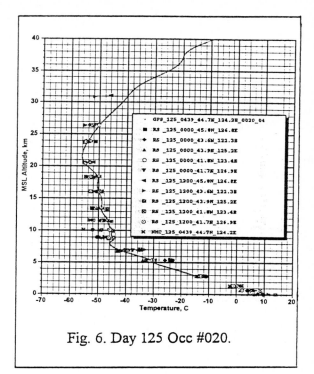

Fig. 6. Day 125 Occ #020.

REFERENCES

Eyre, J. R. (1994). Assimilation of radio occultation measurements into a numerical weather prediction system, *ECMWF Technical Memorandum* 199, 34 pp.

Fjeldbo, G., Kliore, A. J., and Eshleman, Von R. (1971). The neutral atmosphere of Venus as studied with the Mariner V radio occultation experiments, *The Astronomical Journal* 76, 123–140.

Phinney, R. A., and Anderson, D. L. (1968). On the radio occultation method for studying planetary atmospheres, *J. Geophys. Res.* 73, 1819–1827.

Tyler, R. L. (1987). Radio propagation experiments in the outer solar system with Voyager, *Proceedings of the IEEE* 75, 1404–1431.

Zou, X., Kuo, Y.H., and Guo, Y. R. (1995). Assimilation of atmospheric radio refractivity using a mesoscale model, *Mon. Wea. Rev.* 123, 2229–2249.

INITIAL RESULTS OF GPS-LEO OCCULTATION MEASUREMENTS OF EARTH'S ATMOSPHERE OBTAINED WITH THE GPS-MET EXPERIMENT

G. A. Hajj, E. R. Kursinski, W. I. Bertiger, S. S. Leroy, T. K. Meehan, L. J. Romans, and J. T. Schofield

Jet Propulsion Laboratory
California Institute of Technology

ABSTRACT

The radio occultation technique, which has been repeatedly proven for planetary atmospheres, was first utilized to observe Earth's atmosphere by the GPS-MET experiment (launched in April 1995), in which a high performance GPS receiver was placed into a low-Earth orbit. During certain phases of the mission, more than 100 occultations per day are acquired. A subset of this occultation data is analyzed and temperature in the neutral atmosphere and electron profiles in the ionosphere are obtained. Comparing about 100 GPS-MET retrievals to accurate meteorological analyses obtained from the European Center for Medium-range Weather Forecasting at heights between 5-30 km, temperature differences display biases of less than 0.5K and standard deviations of 1-2K in the northern hemisphere, where the model is expected to be most accurate. Furthermore, electron density profiles obtained for different geodetic locations and times show the main features that are expected in the ionosphere.

1. INTRODUCTION

When a signal transmitted by the global positioning system (GPS)[1] and received by a low-Earth orbiter (LEO) passes through the Earth's atmosphere [Fig. 1] its phase and amplitude are affected in ways that are characteristic of the index of refraction of the propagating medium. By applying certain assumptions on the variability of the index of refraction of the propagating media (e.g. spherical symmetry in the locality of the occultation), phase change measurements between the transmitter and the receiver yield refractivity profiles in the ionosphere (~60-1000 km) and neutral atmosphere (0-50 km). The refractivity, in turn, yields electron density in the ionosphere, and temperature and pressure in the neutral atmosphere. In the lower troposphere, where water vapor contribution to refractivity is appreciable, independent knowledge of the temperature can be used to solve for water vapor abundance.

The radio occultation technique has a 30 year tradition in NASA's planetary program and has been a part of the planetary exploration programs to Venus, Mars and the outer planets [see, for example, Tyler, 1987]. However, the application of the technique to sense the Earth's atmosphere using

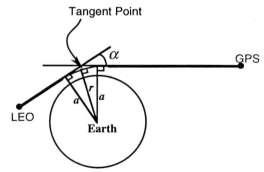

Fig. 1: Occultation geometry defining a, r, α and the tangent point.

---

[1]The GPS consists of 24 satellites in 6 evenly distributed orbital planes, with an inclination of ~55⁰, altitude ~20,200 km, and a revolution period of ~12 h. Each satellite transmits two carrier signals, at 19.0 cm and 24.4 cm wavelength modulated by a Pseudo-random code (P-code). L1 is further modulated by a Coarse-Acquisition code (C/A code).

GPS, first suggested by Melbourne et al. [1988] and Yunck et al. [1988], was tested for the first time with the launch of the GPS-MET mission on April 3, 1995. GPS-MET is an experiment managed by the University Corporation of Atmospheric Research (UCAR) [Ware et al., 1995] and it consists of a 2 kg GPS receiver piggybacked on the MicroLab I satellite which has a circular orbit of 730 km altitude and 60° inclination. The GPS receiver is a space qualified TurboRogue [Meehan et al., 1992] capable of tracking up to 8 GPS satellites simultaneously at both frequencies transmitted by GPS. Under an optimal mode of operation, the GPS receiving antenna boresight is pointed in the negative velocity direction of the LEO and provides 100-120 globally distributed setting occultations per day. By the end of the mission (nominal life time of 6 months), thousands of occultations will have been collected and can be used to assess the accuracy and potential benefit of the GPS radio occultations.

To date, a relatively small fraction of all recorded occultations have been analyzed using the Abel transform approach (presented below). This paper discusses how the GPS-MET data are analyzed and presents some results of temperature retrievals compared to radiosonde measurements and atmospheric analyses obtained from the European Center for Medium-range Weather Forecast (ECMWF). It also presents some preliminary results of electron density profiles obtained in the ionosphere. The paper is structured as follows: Section 2 gives a brief background on the radio occultation technique. The basic features of the technique are presented in section 3. The manner in which the GPS-MET phase data are calibrated to isolate the atmospheric excess phase is described in section 4. Section 5 presents an individual temperature profile and statistics obtained for all occultations available from 2 days during the experiment. These retrievals are compared to atmospheric analysis from ECMWF. In section 6, we show retrievals of ionospheric profiles obtained at different times of day and geographical locations. A conclusion is given in section 7.

2. RADIO OCCULTATION TECHNIQUE

The basic observable for each occultation is the phase change between the transmitter and the receiver as the signal descends through the ionosphere and the neutral atmosphere. After removal of geometrical effects due to the motion of the satellites and proper calibration of the transmitter and receiver clocks, the extra phase change induced by the atmosphere can be isolated. Excess atmospheric Doppler shift is then derived. This extra Doppler shift can be used to derive the atmospheric induced bending, α, as a function of the asymptote miss distance, a, [Fig. 1]. Assuming a spherically symmetric atmosphere, the relation between the bending and excess Doppler shift, Δf, is given by

$$\Delta f = \frac{f}{c}\left[\vec{v}_t \cdot \hat{k}_t - \vec{v}_r \cdot \hat{k}_r + \left(\vec{v}_t - \vec{v}_r \right) \cdot \hat{k} \right],$$ (1)

where f is the operating frequency, c is the speed of light, $\vec{v}_t$ and $\vec{v}_r$ are the transmitter and receiver's velocity respectively, $\hat{k}_t$ and $\hat{k}_r$ are the unit vectors in the direction of the transmitted and received signal respectively, $\hat{k}$ is the unit vector in the direction of the straight line connecting the transmitter to the receiver.

The spherical symmetry assumption can also be used to relate the signal's bending to the medium's index of refraction, n, via the relation

$$\alpha(a) = 2a \int_a^\infty \frac{1}{\sqrt{a'^2 - a^2}} \frac{d \ln(n)}{da'} da',$$ (2)

where $a = nr$ and r is the radius at the tangent point [Fig. 1]. This integral equation can then be inverted by using an Abel integral transform given by

$$\ln (n(a)) = \frac{1}{\pi} \int_a^\infty \frac{\alpha(a')}{\sqrt{a'^2 - a^2}} da'.$$ (3)

The refractivity, N, is related to atmospheric quantities via

$$N = (n-1) \times 10^6 = 77.6 \frac{P}{T} + 3.73 \times 10^5 \frac{P_W}{T^2} - 40.3 \times 10^6 \frac{n_e}{f^2} , \quad (4)$$

$$P = \frac{\rho R T}{m} , \quad (5)$$

$$\frac{\partial P}{\partial h} = -g\rho , \quad (6)$$

where P is total pressure (mbar), T is temperature (K), P_W is water vapor partial pressure (mbar), n_e is electron density (m^{-3}), f is operating frequency (Hz), ρ is density, R is the gas constant, m is the gas effective molecular weight, h is height, g is gravitational acceleration.

When the signal is passing through the ionosphere (tangent point height > 60 km), use of a single GPS frequency is sufficient to estimate α to be used in Eq. (3). Moreover, the first two terms on the right hand side of Eq. (3) are negligible, therefore, knowledge of the index of refraction leads directly to electron density.

When the signal is going through both the neutral atmosphere and the ionosphere (tangent point height < 60 km), a linear combination of the two bending angles, associated with the two GPS frequencies, is used to isolate the *neutral* atmospheric bending and its refractivity profile is derived by use of Eq. (3) [Vorob'ev and Krasil'nikova, 1993]. In the stratosphere and the region of the troposphere where temperature is colder than ~250K, the water vapor term in Eq. (4) is negligible. Therefore, knowledge of refractivity yields the density of the medium by use of the ideal gas law (Eq. 5). The density in turn yields the pressure by assuming hydrostatic equilibrium (Eq. 6) and a boundary condition at some height. Applying the gas law once more, knowledge of density and pressure yields the temperature. In the troposphere, at height where the temperature is larger than 250K, the water vapor term in Eq. (4) becomes significant and it is more efficient to solve for water vapor given some independent knowledge of temperature [Kursinski et al., 1995a].

3. GPS RADIO OCCULTATION FEATURES

Details about vertical and horizontal resolution of the technique, and refractivity, temperature, pressure, water vapor or electron density accuracies as a function of height, are given elsewhere in the literature [Hardy et al., 1993, Kursinski et al., 1993; Hajj et al., 1994]. In this section we quickly summarize the results of these studies.

Due to the nature of the measurement, which is a pencil-like beam of the electromagnetic signal probing the atmosphere, the technique has a much higher vertical and across-beam resolution than horizontal (i.e. along the beam). The vertical resolution of the technique is essentially set by the physical width of the beam where geometrical optics is applicable. This scale is set by the Fresnel diameter which, in vacuum, is given by

$$D_{vacuum} = 2 \sqrt{\frac{\lambda \, R_{GPS} R_{LEO}}{\left(R_{GPS} + R_{LEO} \right)}} , \quad (7)$$

where λ is the signal's wavelength, R_{GPS} and R_{LEO} are the distances of the tangent point (see Fig. 1) to the GPS and LEO respectively. For a LEO, D_{vacuum} is ~1.5 km. In the presence of a medium, due to bending induced on the signal, the Fresnel diameter is ~0.5 near the surface and approaches 1.5 km above 20 km altitude where bending becomes small. When the signal encounters sharp gradients in refractivity due to either water vapor layers near the surface or sharp electron density changes at the bottom of the ionosphere, the Fresnel diameter shrinks to ~200 meters.

A horizontal resolution scale is set by the length of the beam inside a layer with a Fresnel diameter thickness. This length is 160-280 km for a Fresnel diameter of 0.5-1.5 km.

In the ionosphere, the vertical scale is still set by the Fresnel diameter; however, the horizontal scale can extend several thousands of kilometers due to the large vertical extent and scale height of the ionosphere. These features of the ionosphere allow one to use tomographic approaches in order to combine information from neighboring occultations to solve for horizontal and vertical structure [Hajj et al., 1994].

Under ideal conditions, when a LEO tracking GPS has a 360° field of view of the Earth's horizon, about 750 occultations per LEO per day can be obtained. However, side-looking occultations (GPS-LEO link > 45° from velocity or anti-velocity of LEO) sweep across a large horizontal region, and the spherical symmetry assumption described in Sec. 2 becomes inaccurate. Discarding side-looking occultations, one LEO provides up to 500 occultations per day.

In the case of GPS-MET, only an aft-looking antenna was mounted on the satellite, which reduces the viewing geometry to 1/2 the Earth's limb (±90° from boresight). In addition, in order to calibrate the clocks of the occulting transmitter and receiver, one other GPS transmitter and one ground GPS receiver are required (see Fig. 2; the technique of calibration is described in more detail in the next section). This requirement, in addition to some memory limitations inside the flight receiver, limits the number of occultations to about 100 per day.

A high inclination LEO provides a set of occultations that covers the globe fairly uniformly. This feature is particularly advantageous when comparing LEO-GPS occultation coverage to that obtained from balloon launched radiosondes. A total of about 800 radiosondes are launched each 12 hours from sites around the world. The vast majority of these sites are over the northern hemisphere continents, particularly Europe and North America. This creates the need for high resolution temperature/pressure/water vapor profiles in the southern hemisphere and over the oceans. The contribution of radio occultation retrievals to climate and weather modeling should be particularly important in these regions. (Global data provided by spaceborne nadir sounders average over large—3-7 km—vertical distances.)

When compared to infrared spaceborne sounders, the radio occultation technique has the advantage of being an "all-weather" system. Namely, it is insensitive to aerosols, cloud or rain due to the relatively large GPS wavelengths.

Unlike other techniques such as radiosonde or microwave sounders, where instruments need constant calibration, the GPS radio occultation provides a self calibrating system, as will be discussed in more detail below. The long term stability inherent in radio occultation make this an excellent system to keep an accurate record of climate changes.

4. CALIBRATING THE GPS SIGNALS/ISOLATING ATMOSPHERIC EXCESS DELAY

The main observable used in an occultation geometry is the phase change between the transmitter and the receiver as the occulting signal descends through the atmosphere. This phase change is due to (1) the relative motion of the LEO with respect to the GPS, (2) clock drifts of the GPS and LEO and (3) delay induced by the atmosphere. In order to derive the excess atmospheric Doppler shift, one must remove the contribution of the first two effects.

Accurate knowledge of the GPS orbits comes from an overall solution involving all 24 GPS satellites and a global network of ground receivers. The LEO orbit is determined by use of other links tracking the non-occulting GPS satellites.

When the occultation is mostly radial (i.e. GPS-LEO link has no horizontal motion out of the occultation plane), the occultation link descends through the ionosphere and stratosphere at a rate of about 3 km/sec; thus, crossing a Fresnel diameter (see Sec. 3) in about 0.5 seconds. However, in order to investigate sub-Fresnel structure (by examining the diffraction pattern of the received signal's phase and amplitude) and for other purposes (such as eliminating different signals caused by atmospheric multipath in the lower troposphere) the occulting data is taken at a rate of 50 Hz. In order to calibrate the LEO clock, one more GPS transmitter is tracked by the LEO at the same high rate (link 2 in Fig. 2). In addition, in order to calibrate the GPS clocks, a ground receiver tracks both GPS satellites at 1 Hz (links 3 and 4 in Fig. 2). One can interpolate the lower rate GPS clock solutions to 50 Hz, due to the greater clock stability (of order 10^{-12} sec/sec, as opposed to 10^{-9} sec/sec for the LEO clock), and the smoothness of the DoD Selective Availability dithering.

Knowing the position of all four participants (i.e. two GPS satellites, one LEO and one ground receiver), and modeling various physical effects such as light travel time, the three spaceborne clocks can be solved for w.r.t. to the ground clock. The net result of the calibration is the excess phase due to the atmosphere as a function of time (see Fig. 3.a).

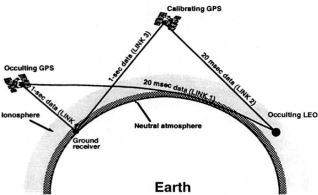

Fig. 2: The occultation geometry involving two GPS transmitters, one ground receiver and one space receiver.

5. DATA ANALYSIS AND TEMPERATURE PROFILES

In this section we show the various steps of processing for a single retrieval in order to understand the basic characteristics of the atmospheric effects on the signal. We then look at statistical differences between two days' worth of occultations and a numerical weather prediction model.

An Individual Retrieval

After applying the calibration described in the previous section, we obtain the atmospherically induced phase delay (up to a constant bias). Fig. 3 shows the L1 delay, Doppler shift and instrumental signal-to-noise ratio for an occultation near Pago Pago, -14 N and 190 E near midnight UT of April 25, 1995. The following features can be observed from these two plots: 1- The phase has a constant bias of about 20 m; this bias is irrelevant for consequent processing since it is the phase time derivative that is used. 2- Near the bottom of the ionosphere, there is a sharp fluctuation of the SNR due to the sharp gradient in refractivity which causes more

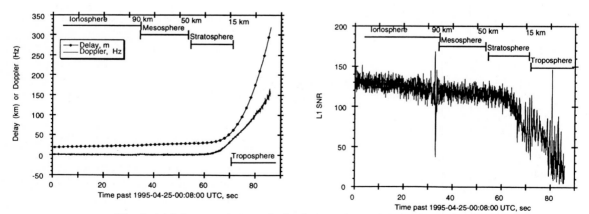

Fig. 3: (a) left: excess atmospheric phase and doppler as a function of time;
(b) right: receiver's signal-to-noise ratio as a function of time

bending and therefore defocusing. This is suggestive of the sensitivity of GPS radio occultation to sense the sharp structure of the bottom of the E-layer in the ionosphere. 3- The rapid increase in excess phase and Doppler shift and the decrease in SNR starting at the lower stratosphere is due to the fact that atmospheric bending is becoming significant. This bending causes the SNR to drop from ~130 volt/volt at the top to ~35 v/v at the bottom (averaged over 1 sec.), corresponding to about 11 dB of signal loss, and finally to lose the signal. 4- The SNR shows a clear oscillation near the tropopause which is indicative of a diffraction pattern caused by the sharp change in temperature lapse rate. 5- The SNR shows a peak (~150 v/v) in the

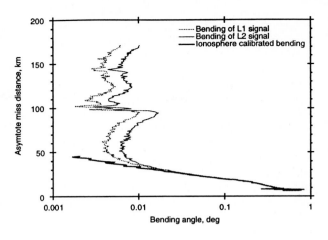

Fig. 4: Bending of GPS-L1 and L2 signals and ionospheric free bending as a function of asymptote miss distance

middle-troposphere which can be caused by signals coming from a large region (relative to a Fresnel zone) and focusing near the receiver. The corresponding L1 and L2 bending for the same occultation are shown in Fig. 4. Again the sharp feature around 90 km is caused by the sharp curtailing of electron density. The L2 bending as a function of asymptote miss distance, $\alpha_2(a_2)$, is interpolated to the L1 asymptote miss distance and the following relation is used to calculate the neutral atmosphere's contribution to bending [procedure suggested by Vorob'ev and Krasil'nikova, 1993]

$$\alpha(a) = 2.54\ \alpha_1(a_1) - 1.54\ \alpha_2(a_1), \qquad\qquad (8)$$

where the first and second coefficients of Eq. 8 corresponds to $f_1^2/(f_1^2 - f_2^2)$ and $f_2^2/(f_1^2 - f_2^2)$ respectively, and f_1, f_2 are the operating frequencies for L1 and L2 respectively. The difference in bending in L1 and L2 frequencies is due to the dispersive nature of the ionosphere (which leads to Eq. 8). Above 40 km, bending due to the ionosphere dominates.

Using the ionosphere free bending, $\alpha(a)$ and Eqs. (3)-(6), temperature is derived in the neutral atmosphere and is shown as a function of pressure in Fig. 5. Also shown on the same figure are temperature profiles obtained from a nearby radiosonde and a stratospheric numerical weather prediction model obtained from the National Meteorological Center (NMC). The GPS-MET profile agrees with the radiosonde and the NMC analysis to about 2K between 450-10 mbar and to the NMC analysis to about 10 K

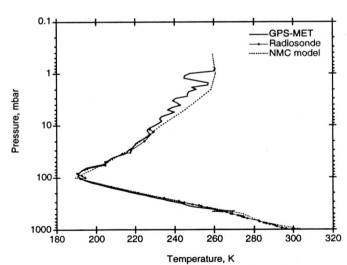

Fig. 5: Temperature profile from GPS-MET, radiosonde and NMC stratospheric model

between 10-1 mbar. The only auxiliary information used in deriving the GPS-MET temperature is an initial condition of temperature at 50 km altitude equal to the NMC analysis temperature. Given the measured density at that height, this initial condition can be translated into a pressure boundary condition which is needed in order to integrate Eq. 6. The oscillation of the GPS-MET temperature above 10 mbar can be attributed to thermal noise in the GPS phase measurement and residual ionospheric effects. The lowest point in the GPS-MET profile corresponds to about 7 km where the signal is lost. This loss of the signal is due to signal defocusing which is exacerbated by the presence of water vapor layers in the lower troposphere.

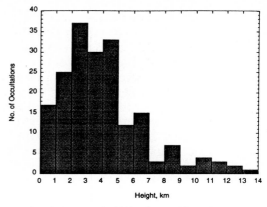

Fig. 6: Lowest height of occulting signal for days April 24,25 and May 4,5 of 1995.

More individual temperature retrievals as well as statistical differences between numerical weather prediction analyses obtained from the European Center for Medium-range Weather Forecasting (ECMWF) and GPS-MET are also discussed by Kursinski et al. [1995b].

Statistical Comparisons

GPS-MET, which is a secondary instrument on MicroLab I, is configured in a favorable geometry (antenna boresight in the anti-velocity direction) only for about two weeks out of each repeat cycle of the satellite (55 days). Thus far, this has happened twice: once between April 22-May 6, 1995, and again between June 17-July 11, 1995. (These periods were interrupted by times of non-ideal viewing geometry, due to attitude control problems.) AS[2] was off during these periods and data for April 24, 25 and May 4, 5 were analyzed. On each of these days respectively, 98, 119, 98 and 69 occultations were recorded, with about half of these successfully inverted, while the rest were automatically discarded, normally due to a data gap in one of the four links discussed above (see Fig. 2). The number of occultations for these four days as a function of the lowest height that an occultation reaches is shown in Fig. 6.

In order to assess the accuracies of retrieved temperature profiles from GPS occultations, we compare with the 6-hour ECMWF analyses. These are among the best available global analyses of atmospheric temperature structure below 10 mbar, and comparison against them has become a standard method for evaluating the accuracy and resolution of observational results [Flobert et al., 1991]. Fig. 7 shows temperature difference statistics for all successfully retrieved profiles for May 4 and 5, 1995. In order to eliminate temperature retrieval errors due to water vapor, tropospheric temperatures exceeding 250 K have been excluded from the comparisons. The three panels in Fig. 7 display temperature difference statistics for the northern high latitudes (30N-90N), the tropics (30S–30N), and the southern high latitudes (30S-90S). Within each latitude zone, retrieved profiles are widely scattered in both location and time.

It is clear from Fig. 7 that agreement between the two data sets in the northern hemisphere is impressive with mean differences of generally less than 0.5 K and difference standard deviations of typically 1 to 2 K. It should also be remembered that these differences include retrieved vertical structure that is not resolved by the ECMWF analysis, especially above 100 mbar. This agreement is particularly significant because the ECMWF analyses are expected to

---

[2]Anti Spoofing (AS) is the DoD encryption of the GPS signal for non-authorized users. This causes the L2 phase data noise to increase by about a factor of 10 for the current GPS-MET receiver.

be most accurate in the northern hemisphere. Although both radiosonde and TOVS (TIROS Operational Vertical Sounder, a space-based sensors with typical 3-7 km vertical resolution) data are assimilated into the ECMWF model, the analyses are expected to be less accurate in some regions of the southern hemisphere due to the sparse distribution of radiosondes. Southern hemisphere radiosondes cluster over a few land masses whereas the occultations fall mostly over the ocean. Fig. 7 shows that in the southern hemisphere, both mean temperature differences and standard deviations increase at lower altitudes. As the occultation retrieval process has little dependence on latitude, the good agreement in the northern hemisphere suggests that the larger systematic and random differences at southern latitudes originate in the analyses rather than in the retrieved profiles. Further inspection of the data shows that this difference feature is produced by a small sub-set of 8 occultation profiles concentrated far from radiosonde ascents in the southern hemisphere storm track and close to the ice edge, where problems in the assimilation of TOVS data are known to arise [Eyre et al., 1993]. Agreement with the remaining 25 profiles is comparable with that achieved in the northern hemisphere.

Temperature differences at tropical latitudes also display distinctive structure in Fig. 7. On average, retrieved profiles are about 1 K colder than the analyses between 300 and 100 mbar whereas above 70 mbar, they are warmer by a similar amount. A statistical comparison between tropical radiosondes and the ECMWF analysis revealed a qualitatively similar temperature difference structure, although the radiosonde temperatures in the upper troposphere are generally not quite as cold as the retrievals. Retrieved temperature gradients are systematically larger than analysis gradients just above the tropopause. These gradients are associated with wave-like structure often seen in the retrievals just above the tropical tropopause, and not resolved in the ECMWF analyses. While tropospheric standard deviations are similar to those in the northern hemisphere, stratospheric values are somewhat larger due, perhaps, to waves above the tropopause. Accurate temperature measurements near the tropical tropopause are needed to understand convection and energy transfer within the atmosphere, troposphere-stratosphere exchange processes, and future climatic variations. Although the temperature retrievals are preliminary, and in spite of the clear need for more data, the tropical results are felt to be reliable because of the excellent agreement achieved in the northern hemisphere.

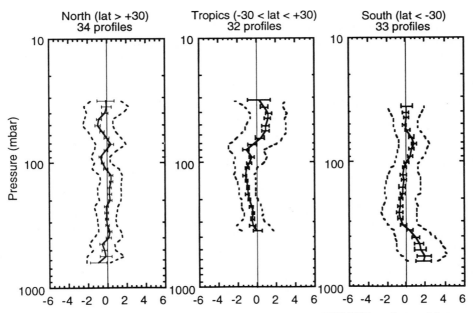

Fig. 7: GPS occultation retrievals from 95/05/04 and 95/05/05 vs. ECMWF analyses. Mean temperature difference with standard error bars (K) (solid lines); One standard deviation about mean difference (shaded lines).

6- IONOSPHERIC PROFILES

Although the purpose of the GPS-MET experiment was mainly to demonstrate the usefulness of the GPS radio occultation for sensing the neutral atmosphere, the same technique can be used to obtain profiles of electron density in the ionosphere. In the ionosphere, the spherical symmetry assumption is not as accurate as in the neutral atmosphere, for reasons that are described in Sec. 3. Nevertheless, in this section we show some representative profiles of electron densities obtained from GPS-MET with the spherical assumption. A first order, but significant, improvement of the spherical symmetry has been proposed elsewhere [Hajj et al., 1994] where global maps of integrated zenith electron density [Mannucci et al., 1994] can be used in order to constrain the horizontal variability.

The nominal design of the GPS-MET receiver was to collect data at three different rates depending on the geometry of the GPS-LEO link. When the link is at positive elevation (i.e. looking above the LEO local horizontal at 730 km) the rate is 0.1 Hz. When the link has a negative elevation (i.e. its tangent point is below the LEO altitude), data is taken at 1 Hz rate. When the tangent point gets as low as ~120 km altitude (30 km above which the neutral atmospheric effect starts to be detectable) the data is taken at 50 Hz. (Data used in deriving Fig. 5, had a high rate starting at 180 km.) Due to complications with the receiver software, however, one second data has not been collected. Instead, 0.1 Hz data was available through the ionosphere down to 120 km below which 50 Hz was taken. The inversions shown below are based on connecting these two data rates which explains the higher density of points below 120 km.

One can readily distinguish numerous prominent features of the ionosphere at day and night time and for different geodetic latitudes. These profiles are obtained around midnight (night-time profiles) and noon (day-time profiles) of May 4, 1995. The main features that are readily observed are the presence of the three distinct layers, E, F1 and F2 in the mid-latitude and equatorial day-time profiles, the higher electron density during the day, the sharp drop of the F1 region at night, the higher F2 peak near the equator and the very low peak at high-latitude night. Normally, one would expect the electron density to drop down to effectively zero around 60 km. The fact that they do not can be attributed to the spherical symmetry assumption used in the

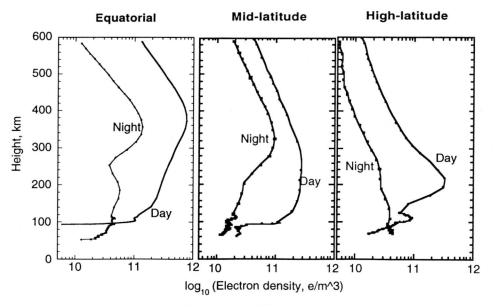

Fig. 8: Ionospheric electron density profiles for three different geographical regions for night and day time obtained with the assumption of spherical symmetry.

retrieval which can create an overall bias in the E-layer electron density, although the point-to-point structure can be accurate. Ways of improving these retrievals are now underway and will be presented in a future work.

7. CONCLUSION

Based on theoretical estimations and simulations [Hardy et al., 1993] atmospheric temperature profiles are expected to be accurate to the sub-Kelvin level between 5-30 km heights. Initial results of GPS-MET are consistent with these predictions. The GPS radio occultation measurements combine accuracy with the vertical resolution necessary to resolve tropopause structure in a way that is well beyond the capabilities of current space-based atmospheric sounders. A single orbiting GPS receiver provides up to 500 globally distributed soundings daily. The density of these measurements exceeds that of high vertical resolution radiosonde soundings by several factors in the southern hemisphere The coverage, robustness, accuracy, vertical resolution, and insensitivity to cloud inherent to GPS radio occultation suggest that it will have a major contribution to global change and weather prediction programs around the globe.

In the ionosphere, GPS radio occultations provide electron density profiles. Spherical symmetry is accurate enough to see the prominent structures in the ionosphere, but improvements over this assumption, such as using information from ground data and/or nearby occultations, can be applied to get more accurate profiles.

ACKNOWLEDGMENTS

We thank M. Exner and R. Ware of UCAR for providing the GPS-MET flight data and Y. Bar-Sever, T. Lockhart, R. Muellerschoen and S. Wu of JPL for their significant contributions. This research was performed at JPL, supported by NASA and the Caltech President's fund.

REFERENCES

Eyer et al., *Quart. J. R. Met. Soc.*, 119:1427-1463, 1993.

Flobert, J. F. et al., *Mon. Weath. Rev.*, 119:1881-1914, 1991.

Hajj G. A. et al., *Int. J. of Imaging Sys. and Tech.*, Vol. 5, 174-184, 1994.

Hardy K. R. et al., *Proc. of the ION-GPS 93 Conf.*, Salt Lake City, pp. 1545-1557, 1993.

Kursinski E. R. et al, *Proc. of the 8th Symp. on Meteorological Observations and Instrumentation*, pp. J153-J158, American Meteorological Society, Anaheim, 1993.

Kursinski E. R.et al, *Geophy. Res. Lett.*, 1995a.

Kursinski E. R. et al, in press, 1995b.

Mannucci, A. J. et al., *Proc. of the Int. Beacon Satellite Symp.*, L. Kersley ed., University of Wales, Aberystwyth, pp 338-341, July 1994.

Meehan et al, *6th Int. Geodetic Symp. on Satellite Positioning*, Columbus Ohio, 1992.

Melbourne W.G. et al., *GPS geoscience instrument for EOS and Space Station*, GGI proposal to NASA, July 15, 1988.

Tyler G. L., *Proc. IEEE*, 75:1404-1431, 1987.

Vorob'ev V.V. and T.G. Krasil'nikova, *Fizica Atmosfery i Okeana*, 29, 626-633, 1993.

Ware R. M. et al, in press, 1995.

Yunck T. P. et al., *Proc. of IEEE position location and navigation symposium*, Orlando, 1988.

THE LASER RETROREFLECTOR EXPERIMENT ON GPS-35 AND 36

E. C. Pavlis
Dept. of Astronomy, University of Maryland,
Space Geodesy Branch, NASA/GSFC 926, Greenbelt, MD 20771

Ronald L. Beard
Space Applications Branch, Naval Center for Space Technology
U.S. Naval Research Laboratory, Washington, D.C. 23075

INTRODUCTION

The purpose of this project is to identify and investigate means of enhancing the Global Positioning System (GPS) system integrity and performance. This project involves installing laser retroreflector arrays on–board Global Positioning System (GPS) satellites, tracking the satellites involved in cooperation with the NASA Satellite Laser Ranging (SLR) network and collecting data for analysis and comparison with GPS phase and pseudorange data. The Laser Retroreflector Experiment (LRE), previously known as the Advanced Clock Ranging Experiment (ACRE), was submitted by the U. S. Naval Research Laboratory (NRL) to the Tri–Service Space Test Program for spacecraft integration funding as a Tri–service space experiment.

THE EXPERIMENT

The objective of such an experiment is to provide an independent high precision measurement to compare and calibrate the GPS pseudoranging signal. This project is a cooperative effort involving the NASA Goddard Space Flight Center SLR group, the NRL and the University of Maryland. Installation of the LRE on the GPS satellite was performed in conjunction with the GPS Joint Program Office and their contractor, Rockwell International, the Air Force Space Command and the Second Satellite Operations Squadron.

The navigation component of GPS is a predicted, real–time, passive ranging system, made up of space, control and user segments. The space and control elements comprise the system proper, and the user segment operates passively utilizing the products of the system transmitted by the space segment. Tracking network data are similar in content to that used by the user segment and is relayed to the Master Control Station (MCS) for computation and prediction of the system states which are uploaded into the satellites for the users. Embedded in the space and control segments are atomic clocks to maintain all elements of the system in synchronization and enable the precise timing of propagation measurements (Pseudoranges) the users make to determine range between themselves and the satellites. Precise and accurate SLR-derived ephemerides are compared to GPS orbits generated by the MCS and the Defense Mapping Agency post-processed precise ephemerides to separate the satellite position and on–board atomic clock errors. This error separation should provide a foundation for better understanding the satellite clock on–orbit performance, error propagation within the MCS data computation process, and an independent calibration of GPS accuracy.

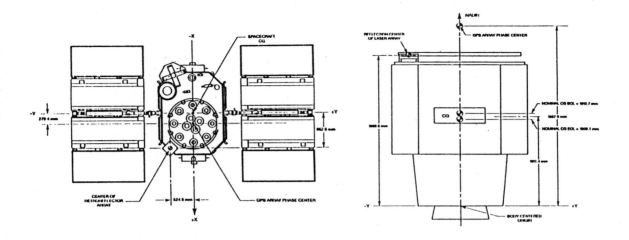

Fig. 1. Location of the LRE with respect to the center of gravity of the spacecraft and the GPS array phase center. (BOL, beginning of lifetime, EOL, end of lifetime).

The LRE is a panel of laser retroreflector cubes, 24 x 19.4 cm. This array consists of 32, 2.7 cm reflectors of the design used on–board GLONASS satellites. These arrays were built and tested by the Russian Institute for Space Device Engineering in a cooperative arrangement with the University of Maryland. The placement on the selected satellites, NAVSTAR 35 and 36 is shown in Fig. 1.

LASER TRACKING

Signal strength of laser returns from the LRE is estimated to be a factor of 36 lower than that for GLONASS and a factor of 3 to 4 lower than Etalon (the Russian laser retroreflector satellite at GLONASS/GPS altitude). Good GLONASS returns to the NASA mobile laser sites (MOBLAS) are roughly equal to that from LAGEOS. With the LRE the ranging returns are estimated to be 10 to 20% of that possible for Etalon. Ranging returns could be increased to about the same level as Etalon if the MOBLAS operated in lunar mode during night–time tracking. Daylight tracking from MOBLAS is more difficult primarily due to the high background noise. Results presented here are for NAVSTAR 35 only since NAVSTAR 36 was launched significantly later and has only been sporadically tracked.

Twelve sites have successfully tracked NAVSTAR 35 with varied frequency. The U.S. systems at Monument Pk., CA, Greenbelt, MD, Quincy, CA, McDonald Obs., TX, Haleakala, HI, Yarragadee, Australia and the international sites at Herstmonceux, U.K., Graz, Austria, Wettzell and Potsdam, Germany, Maidanak, Uzbekistan, and Evpatoria, Ukraine. Tracking is sparse, with a non–uniform spatial distribution, limited to nighttime passes. Only over short periods of a day or so were the sites successful in tracking simultaneously the satellite. On November 18, 1993 ten passes of data were acquired. This was the chosen day to do preliminary comparisons with the GPS–derived orbits.

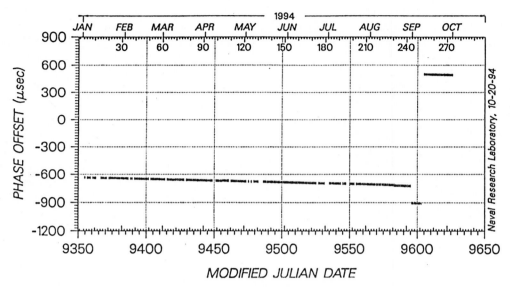

Fig. 2. Raw phase offset between Colorado Springs and USNO PPS receivers.

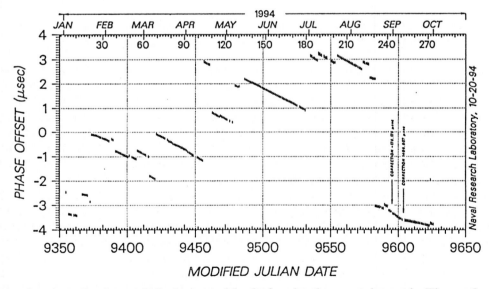

Fig. 3. Linearly detrended phase residuals for the data set shown in Figure 2.

156

GPS TRACKING

NRL collects the GPS data along with the laser tracking data for inter comparison studies. Tracking data from the GPS Control Segment stations, USNO, the broadcast position data and DMA precise ephemerides are collected. These are continuous data over the in–orbit operation of the satellites. To utilize the GPS derived tracking data for inter comparison with the laser derived data, the local clocks at the GPS Monitor Station sites must be accounted for since they are the basis for the GPS tracking measurements. In GPS itself these clocks are accounted for by the use of GPS Time which is a common synchronization time computed at the MCS. However, the GPS ranging measurements are directly related to the local clocks whose performance must be removed if the satellite clock is to be isolated from the satellite orbital position and evaluated. The laser data is independent of this influence on ranging measurements since the local clock is used for time tagging.

To determine the performance of the station clocks to the Master clock at USNO, common view time comparisons with USNO were made to the Colorado Springs, Hawaii and Ascension stations. The data show that large jumps and discontinuities are present as indicated in Figures 2, and 3. These jumps are due to changes in the local clocks or switching necessary for the operation of the system. Navigation users would not be aware of these changes since they use GPS Time which is a computed time accounting for these changes. For this experiment, removal of the local clock and the satellite position error by laser derived positions from the GPS tracking data will leave the satellite clock as the principal error component.

ORBIT ANALYSIS AND RESULTS

Reduction of the SLR data set follows a combination of techniques used in the analysis of LAGEOS laser tracking data and GPS radiometric data. In general we have adopted the IERS Standards (McCarthy, 1992) with minor excursions (e.g. 18x18 JGM-2 gravity field vs. GEM-T3). The model used to describe attitude variations is the abridged version T20 of Rockwell International's "ROCK42" model after (Fliegel et al., 1992). An additional acceleration along the satellite body-fixed Y-axis, the so-called Y-bias, is also adjusted. Due to the length of the arc used, once per revolution accelerations are also included and adjusted over the same intervals that the constant accelerations apply. These parameters along with the state vector at epoch are the only force model parameters that are adjusted.

Measurement modeling accounts for tropospheric refraction, tidal variations of the site including ocean loading, tectonic motions, and occasionally measurement biases. Tectonic motions for the sites are either from the LAGEOS-based solution SL8.3 (Smith et al., 1994) or the NUVEL-1NNR (DeMets et al., 1990). Only simple measurement biases were adjusted on a few occasions for certain sites. A long arc of about 104 days was continuously extended as new data become available. This arc was used to check on the fidelity of the force model. The data fit the arc with an rms of 3cm which is well within the expected levels. The geographical distribution of the data set does not include southern hemisphere tracking and that can introduce significant biases in the orbits.

November 18, 1993 being the best tracking day within this data set was used as a test day to verify orbit quality and gain some insight in the level of agreement with the "radiometric data"-determined orbits that the International GPS Service (IGS) for Geodynamics is routinely distributing (Beutler, 1993). Two fourteen day arcs were fit to the data; one for November 5-18 inclusive and one beginning on November 18. These arcs have only 12 hours worth of data in common: 11:00 UT to 23:00 UT, on November 18. The data fit either arc with an rms residual of 1.9 cm. In both cases, the state vector and one set of accelerations were estimated. The two orbits are based on just over 200 normal points each. For arcs of such length this can hardly be called a sufficient amount of data.

The trajectories from the two adjustments were then compared in terms of radial, cross-track, and along-track differences over their common segment. Statistics from this comparison indicate that the SLR orbits are consistent at the 5 cm level radially.

Despite the fact that the SLR data distribution is not as optimal as would be preferred for a precise orbit determination, it is still worthwhile comparing to the GPS-derived orbits distributed by IGS for geodetic work. The IGS orbit was rotated into the inertial frame and used as "observations" with the GEODYN data analysis software package to restitute a dynamic orbit fitting that data. The converged trajectory was then compared to the SLR-derived orbit in the radial, cross-track, and along-track directions. Statistics of these differences of the IGS orbit from both SLR 14-day arcs are shown in Table 1. The common segment of course is only one day (November 18) in both cases.

Table 1. Trajectory differences between the SLR-derived and GPS-derived orbits.

Orbit Pair :	SLR-1 vs. IGS GPS			SLR-2 vs. IGS GPS		
Direction	Radial	Cross	Along	Radial	Cross	Along
Mean [cm]	8.9	63.3	39.7	3.6	41.5	58.7
RMS [cm]	7.7	56.5	75.1	9.8	90.9	72.9

CONCLUSIONS

Tracking of the LRE is sparse due to its low priority within the SLR data collection schedule. Limited orbit comparisons show at least the level of compatibility of the SLR and IGS orbits at about 10 cm in the radial direction. A more uniform and extended SLR data set will be required before we can reliably determine an orbit at the few centimeter level of accuracy. The complication of removing the local atomic clock offset and drift from the GPS data is being accomplished using the common view technique of simultaneous observations of the satellites at two sites. These comparisons should be of sufficient accuracy to remove these effects from the individual satellite tracking data. With SLR derived positions having sufficient confidence the resulting satellite atomic clock performance should be easily isolated for evaluation.

REFERENCES

Beutler, G. (1993). The 1992 IGS Test Campaign, Epoch '92, and the IGS PILOT Service: An Overview, *1993 IGS Workshop*, G. Beutler and E. Brockmann, (eds.), University of Berne.

DeMets, C., R.G. Gordon, D.F. Argus, and S. Stein (1990). Current Plate Motions, *Geophys. J. Int.*, **101**, pp.425-478.

Fliegel, H.F., T.E. Gallini, and E. Swift (1992). Global Positioning System Radiation Force Models for Geodetic Applications, *J. Geophys. Res.*, **97** (B1), pp.559-568.

McCarthy, D.D. (ed.) (1992). IERS Standards (1992), *IERS Technical Note* **13**, Observatoire de Paris, IERS, Paris.

Smith, D.E., R. Kolenkiewicz, R.S. Nerem, P.J. Dunn, M.H. Torrence, J.W. Robbins, S.M. Klosko, R.G. Williamson, and E.C. Pavlis (1994). Contemporary global horizontal crustal motion, *Geophys. J. Int.*, **119**, pp.511-520.

Chapter 3

Kinematic Applications of the GPS

KINEMATIC GPS TRENDS - EQUIPMENT, METHODOLOGIES AND APPLICATIONS

M. E. Cannon and G. Lachapelle
Department of Geomatics Engineering
The University of Calgary, 2500 University Drive N.W.
Calgary, Alberta T2N 1N4 CANADA

ABSTRACT

Developments in the area of kinematic GPS are accelerating as a broader community requires higher levels of accuracies. This paper first reviews developments in receiver technologies and processing algorithms. These include low C/A code noise receiver developments, receiver multipath reduction techniques (e.g., MET and MEDLL), codeless tracking methods (e.g., P-W), and on-the-fly (OTF) ambiguity resolution techniques. The use of kinematic of GPS is illustrated by land and shipborne applications including precision farming, structural monitoring and shipborne water level profiling. The paper concludes with a discussion on the future of kinematic positioning.

INTRODUCTION

Kinematic positioning has been accepted within the broad GPS community as the case when the precise carrier phase measurement is used in the positioning model. Fig. 1 gives an overview of the various positioning modes of GPS and the associated accuracies when operating under reasonable conditions (i.e. favorable geometry and relatively low multipath). As can be seen, when using the carrier phase measurement, accuracies on the order of a few cm can be obtained when integer ambiguities are resolved, whereas decimetre accuracies are generally achievable if so-called 'floating' ambiguities are estimated. This second case is typically for longer monitor-remote separations (i.e. greater than 25 km) or under significant multipath or unstable atmospheric conditions.

As Fig. 1 indicates, the achievable accuracy when using the code observable in DGPS mode may approach near-decimetre levels of accuracies when using narrow correlator-type technology as well as carrier smoothing. Improvements in code correlation technology are discussed in the following section and encompass improved multipath elimination technologies as well as measurement resolution. Single point GPS performance has also been improved for the post-mission case by utilizing precise orbit and clock information which are computed using International GPS Service (IGS) data typically by federal government agencies (e.g. Natural Resources Canada). In some cases, the availability of precise orbit and clock data has been sufficient to meet accuracy requirements previously met by differential processing.

The following paper describes some of the developments in receiver technologies and their impact on precise GPS positioning. A brief discussion on general considerations for integer ambiguity resolution is included along with the variations in methodologies in which research groups are currently investigating. Subsequently, some kinematic GPS applications are described including precision farming, structural monitoring and water level profiling. The paper concludes with a discussion on the future of kinematic positioning.

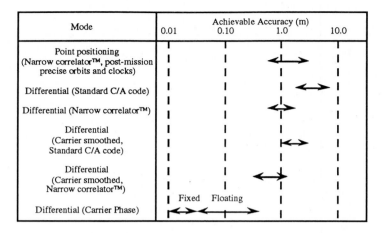

Fig. 1. GPS operating modes and achievable accuracies.

GPS EQUIPMENT DEVELOPMENTS

GPS receivers typically output both the pseudorange (code) observable as well as the carrier phase. Code correlation technologies have been enhanced during the past several years, most notably with the development of Narrow Correlator spacing which improves the resolution of C/A code pseudoranges to the 10 cm level compared with standard C/A code noise levels of 1-2 m (Fenton et al., 1991). Although noise is one component in the position error budget, multipath is generally of a greater concern since it can easily vary from a few decimetres to several metres. In order to reduce multipath, new antenna technologies are being developed as well as improved receiver tracking loop design. In the latter case, two such examples are the Multipath Elimination Technology (MET) and the Multipath Elimination Delay Lock Loop (Townsend and Fenton, 1994).

Codeless and Semi-codeless techniques are used to access L2 when A.S. is on and are important to provide ionospheric corrections or to form a widelane observable. Several methods have been developed to provide this L2 data. These are listed below (see Van Dierendonck, 1994 for further information):

Table 1. Codeless and Semi-codeless techniques and power loss.

Squaring:	30 dB loss, L2 λ = 12 cm
Cross-correlation:	27 dB loss, L2 λ=24cm, L1 - L2 range, benign dynamic applications
P-W correlation:	14 dB loss, L2 λ=24cm, L1 and L2 Y code pseudoranges available

GPS ON-THE-FLY KINEMATIC POSITIONING ALGORITHMS

On-The-Fly (OTF) ambiguity resolution refers to the case when carrier phase integer ambiguities are estimated when the remote GPS antenna is moving. This is in contrast to the case when a static initialization is required to determine the integer values. Virtually all carrier phase processing algorithms that utilize an OTF technique, rely on the double difference carrier phase observable as the primary measurement.

Whether the system is for real-time or post-mission use, the algorithm is generally treated the same. Clearly, with real-time implementations, data communication and hence latency may become significant problems. Data outages and cycle slips can also severally limit the performance of the system and the environment in which it can be used.

The time to ambiguity resolution can range from a few seconds to several minutes depending on some of the following considerations:

- use of L1 vs L1 - L2 (widelane) observable (λ_{L1}= 19 cm, λ_{L1-L2} = 86 cm)
- distance between reference and remote receivers
- number and geometry of satellites
- differential atmospheric conditions
- multipath conditions, code and carrier phase noise
- ambiguity search method used

Intense research activities are currently underway by a number of groups aimed at developing effective OTF methods. These methods include the ambiguity function method, least-squares search, conditional covariance approaches, covariance decorrelation and integer programming. These methods typically assume a known stochastic behavior for unmodelled errors (e.g., noise, multipath, differential atmospheric effects), which if present, will negatively affect the performance of the algorithms.

KINEMATIC APPLICATIONS USING GPS

Many applications have been developed using precise kinematic GPS techniques. The following are a selection of a few applications in the land and marine environments.

Precision Farming

In the past, farmers generally treated an entire field with the same quantity and type of fertilizer, without regard for possible variations in topography, salinity and soil type within the field. By applying a variable amount and type of fertilizer as a function of location within the field, the overall crop productivity can be improved. This is called *precision farming*. The fertilizer application is determined from a prescription map derived from information previously collected. As the fertilizer prescription is applied to smaller parcels of land, the overall economic gain is improved since local conditions are taken into account. The process begins with the harvesting process where the crop yield is monitored as a function of location. The resulting yield map is used, together with other soil information such as salinity, to prepare a prescription map which is used during the fertilizing process to optimize spreading.

The University of Calgary, together with Alberta Agriculture, is currently developing a system which can be used to measure the variability of crop production as a function of location (Lachapelle et al., 1994). The system consists of several components which include GPS receivers for positioning, a yield monitoring system which outputs the instantaneous Bu per acre, an EM conductivity meter for salinity measurements, and soil

163

samples for determination of soil types and nutrients. The DGPS/yield monitoring data can be collected under normal combining operations, see Fig. 1 for the concept. As the fertilizer moves along a pre-determined trajectory, the fertilizer prescription for the current cell is retrieved from the Geographic Information System (GIS).

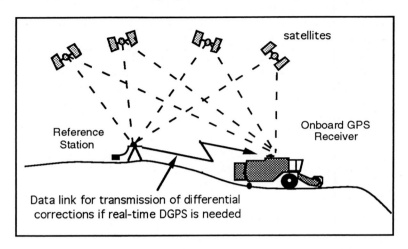

Fig. 1. Positioning of combine harvester with DGPS.

The precise positioning system needed to assign spatial coordinates to the field measurements must have an accuracy of 1 m or better and should be available in real-time on a practically continuous basis (< 1 second). GPS can meet, in differential mode, these requirements using relatively low cost user equipment. Position updates are available at a rate of several times per second if required. Two 10-channel C/A code narrow correlator spacing NovAtel GPSCard™ sensors were selected as they have shown to provide sub-metre accuracy in previous field tests using a robust carrier phase smoothing of the code approach.

Four test fields across Alberta ranging from 80 to 200 acres have been selected for study over a four year period. These areas vary in soil type as well as topography and salinity conditions. The project began during the 1993 harvest and is expected to continue until 1997. Positioning results are presented in Table 3 for one of the test sites which consisted of gently rolling hills. The reference station was installed near the field and the moving platform was operating within a few km from the reference station. The crop was harvested on September 21. On November 9, soil samples were taken at various locations using an All-Terrain-Vehicle (ATV). The ambiguities could be resolved OTF on September 21 and November 9 using typically 10 to 15 minutes of data with a choke ring groundplane at the reference station only. This is to be expected when using single frequency data. On each day, a minimum of six satellites were tracked and the theoretical PDOP was always smaller than 3. Data was collected at a rate of 1 Hz. Although cycle slips were occasionally detected on low satellites, the tracking stability of the GPSCard™ was fully satisfactory.

Table 3 gives comparisons between coordinates at crossover points during each of the test days, as well as between the two test days. The results show that by using carrier smoothing, a repeatable accuracy of about 0.5 m or better can be achieved, whereas accuracies below 15 cm can be reached using carrier phase kinematic processing. Although the OTF results exceed the current positional requirements, the agricultural community is interested in these levels of accuracy to generate topographic information.

Table 2. RMS height differences at crossover points using carrier phase smoothing of the code and OTF solutions.

Date	Carrier Smoothing (m)	OTF (m)
Sept. 21	0.25	0.07
Nov. 9	0.56	0.08
Sept. 21 vs Nov. 9	0.43	0.14

The use of GIS and GPS in agriculture is expanding rapidly and the precision farming case reported here is only one of several applications which include salinity measurements (Cannon et al. 1994), aerial and terrestrial crop spraying and seeding, and animal tracking. In the future, the extension of this technology to unmanned farm vehicles is projected to occur.

Structural Monitoring

The use of precise GPS techniques for structural monitoring has become increasingly widespread as an alternative to measurements using accelerometers, laser interferometers or electronic distance measurement (EDM) instruments. For many applications, kinematic GPS processing algorithms must be used due to the relatively high frequency of the movement. One such example is the deformation monitoring of a tall structure as described in Lovse et al. (1994). Fig. 2 shows the Calgary Tower which is a 160 m structure built in 1968 to a design specification of 165 mm movement in a 160 km/h wind.

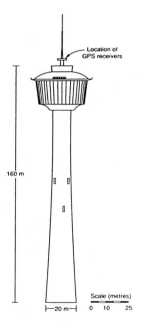

Fig. 2. Calgary Tower.

In order to measure the actual measurement of the tower, a GPS receiver was placed near the base of the main communication antenna above the observation deck. The GPS reference receiver was situated in a stable location on the roof of an apartment building approximately 1 km from the tower. Single frequency receivers were used in the project since the baseline separation was relatively short.

Data was collected for 15 minutes at a 10 Hz rate. A high data rate is required in these types of applications since the deformation may have a frequency of up to several Hz. During the particular session in which the data was recorded, the winds were typically 60 km/h from the west, however gusts of up to 100 km/h were noted. The GDOP varied between 2 and 3, and there were 8 satellites observed above a 10 degree elevation cutoff.

In order to process the data in post-mission, OTF techniques were utilized. About 3 minutes of data was needed to resolve the integer ambiguities after which the double difference measurement residuals were below 5 mm. Fig. 3 shows a representative plot of the tower movement over a 1 minute time interval.

The movement of the tower is larger in the north-south direction even though the wind was from the west. This is due to the eddies that are produced on the north and south sides of the tower which create zones of slightly different pressure which tend to push the higher pressure side and pull on the lower pressure side. As well, the wind is out of phase with the east-west movement of the tower which dampens movement in that direction.

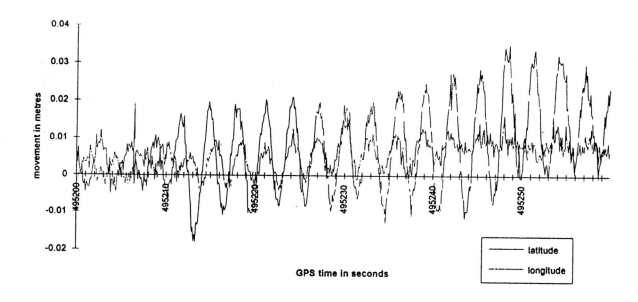

Fig. 3. Movement of the Calgary Tower.

Although there is no external reference to check the motion as depicted in Fig. 3, a visual analysis shows that the movement is regular and within the expected range of such a structure. The north-south motion had an amplitude of approximately ± 15 mm and an east-west component of ± 5 mm.

Water Level Profiling

Precise knowledge of water levels is essential for tidal studies and other hydrographic purposes such as the establishment of chart datums. GPS offers the possibility of determining water level profiles with cm-level accuracy using carrier phase measurements on-the-fly. Accurate Bench Marks (B.M.'s) can also be established along the shores if an accurate geoid model is available. This is an important development since levelling operations may be prohibitive in remote areas such as the McKenzie River in Northern Canada. A feasibility study was conducted by The University of Calgary in conjunction with the Canadian Hydrographic Service (CHS) to obtain water level profiles and establish B.M.'s along an 80 km segment of the Fraser River, British Columbia. A first order levelling line along the river was used to assess the accuracy of the GPS-derived orthometric heights. Fig. 4 shows a map of the area.

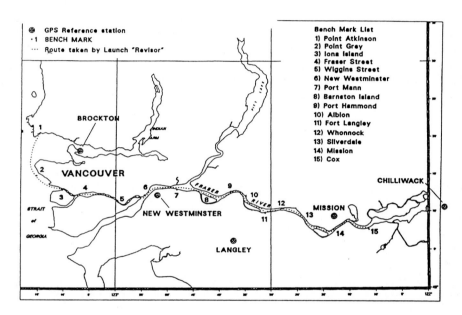

Fig. 4. Fraser River GPS water level profiling survey.

Table 3. Summary statistics of GPS-derived and levelled heights.

RMS agreement between successive GPS height determinations ($\Delta T \leq 200$ s) at each B.M. visit.	5.2 cm
RMS agreement between successive B.M. visits (several hours $< \Delta T < 3$ days).	5.5 cm
RMS agreement between GPS-derived and levelled (B.M.) heights.	9.3 cm
Geoid undulation bias (ΔN) estimated by comparing GPS and levelled heights.	7.1 cm
RMS agreement between GPS-derived and levelled (B.M.) heights (Geoid bias removed)	6.0 cm

Results from the test are given in Lachapelle et al. (1993) and show that an accuracy of about 5.5 cm (RMS) could be achieved between the GPS-derived orthometric heights and the levelled B.M. height when kinematic OTF techniques are utilized. Part of the error budget is due to the height transfer method (accurate to about 2 cm) as well as the geoid. Overall, the feasibility study was successful and the GPS-derived water level profiles are being analysed to determine the extent of tidal and other effects. Table 4 gives a summary of the statistics derived from the comparisons.

FUTURE OF KINEMATIC POSITIONING

The future of kinematic GPS positioning is clearly tied in part to developments in receiver technologies. As L2 codeless tracking capabilities are enhanced, the feasibility of using OTF kinematic processing for a wider range of applications will undoubtedly be shown. However, even with improved technology, the operation of such a system in an urban or forested environment may be prohibitive. The augmentation of GPS with pseudolites or inertial navigation systems (INS) will certainly improve in these cases.

Improvements in processing algorithms can also be expected in the future as a wider research community becomes involved in the development of these algorithms. One of the important components that may not have received sufficient attention as of yet is the characterization of GPS errors with sufficient accuracy so as to allow for extremely reliable integer ambiguity resolution.

New developments in the GPS infrastructure may also assist in kinematic positioning applications. In a recent study mandated by the US Congress and carried out by the National Academy of Public Administration (NAPA) and the National Research Council (NRC) on the future of GPS (NAPA, 1994; NRC, 1994) several proposals have been made to ensure that GPS will be a useful and cost-effective navigation tool for the foreseeable future. Most notably, these proposals include turning Selective Availability (SA) to zero immediately as well as the introduction of a 'civilian L2 signal' on Block IIF satellites (designated the L4 signal). The elimination of SA would have a major impact on real-time kinematic applications since communication link requirements would be loosened and algorithms may be simplified to some extent. A fourth GPS frequency would be particularly critical for long baseline GPS users and also for widelaning to solve integer ambiguities in a relatively short amount of time. A third proposal in the NAPA and NRC reports is to augment the current GPS Control Segment with additional ground monitor stations to improve the quality of the broadcast ephemeris. This would also be beneficial to long baseline real-time GPS users.

The above are some of the areas in which developments may occur to affect GPS kinematic positioning users. No doubt with increased attention paid to this field from areas such as civil aviation, further enhancements of the system may be expected.

REFERENCES

Cannon, M.E., R.C. McKenzie, and G. Lachapelle (1994) Soil Salinity Mapping With Electromagnetic Induction and Satellite-Based Navigation Methods, Can. *Journ. of Soil Science*, Agricultural Institute of Canada, Vol. 74, No. 3, pp. 335-343.

Fenton, P.C., W.H. Falkenberg, T.J. Ford, K.K. Ng and A.J. Van Dierendonck (1991), NovAtel's GPS Receiver - the High Performance OEM Sensor of the Future, *Proc. of IOn GPS-91*, Albuquerque, Sept. 11-13, pp. 49-58.

Lachapelle, G., M.E. Cannon, H. Gehue, T. Goddard and D. Penney (1994), GPS System Integration and Field Approaches in Precision Farming, *Navigation*, Vol 41, No. 3, pp. 323-335.

Lachapelle, G., C. Liu, G. Lu, W. Qiu and R. Hare (1993), Water Level Profiling with GPS, *Proceedings of the ION GPS-93*, Salt Lake City, September 22-24, pp. 1581-1587.

Lovse, J.W., W.F. Teskey, G. Lachapelle and M.E. Cannon (1995), Dynamic Deformation Monitoring of a Tall Structure Using GPS Technology, *Journal of Surveying Engineering*, ASCE, Vol. 121, No. 1, pp. 35-40.

NAPA (1995), *The Global Positioning System - Charting the Future*, National Academy of Public Administration, Washington, D.C.

NRC (1995), *The Global Positioning System - A Shared National Asset*, National Research Council, Washington, D.C.

Schnug E., et al. (1992) Yield Mapping and Application of Yield Maps to Computer Aided Local Resource Management, *Proc. of Workshop on R&D Issues on Soil Specific Crop Management*, Apr 14-16, Minneapolis.

Townsend, B. and P. Fenton (1994), A Practical Approach to the Reduction of Pseudorange Multipath Errors in an L1 GPS Receiver, *Proc. of ION GPS-94*, Salt Lake City, Sept. 20-23, pp. 143-148.

Van Dierondonck, A.J. (1994), Understanding GPS Receiver Terminology: A Tutorial on What those Words Mean, *Proc. of KIS94*, Dept. Geomatics Engineering, Univ. of Calgary, pp. 15-24.

Kinematic And Rapid Static (KARS) GPS Positioning:
Techniques and Recent Experiences

Gerald L. Mader

Geosciences Laboratory

Ocean and Earth Sciences, NOS, NOAA

Silver Spring, Maryland

Since the first demonstrations of airborne kinematic positioning using GPS carrier phase measurements in 1985, the ability to do practical and reliable kinematic GPS solutions in support of operational remote sensing activities has improved significantly. These improvemtns arise from a combination of factors including: GPS receivers able to track 8 or more satellites, more precise pseudorange measurements, the availability of post-processed precise satellite ephemerides, and the development of algorithms to initialize phase bias's while in motion with a minimal amount of phase and pseudorange measurements. These factors have combined to produce an operational kinematic positioning capability accurate to centimeters while extending the range of the technique to hundreds of kilometers. The integer search algorithms used by KARS, that provide on-the-fly bias initialization and cycle slip fixing over extended range operations are described.

This symposium coincidentally marks the tenth anniversary of the first successful airborne kinematic GPS carrier phase solutions. This initial demonstration of the potential of GPS to provide centimeter level positioning to support remote sensing applications was a joint effort between NOAA/NGS Geodetic Research and Development Laboratory (Mader 1985, 1986) and NASA/Wallops Flight Center (Krabill and Martin, 1987). Since these early efforts, kinematic GPS operations and processing have matured and are now used routinely for a wide variety of applications.

Kinematic GPS data processing must address several challenges to be of practical use for remote sensing applications. These include on-the-fly bias resolution or the ability to start data processing without any specific operational procedures, bias resolution and data processing in the presense of significant ionospheric effects, and bias resolution of newly rising satellites.

The KARS program has been written to meet these challenges as well as several additional specific objectives: to apply the same integer search (bias resolution) algorithms to both kinematic and rapid static data processing; to obtain nearly instantaneous bias resolution; to extend the range of kinematic GPS operations to hundreds of kilometers; to provide ionosphere-free operation; and to do this with minimal use of pseudo-range observables.

Knowledge of the double-difference integer biases is what allows the tremendous precision of kinematic carrier phase solutions to be realized. There are several different situations where satellite biases must be determined. These situations are distinguished by the number of biases that may already be known and the kinematic status of the mobile platform. These situations are summarized in Table 1.

The first case would occur when the reference receiver and the mobile receiver, for example, an airplane, are both at the same airport and are tracking for some time prior to any movement. In the second case, the airplane might fly near a reference receiver located in a project area and require initialization with respect to that receiver. The third case might occur after a prior successful initialization on all satellites, but a loss of lock has occurred on all but 1 or 2 satellites perhaps from a steep banking turn where the wing has blocked most of the visible satellites. In these first three cases the position of the mobile receiver cannot be estimated from a bias-fixed carrier phase solution since a sufficient number of biases are not yet known. This is not the situation for the remaining 2 cases where only 1 or 2 satellites are being initialized and there is a sufficient number of other satellites with continuous phase to provide a precise solution.

Table 1. Cases of Bias Initialization

case #	Description	# of known biases	comment
1	full bias initialization	0	rapid-static, on-the-fly (otf)
2	full bias initialization	0	kinematic otf
3	partial initialization	< 3	kinematic
4	initialize rising satellite	> 2	kinematic
5	reinitialize after cycle slip	> 2	kinematic

In each of these cases, the biases are found by an integer search technique. Although this discussion will follow the procedures used in the KARS program written by the author and used by NOAA, practically all integer search techniques share the same basic features.

In this search algorithm, it is recognized that the double-difference biases cannot explicitly be solved. Instead, an initial bias value is estimated for L1 and L2 for each satellite using the best available receiver position. These initial values for satellite j are given by the a priori double difference phase residual:

$$n^j = \delta^2 \Delta \phi^j$$

A search range is then defined for the L1 and L2 bias for each satellite. This search range corresponds to the uncertainty in the initial position estimate and must be broad enough to be sure to include the correct bias values. The search ranges may be found from:

$$\Delta n^j = \frac{f}{c}\left(\left(\frac{x^j - x}{R^j} - \frac{x^{j_o} - x}{R^{j_o}}\right)\Delta x + \left(\frac{y^j - y}{R^j} - \frac{y^{j_o} - y}{R^{j_o}}\right)\Delta y + \left(\frac{z^j - z}{R^j} - \frac{z^{j_o} - z}{R^{j_o}}\right)\Delta z\right)$$

In principle, the correct suite of integer biases could be found by forming all possible permutations of the satellite biases and testing each candidate suite in a least squares solution of the double-difference phase equation. The correct suite should have the lowest residuals, while the incorrect suites of integers will have larger residuals. In practice, this cannot be done because of the enormous number of possible permutations, especially when the initial position uncertainty is large, and the computer time that would be required. Furthermore, many incorrect integer suites will give low residual values, even lower than the correct suite, for data at a particular time just by chance. However, these same integer suites will yield poor residuals using data at different times. The increase in residuals with time as subsequent data is processed using a particular integer suite is a sure indication of a bias error.

The number of possible integer permutations is minimized by examining each satellite separately and eliminating particular integers from further consideration. This may only be done using both the L1 and L2 data. Single frequency L1 data, does not offer the possibility of eliminating possible L1 integers within the L1 search range. Unless the initial position estimate is very good (within about 1 wavelength) and the number of satellites in view is large (generally at least 6), single frequency integer searches may suffer from numerous ambiguous solutions. Dual frequency receivers would also be preferred because of the extended range that is possible when the ionosphere is eliminated.

In KARS the number of possible integers is minimized for each satellite by computing the ionosphere correction for each L1 and L2 integer pair within the search range. These corrections at L1 are given by:

$$I^j = \frac{f_1 f_2^2}{f_1^2 - f_2^2}\left(\frac{\delta^2 \Delta \phi_1^j - n_1^j}{f_1} - \frac{\delta^2 \Delta \phi_2^j - n_2^j}{f_2}\right)$$

The approximate value of these corrections at the zenith are found by simply multiplying by the sine of the elevation. This enables a more uniform comparison between all the satellites. An example of these corrections is shown for a single satellite in Table 2 as a function of the change in the L1 and L2 integers from their initial estimated values. Most of these corrections are unrealistically large. A reasonable estimate of an upper limit for the zenith double-difference ionosphere correction for the 33 km baseline of this test data set is about 0.3 cycles. The integer pairs that yield ionosphere corrections with absolute values less than this cutoff are shaded dark in Table 2. These pairs are saved for further filtering and forming the integer permutations. The remainer are eliminated from further consideration.

Table 2. Estimated L1 Zenith Double-Difference Ionosphere Corrections Over Search Range

N1\N2	-4	-3	-2	-1	0	1	2	3	4
-5	-0.07	0.45	0.98	1.50	2.03	2.55	3.08	3.60	4.13
-4	-0.48	0.04	0.57	1.09	1.62	2.14	2.67	3.19	3.72
-3	-0.89	-0.37	0.16	0.68	1.21	1.73	2.26	2.79	3.31
-2	-1.30	-0.78	-0.25	0.27	0.80	1.32	1.85	2.38	2.90
-1	-1.71	-1.19	-0.66	-0.14	0.39	0.91	1.44	1.97	2.49
0	-2.12	-1.60	-1.07	-0.55	-0.02	0.51	1.03	1.56	2.08
1	-2.53	-2.01	-1.48	-0.96	-0.43	0.10	0.62	1.15	1.67
2	-2.94	-2.42	-1.89	-1.37	-0.84	-0.31	0.21	0.74	1.26
3	-3.35	-2.83	-2.30	-1.77	-1.25	-0.72	-0.20	0.33	0.85
4	-3.76	-3.24	-2.71	-2.18	-1.66	-1.13	-0.61	-0.08	0.44
5	-4.17	-3.64	-3.12	-2.59	-2.07	-1.54	-1.02	-0.49	0.03

In KARS, the integer pairs that pass through this ionosphere filter may next be filtered by their wide-lane value. The wide-lane is the designation for the difference between the L1 and L2 phases. The wide-lane integer is the difference between the L1 and L2 integer biases. It is sometimes possible to independently estimate the wide-lane integer from the L1 and L2 phases and the P1 and P2 pseudoranges from the expression:

$$N_1^j - N_2^j = N_{wl} = \phi_1^j - \phi_2^j - \frac{f_1 - f_2}{f_1 + f_2}\left(\frac{f_1}{c}P_1^j + \frac{f_2}{c}P_2^j\right)$$

This wide-lane expression is independent of the ionosphere, clock errors, and receiver position. However, it is sensitive to multipath and pseudo-range calibration errors which may prevent unambiguous recognition of the wide-lane integer. When it is known, the wide-lane selects a particular diagonal in the N1,N2 matrix as shown by the lightly shaded values in Table 2. Those integer pairs not lying along this diagonal are eliminated from further consideration.

The integer pairs for each satellite that survive these filters are saved and used to create permutations of all possible remaining integers. A least squares solution and the post-fit residuals are found for each integer suite being tested. The integers are now evaluated as complete sets. When the correct integers for some satellites are used with the incorrect integers for other satellites, the residuals should still be large. When the integer combination that contains all the correct integers is tested, the residuals should be minimal. A successful search must meet two conditions. First, the minimum residual value found must be acceptable, indicating a viable solution. Second, the contrast between this minimum value and the next nearest value must be large enough (i.e. the next best solution is bad enough) so that a single solution may be unambiguously identified.

The KARS program can successfully conduct this integer search often using just one or several data epochs. For longer distances where the ionosphere is likely to be a problem, the ionosphere corrections for each integer pair are applied to the phase residuals during the least squares solutions allowing KARS to determine integer bias values out to distances of several hundred kilometers. The typical search time is less than 1 second.

The initial position estimate needed to initiate the search algorithms for cases 1, 2, and 3 in Table 1, comes from a differential pseudorange solution. A search volume centered on this position is proportional to the component errors of that solution. Since no external information or static initialization is required, these solutions are self-starting and this type of technique is described as on-the-fly. Case 1, is distinguished from the other cases primarily by the ability to use alternate techniques to find the initial position when the receiver is static.

Cases 4 and 5 have precise carrier phase solutions available. This allows a much smaller search volume and consequent integer search ranges to be defined. The integer search for these cases is conducted only over those satellites whose biases are not known. All the satellites are used in the least squares solutions with the known biases used for those satellites with continuous phase tracking. The primary distinction between these two cases is the recent ionosphere correction information available for satellites that underwent a cycle slips. This usually allows the ionosphere correction filter to be centered, and a width used, that is more realistic. No such history is available for newly rising satellites although the current satellite constellation may be used to more loosely set these filter values.

Integer search techniques are being used operationally to routinely process kinematic GPS phase data for a wide vareity of applications. Kinematic positions with precisions on the order of centimeters are regularly supporting photogrammetry as well as a number of other remote sensing activities.

The KARS program successfully provides automatic bias resolution including on-the-fly as well as bias resolution at distances of hundreds of kilometers. The program is available for DOS and UNIX environments and includes a user interface program for easy creation of required input files.

REFERENCES

Krabill, W.B., and Martin, C.F., Aircraft Positioning Using Global Positioning System Carrier Phase Data, Journal of the Institute of Navigation, Vol. 34, No. 1, 1987

Mader, G.L., Aircraft Positioning Using GPS Carrier Phase Measurements, Fall AGU Meeting, San Francisco, December 1985.

Mader, G.L., Dynamic Positioning Using GPS Carrier Phase Measurements, Manuscripta Geodaetica, 11, 1986.

DEVELOPMENTS IN AIRBORNE
'HIGH PRECISION' DIGITAL PHOTO FLIGHT NAVIGATION
IN 'REALTIME'

Günter W. Hein, Bernd Eissfeller, Jürgen Pielmeier
Institute of Geodesy and Navigation (IfEN)
University FAF Munich, D-85577 Neubiberg, Germany

ABSTRACT

At the 'Institute of Geodesy and Navigation' a completely integrated digital photo navigation system is under development.

The objective of this system is to support the customer through computer integration from project planning via photoflight guidance to post-mission data evaluation. With this integrated system, we expect a dramatic increase in photogrammetric productivity by saving time and money and improving mission data quality.

The system consists currently of two major parts, the 'off-line' mission planning program 'VAMP' and the 'on-line' mission navigation system 'SANC'. The 'Visual Aided Mission Planning' program is based on Windows 3.1 and integrates the capabilities of project planning with photoflight and VFR-flight planning on an underlying objectoriented GIS-database, running on standard PC hardware. The 'Satellite Aided Navigation Copilot' system enables realtime DGPS position, velocity and attitude determination and navigatiopn in different modi, according to the required precision level, dependent on photogrammetric map scale.

First, the tests have shown the potential for flight guidance and photogrammetric imaging with the availability of attitude information. Second, a lot of benefit results from a completely integrated planning and navigation system with 'state-of-the-art' visualization.

1. INTRODUCTION

Knowing the *position and attitude* of the photo camera at the moment of exposure enables an adjustment technique called 'aerotriangulation'. With aerotriangulation, you need only a few or ideally no ground control points for block adjustment. This allows fast and flexibel planning and execution of photo flights. An additional task in aerotriangulation is the transformation of the WGS 84 positions into the desired local geodetic system.

An algorithm for aerial photography using GPS could be summarized as follows:

- Planning of 'photopoints' along strips in the photo block area according to the photogrammetric parameters as side overlapping, map scale, camera type and so on.
- Flight guidance along the 'photopoints' using GPS.
- Photo camera shutter triggering at the desired 'photopoint'.
- High precision (precision requirements depends on desired map scale) 'photopoint' position and attitude data logging, with camera and GPS time synchronized.

The last two topics are photo flight specific. The first two topics however are generally valid if the term 'photopoint' is replaced with 'waypoint'. Using 'waypoints' enables it to apply mission planning and flight guidance to general flight tasks as 'take off', 'enroute navigation' and 'approach'.

2. SYSTEM OVERVIEW

The 'Visual Aided Mission Planning' ('VAMP') 'off-line' digital mission planning software and the 'Satellite Aided Navigation Copilot' ('SANC') 'on-line' realtime DGPS navigation system are connected at two points.

The first point of contact between the two system parts is the *GIS database*. This database is essentiel for the visual mission planning and the flight guidance positioning visualization. To use the same database is very important for the integrity of mission execution.

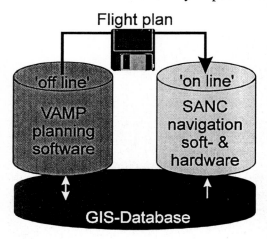

Fig. 1 Abstract system view

The second point of contact between the two main system parts is the *'flight plan'*. The flight plan consists of 'waypoints' describing the desired flight path and additional information for special purposes at the different waypoints. A photopoint for example is a *waypoint plus the instruction* for making an exposure with the photo camera at this waypoint. The flight plan is transported via disk.

At an abstract general system level, the *GIS database* represents a third system component *beside mission planning* and *flight guidance*. In the next sections, these three main topics will be further discussed.

Therefore in the next sections the main topics are 'core concept' and 'system architecture' of 'Visual Aided Mission Planning', 'Satellite Aided Navigation Copilot' and 'GIS Database'.

3. VAMP SYSTEM

Flight planning is the starting point of every flight mission. VAMP should provide the capabilities to plan a complete mission from the beginning to the end in the final version. At the moment, VAMP is able to plan the 'Photo Flight' parts of a mission. To meet the system objectives, VAMP was developed on standard PC hardware using the Windows 3.1 GUI.

The core concept of VAMP consists of three different 'working levels'. These levels are 'Project Management', 'Mission Planning' and 'Task Editing'. Every project document is organized according these three levels. In the project management view, you can notice the document organization.

The highest working level is *'Project Management'* ('PM'). Every project consists of 'mission' and 'task' objects. On the 'Project Management' level you can create and add mission objects to the project and respectively create and add the different task objects to a mission. Furthermore you can change this project structure by deleting, moving or copying the missions and tasks with 'point and click' actions. The 'Project Manager' is the central part of VAMP application. From this level, all project related parts are controlled.

The mediate working level is *'Mission Planning'* ('MP'). Every mission consists of a number of different tasks. The main function of the mission planer is to *generate a flight plan* out of the tasks of the desired mission. This generation process is supported by some other mission planning functions.

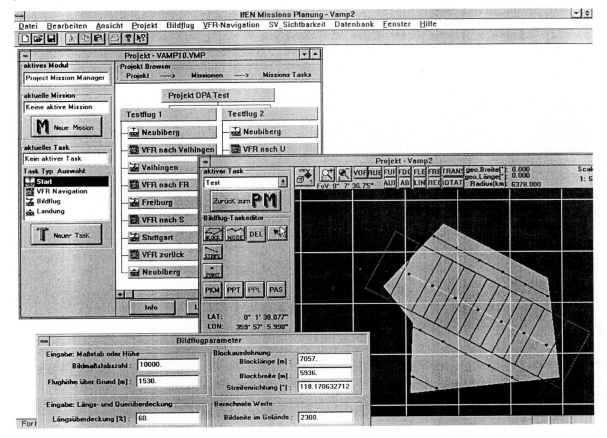

Fig. 2 VAMP Sreen Shot with Project Manager, Task Editor, Photo flight parameter dialog box

The lowest working level is 'Task Editing' ('TE'). In this working level, the different tasks are edited on an underlying map, extracted from the GIS database. Geographic coordinates are entered visual in the actual map coordinate system. In the 'Task Editing' level every task has different features. Only one task of a mission can be active at the same time. The task editor according to the active task type is loaded and then allows visual editing in 'Task Editor View'. The 'Photo Flight Task' editor for example offers the following features:

- Enter/edit photo block area, photo strip, single photo point, ground control points
- Select photo camera and edit photo parameters

In the 'Task Editing Level' all task-specific and geographic related informations are entered with 'point and click'.

As summary, planning with VAMP includes the following steps:

- 'PM': generate missions and tasks of desired type, select active mission, select active task
- 'TE': visual edit all tasks with the loaded task editor on the underlying database map
- 'TE': generate flight plan starts 'MP'
- 'MP': consistency check, generate time correlations, consider GPS constellation
- 'MP': write verified flight plan for a misson to disk

.4. GIS DATABASE

The GIS database is essentiell to both systems. VAMP and SANC.

The database is object oriented, that means it consists of persistent objects. The object oriented approach to the database enables 'real world modeling' with objects and object relations.

The database concept is based on two main object classes.

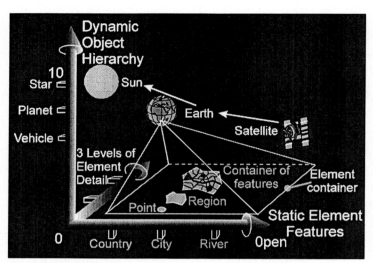

Fig. 3 GIS Database Structure

The first class is the *'Dynamic Object'* class. These objects are hierarchically ordered in the database. Dynamic objects have their own coordinate system. This is very important because this allows transformation of these objects as translation and rotation. Every dynamic object relates to a real world object. For example, the dynamic object class of level 7 in the database hierarchy is of type 'PLANET'. One well known object of this level is the planet 'earth'. The planet 'earth' has coordinates relative to is father object, the 'sun', which is of level type 'STAR'. The 'sun' has an dynamic object container. This object container possesses all objects related to the sun. This relations enables for example the calculation of the 'sun' position at a certain moment relative to the container object 'earth'.

This enables simulation of the spatial and temporal behavior of the dynamic objects in the database, e.g. the orbits of GPS-satellites or flight paths of airplanes relating to their father object 'earth'. Furthermore it's possible to navigate very quickly through the whole database. Consequently every object has a shape, for the type 'PLANET' we would say surface. Every dynamic object has an container of surface objects. This surface objects are of the *'Static Element'* class. This class type is the second main object class of the database. The main difference to the dynamic class is, that static elements have no own coordinate system. Therefore they are not transformable. They have only positions on the surface relative to the coordinate system of their dynamic owner object. Static elements are geographically described as points, lines or regions. Every element has a special feature. Every element type is characterized be the feature, e.g. 'Country', 'City', 'River' and 'Street' for GIS-specific data or 'Runway', 'Navaid', 'Intersection' for NAV-specific data.

5. SANC SYSTEM

The 'SANC' has to execute the mission. The core concept of 'SANC' has three levels of 'Task Execution'.

On every level, maximum mission exeution safety has priority.

The basic level is *'Positioning'*. This level gets the position and attitude information from the dedicated GPS receivers and 3d-inertial sensors. The highprecision, realtime 'Positioning' capability here is considered as elementary for the 'SANC' system. Nevertheless it requires already the complete know-how of realtime differential GPS positioning with carrier smoothed pseudoranges and integration with the INS data. While the GPS system is providing the *absolut* position and attitude, but with only 1Hz positioning updates, the INS provides *relative* position and attitude, but with a much faster update rate. Combining these data increases *performance* and *integrit*y drastically.

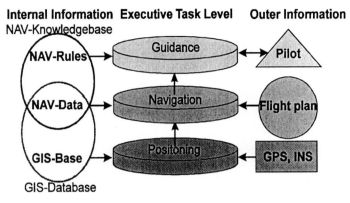

Fig. 4 SANC Core Concept

The mediate concept level is *'Navigation'*. This level compares the actual flight direction with the next waypoint of the external flight plan and calculates the necessary new heading. Consequently, the navigation level computes the new heading to reach the waypoint and control all remote sensors onboard.

The highest system level of 'SANC' is guidance. On this level, there's also the standard 'Go' function besides the normal visualization capabilities, which allows the pilot to decide on-line a new flight target. Using the 'Navigation Rules' of the database, the guidance level calculates the new route to the desired flight target.

6. CONCLUSIONS

The tests of theses systems have shown their potential for future developments. Especially three conclusions can be drawn:

- Integration of different *external, dynamic* (GPS /INS) and *internal, static* (GIS/NAV) data multiplies system performance on all task levels.
- This data integration renders new advanced visualization techniques, increasing *'usability'* of mission planning and flight guidance system drastically. Especially the different flight visualization modi have shown their special advantages:
  - 'Standard mode' for low precision navigation
  - 'Move mode' for medium precison flight guidance tasks
  - 'Tunnel mode' for high precision flight guidance tasks (Approach, Photo flight)
- Distinction between general flight tasks and special mission tasks allows fast adaption of the planning and flight guidance system to new special applicions.

7. REFERENCES

Hein, G.W. (1993): GPS-Bildflugnavigation. Leipziger BILDMESS-TAGE 1993, Institute of Geodesy and Navigation, FAF Munich.
Hein G.W., G. Baustert, B. Eissfeller, H. Landau (1989): High Precision Kinematic GPS Differential Positioning and Integration of GPS with a Ring Laser Strapdown Inertial System. Navigation, 36, 77-98

HIGH PRECISION DEFORMATION MONITORING USING DIFFERENTIAL GPS

Günter W. Hein and Bernhard Riedl
Institute of Geodesy and Navigation (IfEN)
University FAF Munich, D-85577 Neubiberg, Germany

ABSTRACT

Post-mission analysis of observations to the Global Positioning System (GPS) have proven in the past their capability to derive sub-centimeter to millimeter level accuracies for baselines up to 5-10 km.

Considering that we are interested (1) in monitoring three-dimensional relative position changes in the millimeter range over short time spans, and (2) that we want to have the information in an economical way in or near real-time, (3) that we operate only in a radius of up to 7-10 km around a master station, we have built up a continuous real-time DGPS deformation monitoring system. The high portability as well as the high data rates (up to 20 Hz) recommend such a system for its use in monitoring of all kinds of man-made structures where the early detection of possible movements may prevent disasters.

The paper presents this new sophisticated system, outlines the algorithms, the software and hardware architecture as well as reports from first experiences in the field.

1. INTRODUCTION

This paper discusses the developments for a real-time continuous deformation monitoring system in the millimeter range based on satellite observations to the American Global Positioning System (GPS). This monitoring system is suited especially for disaster research (earthquakes, volcano eruptions) as well as for deformation monitoring of engineering objects.

When monitoring engineering objects like dams, bridges, etc. terrestrial observations are commonly used. These techniques include automatic theodolite systems and levelling instruments as well as laser rangefinders and terrestrial photogrammetry. The objects under consideration are generally observed with time intervals of one or more days or even months in between. All of the above mentioned systems have one major disadvantage: they can not gain our insights into the short-time behaviour of deformation objects, since they have not the capability to provide results in real-time. This is also why, for example, the International Commission for Large Dams [*ICOLD, 1988*] considered those contemporary surveying techniques to be of only limited importance for monitoring purposes.

The DGPS real-time prototype deformation system, presented and discussed here, is a system which is able to track satellite observations and to derive position solutions with a frequency of up to 20 Hz.

Since the system is built up with single-frequency GPS receivers, it is very cost-effective.

However, the radius of application is limited to a maximum distance of less than, say 7-10 km, between reference (master) and monitor station due to ionospheric considerations unless dual-frequency instruments are used. It is easily transportable, and therefore especially suited for disaster research.

2. ALGORITHMIC CONSIDERATIONS

In developing the algorithms for such a deformation monitoring system and by using GPS observations we can take into account the following points:

- The coordinates of the network reference station as well as of all the monitor stations are approximately known by the time of system startup. Thus, the carrier phase ambiguities can be fixed immediately. If only the coordinates of the reference station are approximately known, the positions of the monitor stations and, consequently, the carrier phase ambiguities can be resolved and fixed during an initialization period where carrier phase observations of, say half an hour, are processed.
- Since we can further assume that our position changes are smaller than one GPS wavelength, we can always determine our phase ambiguities from the known baselines between master and monitor station. Insofar, loss-of-lock and cycle slips can be easily repaired in real-time.
- When applying low-cost single-frequency receivers we have the full accuracy potential, however, our range of application is limited to less than 7-10 km assuming that the ionospheric impact on the observations is zero. Tropospheric models can be applied taking into consideration possible height differences between the stations.
- According to the latest state-of-the-art in deformation analysis using GPS satellite observations the achievable accuracy might be in the range of 1 to 3 mm [*Frei et al, 1993; Table1*]. In order to smooth out possible oscillations we can incorporate appropriate filtering already in the Kalman filter model, taking advantage of a possible knowledge of the type of dynamics of movements in the data.

The following algorithms are implemented:
- double difference model using carrier phase observations
- appropriate fixing of the integer ambiguities (two possibilities as mentioned above)
- range- and carrier-phase corrections (at the monitor stations)
- low-pass filtering

3. SYSTEM DESIGN

3.1. System overview

The whole set-up of the monitoring network consists of two major components:
- The network master station
- The monitor stations

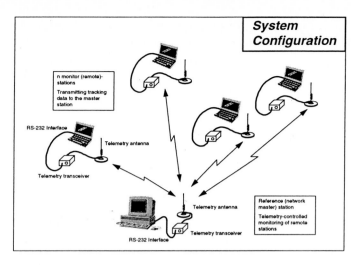

The *network master station* is responsible for the collecting and storing of data received from the network, for position computations and visualizing of the computed results, especially of the distances between all stations. The *monitor stations* are mainly used as data logger and pre-processing units.

In contrast to common reference stations which transmit data to the monitor stations (and, consequently, do not need to carry out calculations for each monitor station), this system just requires the inverse approach. Data is collected by the monitor stations, then it is transmitted to the

Fig. 1. System set-up

network master station. There the data is evaluated and the results are visualized.

Fig. 1 shows the overall set-up of the whole system configuration.

3.2. Network master station (Reference station)

Hardware design. Since the calculations and the data management need a considerable effort, appropriate hardware has to be chosen. Extensive use of platform independent software design (see next section) enables the deployment of PCs (preferably type 80486 and better) in conjunction with MS-DOS and MS-Windows or with workstations using UNIX and OSF/Motif. On a graphical colour display the changes in 3D-positions of the monitor stations can be visualized.

In order to keep the whole set-up cost-effective, only L1 phase receivers are used. At present we are using receiver boards of the NovAtel OEM performance series. They have the capability of tracking 12 channels simultaneously. For the reduction of multipath a choke ring antenna has been used, which usually leads to a 50% decrease of those effects compared to a standard antenna.

For data transmission from and to the monitor stations, there are different possible solutions (e.g. direct links by wire (serial port), telemetry links, cellular phones, modems). At the moment direct links over the serial port and telemetry links have been realized.

Software design. Most important concerning the software is that it has been developed using an object-oriented system design. There are three distinct different groups: (1) I/O operations, (2) computations and data management, and (3) user interface and visualization. This concept is supported using the object-oriented programming language C++. Thus, it is possible to adapt the software easily to other platforms even with different operating systems.

As already mentioned above, most of the calculations take place at the network master station. In addition to the calculations, the graphical output is shown on the screen. On the basis of the object-oriented structure the implementation of the graphical user interface is separated from the calculations and the data I/O operations. In the prototype version of the program runs under MS-DOS and MS Windows.

3.3. Monitor station

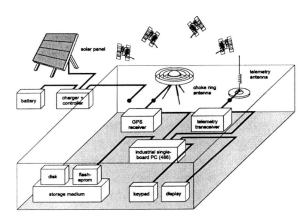

Fig. 2. Monitor station (Hardware design)

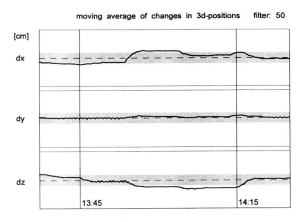

Fig. 3. Noise level of measurements (x, y, z coordinates) *(Shaded areas indicate the millimeter noise level)*

Hardware design. Here again – even more important than for the network master station – it was necessary to find an inexpensive, easy-to-install, transportable solution. Therefore many miniaturized parts are used which make it possible to build very compact stations. All parts are put in a water-proof sealing protecting them against harsh environmental conditions.

The hardware design of the monitor station is shown in Fig. 2.

Software design. At the monitor station the number of calculations is considerably less than at the network master station. Here the raw data is gathered from the GPS receiver. This is screened for outliers, goodness of observation, etc. and structured for storage. Finally, pre-calculated data in the form of carrier phase corrections [*Blomenhofer Taveira and Hein, 1993*] are sent to the network master station .

4. FIRST EXPERIENCES

During a project study a simple prototype system based only on GPS has been developed in order to verify the concept and to show the whole system in operation. For this project receivers of the Magnavox 4200D type were used which were directly linked to the master station (PC). In order to simulate a deformation, the receiver of the monitor station was fixed on a slide allowing us to generate precise three-dimensional movements. Thereby, some insights in the behaviour of the system could be gained, especially with respect to the detection and verification of the correct position change of the receiver at the monitor station. Fig. 3 shows the typical noise level of DGPS derived coordinate components. The time series typically oscillates within a 2 to 3 mm range around the true position. Use of newer GPS hardware (single frequency

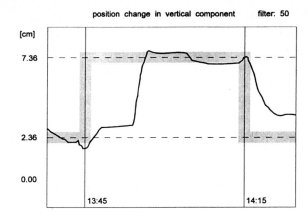

position change in vertical component filter: 50

Fig. 4. Position change in vertical component

cards with narrow correlator) will probably decrease the level down to 1. Fig. 4 shows the observed position change in vertical component in relation to the generated one. The time delay of the response of the monitoring system on the generated antenna position change is due to characteristics of the used low-pass filter which in this case was a simple moving average filter over 50 epochs.

5. CONCLUSIONS

A new real-time deformation monitoring system based on differential GPS and possibly, GLONASS is proposed which is especially suited for disaster research. First trials with a simple prototype system based solely on GPS indicate that it is possible to monitor position changes in the millimeter level with a frequency of 1 to 20 Hz. The new system overcomes the disadvantages of the post-mission static GPS analysis concept commonly used and might offer new geophysical insights into the short-term behaviour of volcano eruptions and earthquakes as well as providing an economical real-time deformation analysis.

Acknowledgment. This project is funded by the German Research Foundation (Deutsche Forschungsgemeinschaft).

6. REFERENCES

Fraile, J., G. W. Hein, H. Landau, B. Eissfeller, A. Jansche, N. Balteas, V. Liebig 1992: „First Experiences with Differential GLONASS/GPS Positioning", Proc. ION GPS-92, Sept. 16-18, Albuquerque, NM, USA, pp 153-158

Frei, E., A. Ryf and R. Scherrer 1993: „Use of the Global Positioning System in Dam Deformation and Engineering Surveys", SPN Zeitschrift für Satellitengestützte Positionierung, Navigation und Kommunikation 2, 42-48, Wichmann Verlag

ICOLD 1988: „International Commission for Large Dams Bulletin No. 60: Dam Monitoring - General Considerations"

Okada, Y. 1985: „Surface Deformation Due to Shear and Tensile Faults in a Half Space", Bull. Seism. Soc. Am. 75, pp 1135-1154

Schneider, G. 1975: „Erdbeben. Entstehung, Ausbreitung, Wirkung", Ferdinand Enke Verlag Stuttgart

Taveira Blomenhofer, E. and G. W. Hein 1993: „Investigations on Carrier Phase Corrections for High-Precision DGPS Navigation", Proc. ION GPS-93, Sept. 21-23, 1993, Salt Lake City, Utah, pp 1461-1468

Multi-Sensor Arrays for Mapping from Moving Vehicles

K. P. Schwarz and N. El-Sheimy
Geomatics Engineering Department
The University of Calgary
Calgary, AB, Canada T2N 1N4
Tel : (403) 220-7377   Fax : (403) 284-1980
e-mail: schwarz@ensu.ucalgary.ca
elsheimy@acs.ucalgary.ca

The field testing of VISAT, a prototype mobile highway mapping system which has been developed in cooperation between the University of Calgary and Geofit Inc. in Laval, Quebec, is described in this paper. The integration of inertial technology and GPS satellite receivers with an array of CCD cameras has resulted in a system that is capable of mapping all visible objects within a 50 m radius of the moving vehicle. Extensive tests in city centers, suburban areas, and rural areas have shown that the system works very reliably and that the expected positioning accuracy of 30 cm is surpassed in many cases. A more detailed analysis of the results shows that the RMS positioning error in across-track direction and height is actually between 5 and 10 cm, while the along-track error is usually larger. Due the restrictions in the length of the paper, only some of the major results can be highlighted.

INTRODUCTION

With the continuing growth of urban centers on a world-wide scale, the demand of city planners for up-to-date information is increasing at a rapid rate. The information needed is expensive to establish by conventional methods and is therefore not well suited for rapid updating. In addition, conventional methods often supply only pointwise information and are therefore not suited to answer the increasingly complex questions concerning the interaction of different factors in urban centers and their time dependencies. To address this problem, the University of Calgary and Geofit Inc., Laval, Quebec, have developed a precise multi-sensor mobile mapping system for operation in urban centers. The total system consists of a data acquisition system, called VISAT, and a measuring and processing system, called GEOSTATION. It implements the idea of storing georeferenced digital images as the basic unit and of combining an arbitrary number of such units, which may be from different time periods, to obtain the specific information required. The system can be used to selectively update GIS data bases very quickly and inexpensively, see also Bossler and Novak (1993) for a discussion.

The system integrates a cluster of video cameras, an Inertial Navigation System (INS), and satellite receivers of the Global Positioning System (GPS). The system carrier is currently a van, but airborne or marine applications can be realized in a similar way. The overall objective of the VISAT development was to build a precise multi-sensor mobile mapping system that could be operated at speeds of up to 60 km per hour and would achieve an accuracy of 0.3 m (RMS) with respect to the given control and a relative accuracy of 0.1 m (RMS) for points within a 50 m radius. This accuracy is required in all environments including inner cities, where stand-alone GPS is not reliable. Accuracy optimization must therefore be done with a view to total system performance, while system design has to be optimized with a view to isolating the error contribution of each sensor. The data flow has to be streamlined to facilitate the subsequent feature extraction process and transfer into a GIS system. For further details system, see Schwarz et al (1993), El-Sheimy and Schwarz (1993), and Li et al (1994).

Figure 1 shows the current VISAT van and its major components consisting of a GPS receiver, a strapdown inertial system, and a cluster of video cameras. A second receiver is needed as a master station for DGPS operation. The hardware components used in the VISAT prototype system include two Ashtech Z-XII GPS receivers, a Litton LTN 90/100 strapdown inertial system, and three COHU 4912 CCD cameras. In the vehicle, the three sensors are interfaced to a regular PC-AT, which controls different tasks through programmed interrupt processes. All components are precisely time-tagged to minimize errors from insufficient sensor synchronization. A navigator in the back seat of the van controls the system through appropriate commands on the computer keyboards.

TESTS AND ANALYSIS OF RESULTS

Test Description

The system was tested in Laval, Quebec, and Quebec City in November 1994. The test areas included open areas, urban centers with narrow roads, and minor and major highways with a number of overpasses. The Quebec City test results will be presented in this paper. They are in four sectors surveyed in three days. Table 1 gives a summary of the surveys made. They show that results in all sectors can be used to evaluate survey repeatability in forward and backward runs on the same day. Data from sector 1 can also be used to assess day-to-day repeatability. Data from sector 3 can also be used to evaluate survey repeatability in forward and backward runs on different days.

Table 1. Survey summary.

Day	Forward	Backward
Day 1	Sector 1- Sector 3	Sector 1
Day 2	Sector 1 - Sector 2	Sector 1 - Sector 2
Day 3	Sector 4	Sector 4 - Sector 3

Fig 1. The VISAT van

In most tests, static initialization was carried out, whenever possible, 20 to 50 m away from a Ground Control Point (GCP) which was used as a GPS master station before and after the survey. During these static initialization periods, images of the GCP were taken.

Accuracy

Because the contractor wanted a 'blind' test, control points were not available along the test course to estimate the absolute positional accuracy of the VISAT system. However, the day-to-day comparisons provided a good test for absolute position accuracy because system performance between days is almost completely independent. Only the use of the same GPS master station and residuals errors in the camera calibration may introduce some correlation between days.

The tests were designed to assess system repeatability under varying conditions. Well-defined landmarks along the test course were used for comparison. Figures 2a to 2c show some of the results.

- **Figure 2.a** shows the relative accuracy of some well-defined landmarks in sector (1), using results of day (1) runs in both forward and backward directions, taking the forward run as the reference

- **Figure 2.b** shows the relative accuracy of the same landmarks for day-to-day repeatability. Sector (1) was resurveyed on the second day to test the system performance in this respect. The figure gives results of a comparison of day (2) and day (1) runs in forward direction, the second run is taken as the reference.

- **Figure 2.c** shows the relative accuracy of some well-defined landmarks in sector (3). The figure gives results of a comparison of the day (3) backward run and the day (1) forward run, with the latter taken as reference. Table 2 summarizes the results of the tests described above.

The conclusions at this point are only tentative because they are based on scarce statistical material and no external control. It appears, however, that an RMS of 0.3 m in the horizontal and a few centimeters in height are achievable for distances of up to 35 m under normal conditions.

The results in height indicate that the GPS/INS positioning component is working at the centimeter level. Since the height component in GPS is the weakest, it can be expected that the X and Y components are at least of the same accuracy. The increase in errors for the horizontal components must therefore be due to the camera array. The most likely explanation is that the increase in RMS(x,y) as compared to RMS(h) is due to the along track error. The across-track error should be small and comparable in size to the height error. Repeatability in the same direction, even on different days, seems to be consistently better than repeatability between forward and backward runs. This indicates remaining calibration errors or improper lens correction. Frequent signal blockage due to overpasses in sector (2) introduce more errors due to INS position reset and ambiguity drift. But the RMS of 0.42 m in horizontal and 0.09 m vertical are still reasonable considering that a total of 28 overpasses and the same number of GPS outages had to be accounted for in this sector, for more details see El-Sheimy et al (1995).

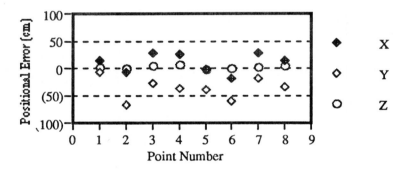

2.a. Repeatability of Forward-Backward Runs on the Same Day (Sector 1)

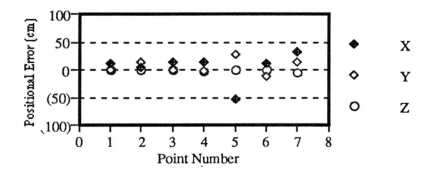

2.b. Repeatability of Forward Runs on Different Days (Sector 1/2)

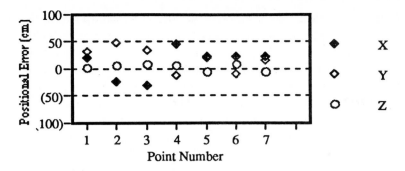

2. c. Repeatability of Forward-Backward Runs on Different Days (Sector 3)

Table 2. Statistical summary of the system repeatability

Repeatability	RMS (x,y) (cm)	RMS (h) (cm)
Forward-backward, same day	± 25 & ± 35	± 3 & ± 4
Forward-backward, same day "Frequent signal blockage"	± 43	±9
Forward-one day, backward-diff. day	±29	±7
Same Direction, Different days	± 20 & ±27	± 2 & ±5

CONCLUSIONS

The VISAT system presented briefly in this paper is a mobile mapping system which can be operated continuously under diverse operational conditions to generate georeferenced digital images. The GEOSTATION environment permits user-friendly viewing of the imagery which contains recent information for user prescribed area. The image database provides a flexible information media which can be used in decision making, answers to queries, and more generally, updating maps and GIS systems.

Pilot project results indicate that highway velocities of 60 km/h can be maintained with adequate data transfer and target positioning in a post-processing mode at the GEOSTATION. Run-to-run and day-to-day repeatability achieved in the pilot project is about 25 cm (RMS) in horizontal and about 5 cm (RMS) in height. There is no significant difference between run-to-run and day-to-day results. This indicates that the GPS/INS component works at a consistent level from day to day. In general, the results of system testing show that the relative accuracy surpasses the development objectives for the prototype system. There is considerable room for improvements which will be implemented in the production system, due for delivery in summer 1995.

Acknowledgment The VISAT pilot project represents the combined effort of a research team at the Department of Geomatics Engineering and Geofit Inc., Laval, Quebec.

References

Bossler, J.D. and Novak, K. . Mobile Mapping System: New Tools for the Fast Collection of GIS Information, GIS'93, Ottawa, Canada, pp. 306-315, March 23-25, 1993.

El-Sheimy, N. and Schwarz K.P. . Kinematic Positioning In Three Dimension Using CCD Technology, VNIS93 Conference, Ottawa, October 12-15 1993.

El-Sheimy, N., Schwarz, K.P., and Gravel, M. . Mobile 3-D Positioning Using GPS/INS/Video Cameras, Mobile Mapping Symposium, Columbus, Ohio, May 24-26, 1995.

Li, R., M. A. Chapman, Qian, L., Xin. Y., K. P. Schwarz. Rapid GIS Database Generation Using GPS/INS Controlled CCD Images, ISPRS94 GIS/SIG, June 6-10, 1994, Ottawa, Canada.

Schwarz, K. P., Martell, H., El-Sheimy, N., Li, R., Chapman, M., Cosandier, D. . VISAT- A Mobile Highway Survey System of High Accuracy, VNIS Conference '93 Conference, Ottawa, October 12-15, 1993.

TESTING OF EPOCH-BY-EPOCH ATTITUDE DETERMINATION AND AMBIGUITY RESOLUTION IN AIRBORNE MODE

Klaus-Peter Schwarz and Ahmed El-Mowafy
Department of Geomatics Engineering
The University of Calgary
2500 University Drive N.W.
Calgary, AB   T2N 1N4   Canada

ABSTRACT

A test for GPS attitude determination in airborne mode was performed in fall 1994 to investigate the performance of an epoch-by-epoch ambiguity resolution technique and to assess the overall performance and accuracy of attitude determination in this environment. The main advantage of an epoch-by-epoch ambiguity resolution is its usefulness for real-time applications. An Ashtech 3DF multi-antenna system was used on a KING AIR 90 airplane with twin turbo engines. The attitude of the 3DF was compared to an accurate inertial system (LTN 90/100) to assess the correctness of ambiguity resolution and accuracy of the 3DF attitude results. Only the heading and pitch were determined in this test. Test results show that with a baseline length of 2.5 metres, the correct ambiguities were determined in 96% of the cases. Heading and pitch accuracies were obtained with a standard deviation of 0.1 to 0.15 degrees.

ATTITUDE DETERMINATION

Attitude determination of a moving body by a GPS multi-antenna system is achieved by measuring interferometric ranges to the satellites and determining the antenna baseline vectors from the range differences. The orientation of the antenna baselines in the WGS 84 system can then be determined from these vectors. The body frame of the aircraft is defined by the phase centres of the antennas mounted on the aircraft. They are usually chosen in such a way that they are approximately parallel to the main axes of the aircraft. The aircraft attitude is defined as the rotation of this body-fixed frame with respect to a well-defined reference coordinate frame, such as the WGS 84 system. If the reference coordinate frame is the local-level frame, the attitude components are azimuth, pitch and roll.

To determine the attitude from antenna vector components in the local-level frame, simple trigonometric relations can be used. If antennas 1 and 2 are mounted on the airplane, such that the baseline 1-2 is assumed parallel to the airplane's primary axis (the longitudinal axis), the heading and pitch of the platform are directly estimated as:

$$\text{Heading} = \tan^{-1} \frac{\Delta E_{12}}{\Delta N_{12}} \tag{1}$$

$$\text{Pitch} = \tan^{-1} \frac{\Delta U_{12}}{\sqrt{(\Delta E_{12})^2 + (\Delta N_{12})^2}} \tag{2}$$

Roll can be determined in a similar fashion as pitch, using a baseline that is orthogonal to the baseline 1-2. For exact definitions and details on general antenna configurations, reference is made to El-Mowafy (1994).

AMBIGUITY RESOLUTION FROM DATA OF A SINGLE EPOCH

For real-time attitude determination, two major problems have to be solved. One is the instantaneous re-initialization of the ambiguities due to loss of lock which typically happens during aircraft turns, the other is cycle slip detection and fixing. A straightforward way to solve both problems is phase ambiguity resolution from data of a single epoch. The paper therefore focuses on this problem, discussing the solution concept and assessing the accuracy of the resulting attitude estimates.

Ambiguities are usually determined using a search procedure. One important step in any search approach is the determination of an approximate solution for the antenna baselines. The search approach can be accelerated significantly by using approximate values of the baselines as close as possible to the correct ones. The number and distribution of the antennas can play a vital role in this process. The minimum number of antennas required to estimate the direction of a line (heading and pitch) is two, while for 3-D attitude determination, a minimum of three non-collinear antennas is required. Instead of using the minimum number of antennas, additional antennas can be used to speed up ambiguity resolution. For instance, Hatch (1989), and Jurgens et al. (1991) have discussed the benefit of placing three collinear antennas along the target line when determining azimuth. Two of the antennas are placed within a distance of less than half a cycle. The azimuth and pitch of the two farthest antennas are roughly estimated from the azimuth and pitch of the two closely spaced antennas. This basic idea has been extended to 3-D attitude determination. El-Mowafy and Schwarz (1995) discuss antenna configurations for different application scenarios. Because the use of additional antennas gives a good approximate baseline estimation from data of a single epoch, the ambiguity search window is reduced to $\pm 2 - \pm 4$ cycles depending on the baseline length used. Thus, ambiguity resolution from data of a single epoch becomes possible and real-time attitude determination becomes feasible.

If data from only one epoch are considered, reliability of the estimate is a problem because the number of measurements used is always small. As a result, ambiguities can be incorrectly determined. However, the incorrect result will only affect a single epoch because the ambiguity determination in the next epoch is completely independent. In addition, an ambiguity moving average can be used to increase the probability of correct ambiguity determination, see (EL-Mowafy, 1994).

TEST DESCRIPTION

A KING AIR 90 airplane with twin turbo engines was used in the test. The multi-antenna system used for attitude determination was the Ashtech Three-Dimensional Direction Finding (3DF) system, which consists of one receiver connected to four antennas. To check results of attitude determination, the output is compared to that of a Litton LTN

90/100 strapdown inertial system, which has an accuracy of 0.02 to 0.03 degrees per hour. The correctness of the determined ambiguities was checked by comparing attitude results of both systems.

No antennas were permitted to be mounted on the wings of the aircraft. Since the fuselage is curved and narrow in transverse direction, it was not practical to mount antennas for a transverse baseline. Consequently, only the heading and pitch were determined in this test. Three collinear antennas were used. Figure 1 illustrates the antenna configuration which allowed fast ambiguity determination because results of the short baseline could be used to limit the search region for the longer baseline.

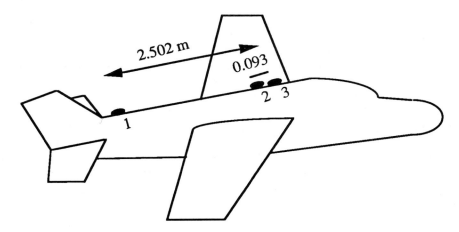

Fig. 1 Antenna configuration on the airplane

The test was conducted on March 29, 1994 close to Calgary. The test began with a static run of 26 minutes, followed by a kinematic one of about 42 minutes. The static test was performed on the runway with engines switched on. The kinematic test began as soon as the aircraft started to move . During the static test the number of satellites observed was 5 to 6, giving PDOPs between 2 and 4. During the kinematic test, the number of satellites observed changed from 3 to 6, due to signal blockage of some satellites during flight maneuvers. Thus, the satellite configuration change did not occur in a typical descending or ascending order. For approximately 42 minutes of flight, between 3 and 6 satellites were observed, allowing ambiguity resolution and attitude determination during that period. The PDOPs values were between 4 and 8 when tracking more than 3 satellites and increased up to 30 when the number of satellites dropped to 3.

The aircraft experienced strong vibrations with accelerations reaching up to 9 m/s² during some periods, compared to less than 1.5 m/s² under normal flight conditions. In addition, satellite tracking was not consistent. Some satellites were received at one antenna, but not at another. This reduced the number of useful data to 2554 epochs. However, the number is sufficient to test the ambiguity resolution technique and assess attitude determination in a statistically meaningful way.

RESULTS OF AMBIGUITY RESOLUTION

During the flight, satellite blockage resulted in a number of cycle slips. In these cases, the ambiguity moving average technique proved to be a powerful approach in finding cycle slips above a threshold of four cycles. Table 1 shows results of instantaneous ambiguity resolution for both static and kinematic tests. Out of 1555 epochs in the static case, only two epochs had wrong ambiguity estimation. In the kinematic case, the epoch-by-epoch ambiguity resolution gave very good results in periods of good satellite visibility, the success rate was about 96%. This result shows that the adopted technique is reliable, especially when considering the high level of dynamics experienced by the aircraft. During the period of poor satellite geometry with 3 satellites, the success rate dropped to 74%. This is expected as there is no redundant measurement available in this case, and poor satellite geometry has a profound impact on ambiguity resolution. It should be noted that these results are specific to the multi-antenna system used and may change when other receivers are used.

Test	Static Test	Kinematic using 4-6 satellites	Kinematic using 3 satellites
number of epochs	1555	2002	552
correct ambiguity resolution	1553	1915	408
percentage of correct resolution	99.9%	95.7%	73.9%

Table 1 Performance of ambiguity resolution technique in airborne mode

ANALYSIS OF ATTITUDE RESULTS

Table 2 summarizes results of attitude determination. Static results are given for comparison and as an indicator of the accuracy achievable under optimal conditions. The static heading accuracy is unusually good, due to the almost total absence of multipath in this component. The static pitch accuracy is more typical for what can be achieved under average conditions. Comparing columns 1 and 2 in Table 2 indicates that for good satellite geometry, the static and kinematic results have comparable accuracy. For poor satellite geometry, kinematic accuracy deteriorates by a factor of 2 to 3.

Attitude Component	Static Test	Kinematic using 4-6 satellites	Kinematic Including 3 satellites
Heading	0.032	0.118	0.273
Pitch	0.127	0.138	0.433

Table 2 RMS of attitude components in static and kinematic tests (deg)

Figures 2 and 3 show the differences between the output of the 3DF and the INS for heading and pitch, respectively. In contrast to the static case, there is no clear evidence of multipath errors in the kinematic test. This is most likely due to randomization of these errors due to orientation changes of the aircraft as well as aircraft vibration.

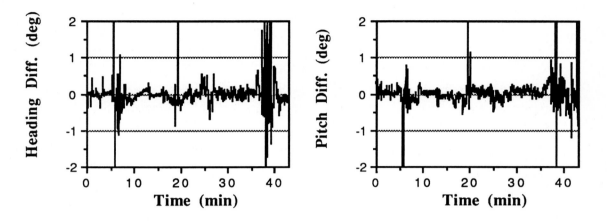

Fig. 2 Kinematic (INS-3DF) heading   **Fig. 3** Kinematic (INS-3DF) pitch

Changes in satellite visibility and geometry described before have a direct effect on the final output. Periods with good satellite geometry showed more stable and consistent errors, while those with poor geometry showed large errors of inconsistent pattern. This is quite obvious when comparing the first three quarters of the data to the fourth. In the former, 5 to 6 satellites with good geometry were observed, resulting in errors bounded in a range usually expected under these favorable conditions. In the fourth quarter poor satellite visibility was encountered with usually not more than 3 satellites available. During this period, the error level is magnified considerably. As shown before, the RMS values were magnified by a factor of two to three.

The results extracted from the periods of normal to good satellite geometry compare well with the ones given in Cohen et al. (1994) and Cannon et al. (1994), when taking the baseline length ratio into consideration. In both studies, the baseline lengths were considerably longer than the 2.5 m used in this test. Cohen et al (1994) achieved an RMS of about 3 arcminutes in attitude for all components using a baseline length of 7.92 m. Cannon et al (1994) achieved 3.5 and 5.8 arcminutes for heading and pitch, respectively, using a baseline length of 6.93 m.

CONCLUSIONS

Test results show that achieving ambiguity resolution from data of a single epoch in an epoch by epoch solution mode is possible in airborne mode. Satellite geometry is a governing factor in achieving reliable results. Under good satellite geometry, the correct ambiguities were found in 96% of the cases. This ratio dropped to 74% when only 3 satellites were visible. The ambiguity moving average approach considerably increased the probability of finding the correct ambiguities. Test results also showed that with a baseline of 2.5 m, the achievable accuracy in heading and pitch is 0.1 to 0.15 degrees. Accuracies can be considerably improved by increasing the baseline length.

ACKNOWLEDGMENTS

This research was supported by a grant from the Natural Sciences and Engineering Research Council of Canada (NSERC). Canagrav Research Ltd. is acknowledged for providing data used in this test.

REFERENCES

Cannon, M.E., H. Sun, T.E. Owen and M.A. Meindl (1994). Assessment of a Non-Dedicated GPS Receiver System for Precise Airborne Attitude Determination, *Proc. ION GPS-94*, Salt Lake City, Utah, September 20-23, 1994.

Cohen, C.E., B.W. Parkinson and B.D. McNally (1994). Flight Tests of Attitude Determination Using GPS Compared Against an Inertial Navigation Unit, *Navigation, Journal of The Institute of Navigation, Vol. 41*, No. 1, Spring, 1994.

El-Mowafy, A. (1994). Kinematic Attitude Determination from GPS, Ph.D. Dissertation, *UCGE Report Number 20074, Department of Geomatics Engineering, The University of Calgary,* 1994.

El-Mowafy, A. and K.P. Schwarz (1995). Epoch by Epoch Ambiguity Resolution for Real-Time Attitude Determination Using a GPS Multi-Antenna System, Accepted for Publication in *Navigation, Journal of the US Institute of Navigation*, 1995.

Hatch, R. (1989). Ambiguity Resolution in the Fast Lane, *Proc. ION GPS-89*, Colorado Springs, Colorado, September 27-29, 1989.

Jurgens, R., C.E. Rodgers and L.C. Fan (1991). Advances in GPS Attitude Determining Technology as Developed for The Strategic Defense Command, *Proc. ION GPS-91,* Albuquerque, New Mexico, September 11-13, 1991.

EXPERIENCES IN DGPS / DGLONASS COMBINATION

Udo Roßbach, Günter W. Hein and Bernd Eissfeller
Institute of Geodesy and Navigation (IfEN)
University FAF Munich, D-85577 Neubiberg, Germany

ABSTRACT

The Institute of Geodesy and Navigation (IfEN) has recently developed a high-precision real-time DGPS navigation system including carrier phase ambiguity resolution on-the-fly for different applications. Although the accuracy is in the centimeter level, the integrity requirements for some applications, for example on-line quality assurance of GPS data, can only be met by combination with other sensors.

Among possible candidates, research work is being done on the GLONASS system, being under deployment by the Russian Federation. The combination of GPS with GLONASS also provides an increased number of available satellites with all its advantages on positioning accuracy.

IfEN has acquired two 3S Navigation R-100/R-101 receivers, enabling the user to observe GPS on L1 C/A-code and GLONASS L1 and L2, C/A- and P-code signals simultaneously. Furthermore, IfEN's differential GPS software was modified to include GLONASS L1 and L2, both C/A- and P-code observations.

This paper reports about first experiences using differential GLONASS observations in combination with DGPS.

1. INTRODUCTION

GPS becomes more and more accepted both for navigation and for geodetic applications. But whereas, by using phase observations and differential measurements, GPS is able to match geodetical accuracy requirements, the integrity requirements for some applications cannot be fulfilled by GPS alone. For this reason, research work is being done on the combination of GPS with other sensors. At the Institute of Geodesy and Navigation of the University FAF Munich, research has been started on the combination of GPS with GLO-NASS three years ago, initially using Russian GLONASS receivers like the ASN-16 [1]. This combination offers the advantage of not having to integrate the GPS receiver with a completely different sensor, as is the case e.g. with INS. Suitable receivers, such as the 3S Navigation R-100/R-101, can be used for GPS and GLONASS observations in parallel. Existing differential GPS software, based on L1 C/A-code pseudorange corrections, was modified to include GLONASS observations on L1 and L2, C/A- and P-code.

2. THE GLONASS NAVIGATION SYSTEM

2.1. Space Segment

GLONASS (Global Navigation Satellite System) is a navigation system similar to the Global Positioning System. It was developed by the former Soviet Union as a counter-piece to the American GPS, and is now put into orbit and run by the Russian Federation.

When completed by the end of 1995, GLONASS will consist of 24 satellites in nearly circular orbits of 19130 km altitude and 64.8° inclination. The satellites are arranged in three orbital planes with eight satellites per plane. Currently, 19 satellites are active.

In contrast to GPS, whose satellites all transmit their signals on the same frequency, but use different codes to distinguish between satellites, all GLONASS satellites use the same code sequence, but transmit on different frequencies in the L1 and L2 bands. These frequencies are $f_{k,L1} = 1602 + k \times 0.5625$ MHz and $f_{k,L2} = 1246 + k \times 0.4375$ MHz, where k means the satellite frequency number (1 - 24).

Unlike GPS, GLONASS satellites broadcast their orbital information as the ECEF components of their coordinate, velocity and Luni-Solar acceleration vectors at a reference epoch. The satellite position to any other epoch is then obtained by integrating the satellite's equations of motion, using these values as initial values [6].

A further difference between GPS and GLONASS is the absence of S/A or other system assurance techniques on GLONASS, which makes a GLONASS positioning solution more coherent than a GPS solution. Positioning accuracy of GLONASS is comparable to that of GPS with S/A turned off.

2.2. Reference frames

Another difference between GPS and GLONASS is the reference frame used to express satellite (and user) positions and time.

GPS uses the Wold Geodetic System 1984 (WGS 84) as coordinate system [7], whereas GLONASS employs the Soviet Geodetic System 1990 (SGS 90), which is nearly identical to the SGS 85 [6]. Although the definitions of these reference frames are nearly compatible, in practice offsets exist between their origin and orientation parameters. For the purpose of combining GPS and GLONASS measurements, the relation between these coordinate frames must be known. According to [4], the x-axis of SGS 85 is 0.6" east of that of WGS 84 and the XOY plane of SGS 85 is 4 m north of that of WGS 84.

Besides the coordinate frame, GLONASS also uses a different time scale. GPS time is a uniform time, whereas GLONASS time is tightly coupled to Moscow time UTC_{SU}. Thus, unlike GPS, GLONASS considers leap seconds. The difference between GPS time and GLONASS time meanwhile is about 10 seconds.

Further, GPS broadcast signals contain parameters to transform GPS time to UTC_{USNO}. In the GLONASS navigation message the difference between GLONASS time and UTC_{SU} is broadcast. As the difference between these two realizations of UTC is usually not available in real-time, in combined GPS/GLONASS measurement scenarios at least five satellites must be observed simultaneously to determine the additional unknown.

3. THE 3S NAVIGATION R-100/R-101 GPS/GLONASS RECEIVER

The 3S R-100/R-101 receivers are capable of tracking GPS and GLONASS satellites simultaneously. GPS satellites can be received on L1 C/A-code, GLONASS satellites can be received on L1 and L2, C/A- and P-code. The receiver entirely provides 20 channels. Eight of these channels (the R-101 channels) are reserved for GLONASS satellites on either L1 or L2, C/A- or P-code. The twelve R-100 channels can receive GPS or GLONASS satellites on L1 C/A-code. Every channel is able to receive one GLONASS satellite, whereas GPS satellites must be tracked on more than one channel. The twelve C/A-channels can receive either twelve GLONASS satellites, seven GPS satellites, or a combination of GPS and GLONASS satellites according to the equation $a + 2b \leq 15$, where a means the number of GLONASS (max. 12) and b the number of GPS satellites (max. 7). Thus, 15 to 20 satellites can be tracked simultaneously, depending on the configuration.

4. COMBINING GPS AND GLONASS MEASUREMENTS

The output data of the 3S receivers are stored in a receiver-specific format. These data are then decoded into RINEX2 format files for input into the differential software. RINEX2 [2] offers provisions for the use of observations to other than GPS satellites, but these features are not consequently implemented in the RINEX concept. For this reason, slight modifications of the RINEX2 format turned out to be necessary.

RINEX2 observation files provide one character and a two-digit number to identify the satellite system and the number of a satellite within the system, whereas in navigation files only the number is used, without a character. Thus, ephemeris data of GPS and GLONASS satellites must be distinguished in a different way. As GPS uses PRN numbers 1 - 32, GLONASS satellites are identified with numbers 33 - 56 (frequency letter + 32) in the navigation files.

As measurements to GPS and GLONASS satellites are taken simultaneously, but on the other hand GPS and GLONASS system times differ from each other, a common time scale must be used in observation files. RINEX, as originally developed for the exchange of GPS data, defines GPS system time to be mandatory. Because for the computation of satellite coordinates based on the Keplerian type ephemeris data the system time is necessary, the differential software must also get access to GLONASS system time. This can be accomplished via the difference between GLONASS time and UTC, along with the difference between UTC and GPS time. Parameters to compute the latter difference are provided in RINEX2 navigation files, but not for the difference between GLONASS time and UTC. Therefore, an additional header record containing GLONASS UTC parameters tentatively was introduced in mixed navigation files.

For reasons of compatibility of the routines for computation of the satellite positions and compatibility with the RINEX format, the GPS orbit description was also used for GLONASS satellites. Satellite positions and velocities at the refence time, as transmitted in the GLONASS ephemerides, are transformed from SGS 90 to WGS 84 coordinates, using the transformation determined in [4]. Then the six Kepler elements of an ellipse fitting the instantaneous satellite

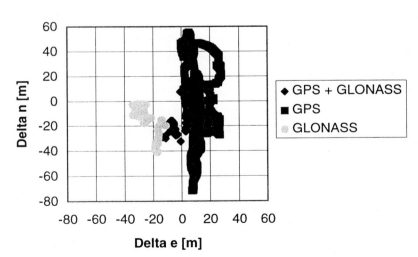

Figure 1: Single point positioning: Deviations from known position in northern and eastern directions. GPS L1 C/A-code, GLONASS L1 P-code.

position and velocity are computed. The correction elements used with GPS ephemerides $(c_{rc}, c_{rs}, c_{ic}, c_{is}, c_{uc}, c_{us}, \Delta n, \dot{\Omega}$ and i) are set to zero, resulting in less accurate ephemeris data for GLONASS satellites than for GPS. This may be significant for single point positioning, as can be seen in Figure 1, but using differential techniques, these orbital errors should cancel out.

Figure 1 further shows clearly the effects of S/A on the GPS positioning solution. In contrast to GPS, the positions derived from GLONASS (without S/A) are much closer together. The mixed GPS/GLONASS solution is also affected by GPS S/A, but these effects are reduced to a great extent.

5. SOFTWARE DESCRIPTION

The used post-processing differential software consists of two programs. One program computes pseudorange corrections for all satellites in view, while the other applies these corrections and computes the user position. Both programs require measurement data in form of (modified) RINEX2 files. Additionally, the reference station software needs the known station coordinates as input, the user software the pseudorange corrections computed by the reference station.

The reference station software reads the observation data epochwise and applies carrier smoothing using a Hatch filter [3], if desired. The difference between the geometric and the measured range from the reference station to a satellite is determined. Before pseudorange corrections are computed, the clock offset has to be compensated. Therefore, the average of the computed range differences as a measure of the clock offset is subtracted from the range differences, yielding the pseudorange corrections. According to [5], the clock correction can be any number, as long as it is applied equally to all differential corrections. But as the receiver can have different clock offsets with respect to GPS and GLONASS time, the average values have to be computed for both systems separately.

The user software reads the observations and generated range corrections for each epoch. Pseudoranges are smoothed, if desired, then the range corrections are added. Assuming short distances between reference station and user station, correction of ionosperic delay will be performed by differential operation. As the tropospheric delay can be different for reference station and user station even for short baselines, if the heights of both stations are different, tropospheric delays are corrected using a Modified Hopfield model.

The position of the user station is then computed using three different Kalman filters in parallel. To compare different solutions, one filter is used for GPS measurements only, one for GLONASS and one for all measurements. Each of the single system filters works on the four unknowns x, y, z and clock offset to system time. The combined filter works on the five unknowns for the position and two clock offsets, as explained earlier.

6. TEST RESULTS

Figure 2 shows test results using GPS L1 C/A-code and GLONASS L1 P-code measurements. Antennae of reference and user stations were located at surveyed locations. Deviations of the computed user position to the known position are shown.

It can be clearly seen that the position derived from GLONASS measurements is better than the one derived from GPS. But with decreasing number of available GLONASS

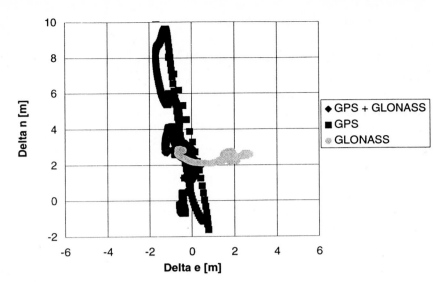

Figure 2: Differential positioning: Deviations from known position in northern and eastern directions. GPS L1 C/A-code, GLONASS L1 P-code.

satellites and increasing number of available GPS satellites, the GPS position becomes more accurate.

Further, it can be seen that the combined solution usually is more precise than GPS alone. Also, the combined solution is more precise than the standalone GLONASS solution.

The drift in the GLONASS solution may be caused by multipath effects.

7. CONCLUSIONS

The combination of GPS and GLONASS measurements in differential observations was tested and proved not only to be feasible, but also to be generally more accurate than measurements to one system alone.

Some details, however, deserve further investigation. Among those is the question of a proper representation of GLONASS measurement and ephemeris data in mixed files.

As the GLONASS system is not yet fully deployed, it suffers from a temporarily low number of available satellites and bad DOP values. With the full constellation expected to be accomplished by the end of 1995, these troubles should be overcome.

APPENDIX A: REFERENCES

[1] Fraile Ordonez, J. M., G. W. Hein, H. Landau, B. Eissfeller, A. Jansche, N. Balteas and V. Liebig, First Experience with Differential GLONASS/GPS Positioning, Proceedings of ION GPS-92, Albuquerque, Sep. 1992

[2] Gurtner, W., G. Mader, Receiver Independent Exchange Format Version 2, GPS Bulletin 3 (3)

[3] Hatch, R., The Synergism of GPS Code and Carrier Measurements, Proceedings of the 3rd Int. Geodetic Symposium on Satellite Doppler Positioning, Las Cruces, Feb. 1982

[4] Misra, P. N. and R. I. Abbot., SGS85 - WGS84 Transformation, in: manuscripta geodaetica, 1994, No. 19, pp. 300 - 308

[5] Design Document for Differential GPS Ground Reference Station Pseudorange Correction Generation Algorithm, U.S. Coast Guard Report No. DOT-TSC-CG-86-2, Dec. 1986

[6] GLONASS Interface Control Document (2nd Rev.), GLAVKOSMOS, USSR, 1991

[7] GPS Interface Control Document (ICD-GPS-200), ARINC Research Corporation, July 1991

PRECISE DGPS POSITIONING
IN MARINE AND AIRBORNE APPLICATIONS

Günter Seeber, Volker Böder, Hans-Jürgen Goldan, Martin Schmitz
Institut für Erdmessung, University of Hannover, 30167 Hannover, Germany

Gerhard Wübbena
Geo++ GmbH, 30827 Garbsen, Germany

INTRODUCTION

Precise DGPS Positioning (PDGPS) is required in a wide range of applications, but ambiguity resolution can still be problematic in a kinematic environment. Pure single frequency processing is limited by the ionosphere to distances up to 15 km and cannot benefit from the use of carrier phase linear combinations. The recent progress in receiver hardware technology enables adequate dual-frequency GPS processing by providing the full wavelength on L2 even under A-S conditions. Different strategies using L1 and L2 carrier phase combinations, ambiguity search algorithms and combined methods to resolve ambiguities have been developed.

For real-time applications it is essential to resolve the ambiguities „on the way" (OTW) while the antenna is moving. Some of the key questions, related to this approach, are
- how many epochs are necessary for the recovery of ambiguities after signal loss
- how reliable are the OTW solutions
- how large is the maximum distance between the reference station and the moving user.

The Institut für Erdmessung (IfE) uses the GEONAP software package for GPS post-processing, which is capable to handle and combine a number of ambiguity resolution techniques. One of the latest features is the possibility of a simultaneous adjustment of both carrier phase observables, offering fast and reliable ambiguity resolution. The identical approach is also used for real-time applications (Wübbena et al., 1995). Hence the results of investigations on reliability and performance in the post-processing mode are also valid for real-time use. The GEONAP software package has been originally developed at the Institut für Erdmessung (Wübbena, 1989, 1991), and it is now further developed and supported by Geo++ GmbH.

This paper describes kinematic dual frequency GPS processing with GEONAP and focuses on PDGPS results obtained in different marine and airborne experiments.

GEONAP SOFTWARE

GEONAP is a software package for post-processing of static (GEONAP-S) and kinematic (GEONAP-K) GPS measurements. The main features of the software are:
- *Multi-station adjustment*: The software is not limited to the processing of single baselines. It allows the simultaneous adjustment of multiple static (reference) and kinematic (mobile) stations. Together with the parameter estimation model (see below)

this feature allows the determination of precise static as well as kinematic coordinates even over long distances.

- *Non-differenced observables*: The adjustment model is based on non-differenced observables. The main advantages are the independency of gaps in observation data sets and the ability to model low dynamic processes as ionospheric and tropospheric delays and interfrequency (intersignal) biases in satellite and receiver hardware.

- *Simultaneous adjustment*: The model allows the simultaneous adjustment of all relevant GPS-observables, i.e. L1 and L2 carrier phase measurements as well as P-(Y-) and C/A-code pseudorange measurements. The parameter model simultaneously estimates geometric and ionospheric parameters. A simultaneous adjustment of single and dual frequency measurements provides ionospheric corrections for the single frequency receivers.

- *Optimized ambiguity search and fixing algorithms*: The ambiguity algorithms take advantage of all available information. Different search and fixing strategies allow a flexible adjustment process even for critical data sets. With simultaneous adjustment of dual frequency carrier phases an automatic wide- and extra-wide-laning procedure yields optimum performance of the ambiguity search and fixing process.

- *Parameter model*: A complete model for all relevant parameters is implemented. All parameters may be simultaneously estimated. Parameter sets are:

  - *Receiver coordinates*: A set of 3 coordinates is estimated for every receiver, every epoch for kinematic observations and every occupied station (static point). All different methods of GPS-surveying, like static, kinematic, stop and go, pseudokinematic, antenna swap etc. are supported and automatically handled. The coordinates of static points may be processed with a network adjustment program which allows the combination of multiple sessions into a network using a complete variance-covariance model.

  - *Receiver clocks*: Modeling ranges from simple integrated white noise models to complicated functional and stochastic models for atomic clocks.

  - *Short arc satellite orbits*: Short arc keplerian parameters may be estimated for every satellite.

  - *Satellite clocks*: Satellite clocks are modeled similar to receiver clocks.

  - *Intersignal hardware delays*: Different delays for different signals or signal components in satellite and receiver hardware may be modeled as low dynamic processes. This inputs more information into the system compared to differencing techniques.

  - *Ionospheric model*: The ionospheric model includes functional and stochastic parameters.

  - *Tropospheric model*: The tropospheric model allows the estimation of a stochastic scaling parameter. Stochastic modeling of the residual tropospheric range error is essential for high precise static control networks, which are observed in long sessions.

  - *Ambiguities*: Ambiguities are estimated for every satellite and every carrier signal. New ambiguities are initialized every time an unrecoverd cycle slip occurs. Due to the non-differenced approach the ambiguity model is not limited to signals with the same nominal wavelength, i.e. the model can be applied to different satellite systems like GLONASS. Ambiguities remaining unresolved from the session solution may be fixed in the network adjustment.

RELIABILITY OF OTW ALGORITHMS

For real-time application it is essential that the ambiguities can be recovered quickly and reliable after any signal loss. In order to study the reliability and velocity of the OTW algorithms in GEONAP, three data sets obtained from two ship experiments near the island of Norderney were analyzed. Norderney is situated about 60 km off the German coast in the North Sea. The first experiment was performed in 1993 using dual frequency P-code Trimble 4000 SSE equipment under SA conditions. The second experiment was carried out in 1994 using Trimble 4000 SSE and Ashtech Z12, both dual frequency receivers with the capability to obtain precise code measurements on both frequencies, even under A-S conditions. The distances between the reference station and the moving receiver reach up to 4 km and the data rate is 1 sec.

The complete data sets were divided into subsets of 2, 3, 5, 10, 30, 60, 120 and 300 seconds each. Reference solutions were computed from the complete data sets with all ambiguities correctly solved using GEONAP. It was then investigated, in how many cases the correct coordinates and hence the correct ambiguities could be recovered by using only the subsets.

Fig. 1 shows the results of five approaches to solve the ambiguities, using

- L1 single frequency measurements with P-code observations (1993)
- extra-wide-laning technique (1993), which solves the ambiguities of different linear combinations step by step
- sequential simultaneous dual frequency processing, GEONAP option ´+L´ (first step in the development of option ´+X´) (1993)
- rigorous (closed) simultaneous dual frequency processing (option ´+X´), using Ashtech Z12 (1994)
- rigorous (closed) simultaneous dual frequency processing (option ´+X´), using Trimble SSE (1994).

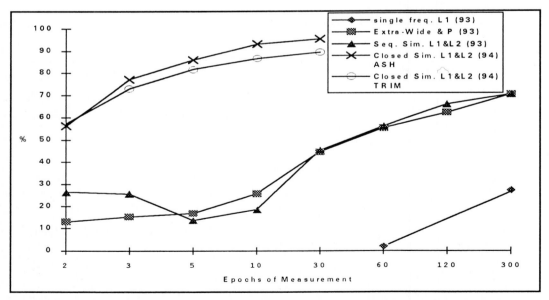

Fig. 1: Reliability of OTW algorithms in GEONAP using different approaches; success rate of ambiguity resolution for a different number of measurement epochs (rate of data : 1 sec)

It can be stated, that in 1993 in about 25 % of all cases the ambiguities could be resolved correctly with only two epochs, and that the success rate improves to about 40 % with 30 epochs of measurement. Using only L1 single frequency data it was possible to solve the ambiguities in about 25 % of all cases with 5 min of epochs. The results were obtained in 1993 with an incomplete satellite coverage, and with a preliminary version of the GEONAP software. In 1994, with the full constellation and improved software models using the new option ´+X´, the success rate approximates to 55 % with only three epochs and up to 90 % with 30 epochs. It should be noted, that the different success rates for Trimble and Ashtech receivers are not an indication for any hardware quality difference.

The results above make no distinction between different satellite constellations. Fig. 2 gives an overview of the number of satellites, cycle slips and loss of data for each satellite in the experiment with the Ashtech Z12 in 1994. The data set is separated into three parts with more than 8, 6 and 5 satellites. The success rate was calculated for each portion of the data set and for different observations lengths.

RXPLOT	LAT : N 53 42 10.0	RECV : Ashtech Z-XII3		ELV : 15
IFE/94	LON : E 7 9 48.0	STAT : asny		S/N : 4
V. 5.3	ALT : 42.8	MARK : 0		SIG : LO

30 seconds:	100 % (from 371 trials)	95,8 % (165)	77,8 % (99)
10 seconds:	100 % (from 462 trials)	90.3 % (206)	56.2 % (119)
5 seconds:	100 % (from 474 trials)	78.5 % (205)	43.8 % (119)
3 seconds:	98.0 % (from 537 trials)	63.2 % (228)	17.4 % (132)
2 seconds:	89.8 % (from 537 trials)	10.1 % (228)	0.0 % (132)

OBSERVATION TIME (GPS): 26.05.94/09:38:02 - 26.05.94/11:56:17

Fig. 2: Reliability of the OTW algorithms in GEONAP with option ´+X´ (1994) for Ashtech Z12; success rates of ambiguity resolution for different numbers of epochs; satellite constellation and data quality indicator from program RXPLOT (IfE)

With more than eight satellites a reliable ambiguity resolution is obviously possible with only five epochs. With six or seven satellites the success rate goes down to nearly 80 %, and with only five satellites to 45 %. However, a successful solution is possible using five satellites and 30 seconds epochs in 78 % of all cases.

The results above were achieved over short distances with a maximum range of 4 km. The objective of another experiment onboard the hydrographic survey vessel VWFS WEGA in 1995 off the German Coast was to examine the use of the PDGPS over larger distances. Two reference stations on the island of Helgoland and on the mainland in Cuxhaven as well as the ship were equipped with Ashtech Z12 receivers.

From this experiment a 20 km trail from Helgoland in direction to Cuxhaven was examined. The distance between Helgoland and Cuxhaven is about 60 km, hence the distances from the reference station Cuxhaven to the vessel range between 40 to 60 km. The reference coordinates were derived from earlier GPS-campaigns with the objective to relate the tide gauge of Helgoland to the tide gauges at the German Coast with centimeter accuracy (Goldan et al., 1994).

The investigations are not yet finished, but preliminary results indicate, that a reliable resolution of ambiguities is also possible over a distance of 20 km with five seconds of data. The success rate using eight satellites over a distance of about 10 to 20 km was close to 90 % with the option '+X'; over 40 to 60 km the first successful resolution of ambiguities was obtained with 120 epochs of 1 sec data.

In order to study the accuracy of the kinematic PDGPS position over a large distance the ambiguities of the trajectory were resolved twice, using Helgoland and Cuxhaven independently as a reference station.

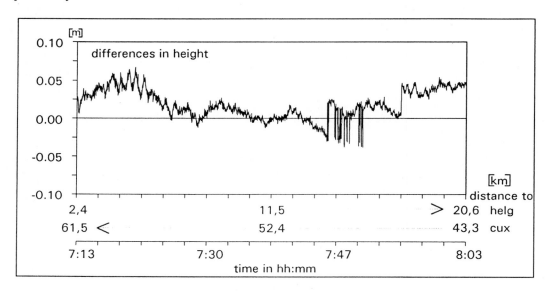

Fig. 3: Differences in height using different reference stations Cuxhaven (40-60 km) and Helgoland (2-20 km); GEONAP option '+X'; VWFS WEGA, Ashtech Z12, rate of data: 1 sec, 28.04.1995.

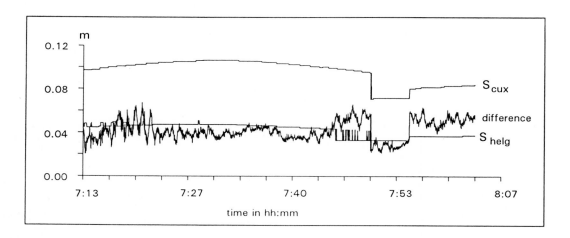

Fig. 4: Standard deviations of the position determination computed by GEONAP using reference stations Helgoland (2-20 km) and Cuxhaven (40-60 km) and total differences between the two solutions; VWFS WEGA, Ashtech Z12, rate of data: 1 sec, 28.04.1995

Fig. 3 shows the differences in the height component between the two solutions. The mean deviation is about 1.5 cm. The maximum values of the differences are less than 6 cm. The reason for the behavior of the curve between 7:45 and 7:55 is found in the rise and set of two satellites at these times.

The root of the square sum of the differences and of the standard deviations in all three position components derived from the GEONAP processing is shown in **Fig. 4**. The mean of the total differences is about 4.2 cm, the mean of the standard deviations amounts to 4.2 cm for the shorter ranges and to 9.6 cm for the longer ranges. The two solutions are highly correlated because of the same data set collected on the VWFS WEGA.

Nevertheless, it can be stated, that the positioning of a survey vessel is possible with sub-decimeter accuracy over large distances up to 60 km. The reacquisition of the lost phase ambiguities becomes more difficult, but the problem can be solved using dual frequency data and a powerful post-processing software.

TIDE OBSERVATIONS ON MOVING PLATFORMS

The next example refers to the determination of two complete tidal cycles in the North Sea (German Bight) at four selected places in the estuary of the river Elbe. The height reference system for this determination was the normal-orthometric height system 'Normal-Null' (NN). Four surveying vessels of different size were anchored on the spot, and considered to be a proof mass in the tidal uplift.

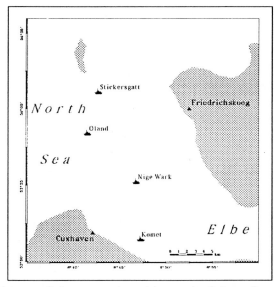

Fig. 5: Positions of the surveying ships and of the reference stations.

The height variations of the ships were observed with kinematic GPS using one Trimble 4000 SSE GPS receiver on each vessel. The biggest vessel was the KOMET with a length of about 68 m, the smallest one was the STICKERSGATT with 14 m. A site in 'Friedrichskoog' and the permanent GPS monitor station 'Cuxhaven' from the State Survey Department of Lower Saxony were used as reference stations. They also were equipped with Trimble 4000 SSE receivers. Fig. 5 shows the definite positions of the four ships in the estuary of the river Elbe, and of the two reference stations.

It is evident that the exact height of the GPS antenna above the sea level has to be determined for each vehicle. To this respect the four ships were towed in the harbor 'Americahafen' of Cuxhaven. In this harbor a registrating tide gauge with connection to the NN height system is available, and the actual height of the sea level in the NN height system can hence be determined. Up to two reference stations were operated nearby at stations with NN heights. At these stations the relationship between NN and the ellipsoidal GPS height in the WGS84 can easily be computed. With these information at hand the exact elevations of the ship antennas above the sea level can be determined (Goldan, 1994).

The ships were moored in the harbor for about one hour before and after the tide measurements in the estuary, and GPS data were gathered. The adjustment of the GPS data was done with the GEONAP software package using an ambiguity free solution from ionospheric free linear combinations of the carrier signals. The final results are formed from a mean value over both observation periods before and after the tide measurement in the estuary.

The precision of a single determination of the antenna height above the water level was found from adjustment to be as small as 2...4 mm. The differences between both measurements show a somehow more realistic value for the accuracy of the antenna height calibration in the order of 1...2 cm. This value is affected by the expected rise of the individual ships caused by fuel consumption.

In between of the two antenna calibration periods the four ships were moored for about 26 hours at the places indicated in Fig. 5. This time span includes two tidal periods. The NN height of the sea level can be computed with the predetermined antenna height over the sea level and a good quasigeoid model (Denker, 1988).

Dual frequency GPS data with a data rate of 5 seconds were recorded. The adjustment was performed again with the GEONAP software package, and with the stations 'Cuxhaven' and 'Friedrichskoog' as static reference stations. All ambiguities could be solved, and the final ellipsoidal antenna positions were determined with the ionospheric-free signal. The height components were smoothed with a simple filter algorithm (weighted moving mean value over 7.5 minutes forward and backward).

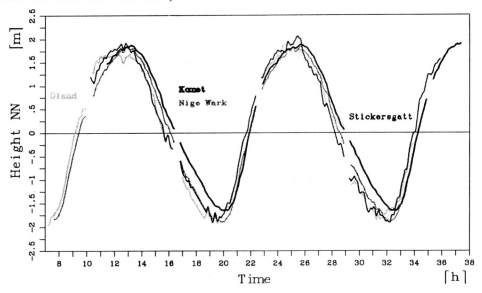

Fig. 6: Tide observations with GPS on four ships, May, 25th 1994

The GPS results were reduced to the water level in the NN-system and resulted into the graphical representation in Fig. 6 for all four ships. Because of the large amount of data a download was necessary every 6 hours. The data gaps are visible in the plots, and they were intentionally located at times of mean water level in between of the low and high tides. The much smoother curves for the larger ships are evident. Accordingly, the smallest ship shows the roughest results. The RMS of a single measurement vary for each ship and is in the order of 3...7 cm. Remaining systematic errors may arise from the geoid model and from errors in the connection between the reference stations and the NN-system. Further, the signals are corrupted by the not compensated ships motion. These movements are in particular present for the smaller ships. Some variations in the results may be caused by multipath or signal interruptions.

It could be demonstrated that ships carrying GPS equipment can be used as floating tide gauges with an accuracy level of a few centimeters. As has been stated above the distance to the reference stations should not be too large because otherwise the solution of the phase ambiguities may be corrupted. Also the systematic influence of the troposphere increases with increasing station separation. At least 5 GPS satellites should be available.

AIRBORNE APPLICATION OF KINEMATIC GPS

The GPS supported bundle block triangulation is widely used and has today become an operational application field of kinematic GPS. The GPS position information is introduced into the bundle block adjustment to support the conventional aerial triangulation and reduces the number of required control points at the ground dramatically (Seeber, 1993, Jacobsen, 1993, 1994).

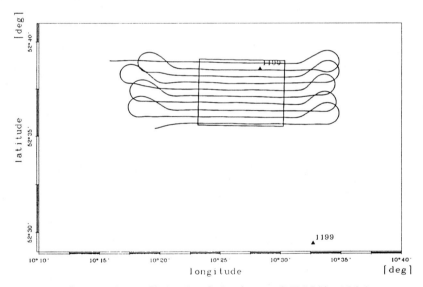

Fig. 7: Photo flight Groß Oesingen (NLVA), 1994

Correct ambiguity resolution is improved by the OTW approaches, but nevertheless, sometimes a few ambiguities remain unsolved. Generally, these ambiguities are then forced to the nearest integer number and the introduced systematic effects are modeled in the bundle adjustment. Commonly, the systematic effects are modeled by shift and/or drift parameters using the information of the projection center from conventional aerial triangulation. This is done in most cases for each strip independently.

Comparing the projection centers derived from a conventional aerial triangulation without GPS allows an independent check of the GPS positions. However, some additional error sources must be taken into consideration, namely the interpolation error of the GPS position on the time event of image exposure, remaining errors due to the transformation of WGS84 coordinates to the national datum, errors due to unrecorded drift compensation of the camera, and eccentricity of GPS antenna and actual projection center.

The following photo flight has been performed by the survey administration of the federal state of Lower Saxony (Niedersächsisches Landesvermessungsamt, NLVA) and the final bundle adjustment consists of 10 strips in east-west direction covering an area of 6×8 km (Fig. 7). The photo scale is approximately 1:7500 and the flying height was 1300 m. The sidelap of the flight was 60 % (Elsässer, 1995). The bundle block adjustment has been performed with the Hannover program system BLUH using 43 equally distributed horizontal and vertical photogrammetric control points as well as selfcalibration. Within the bundle block adjustment the standard deviation of the projection center is estimated for the height component to 2.7 cm and for the horizontal position worse by a factor of 2 to 5.1 cm, and 5.6 cm in east and north component.

The kinematic GPS survey was carried out with Trimble SSE receivers, operating two reference stations during the photo flight. One was located in the block area and the second station approximately 18 km away (Katasteramt 'Gifhorn'). Unfortunately the complete L2 measurements of the kinematic station in the aircraft were lost. Therefore only a single frequency L1 solution could be obtained with the GEONAP processing software. The combined adjustment of single and dual-frequency receivers using the rigorous simultaneous adjustment of L1 and L2 phases did not improve the ambiguity resolution for this particular data set. After fixing as much ambiguities as possible, all remaining ambiguities were forced to the nearest integer. The GPS positions were estimated for every strip using a constant satellite geometry.

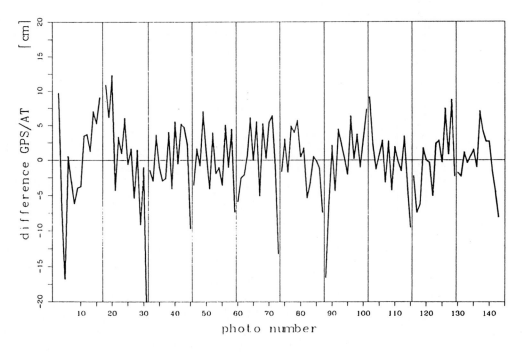

Fig. 8: Height difference of projection centers derived from GPS and Photogrammetry

Fig. 8 shows the differences between the height components of the projection centers derived from GPS and independently from conventional block triangulation. The datum transformation parameters were derived from surrounding well known identical points of the national GPS network (DREF). Spline functions have been used for the interpolation of the projection centers of GPS to the time of exposure. The precisely known eccentricity of the GPS antenna was reduced using the orientation angles from the conventional bundle adjustment. Remaining discrepancies are modeled together with systematic effects due to possible false ambiguity fixing in the shift and drift approaches.

In the case of Fig. 8 only shift corrections have been applied individually for each strip. The differences over all strips amount to ±5.3 cm, or, if additionally a time dependent drift is estimated to ±4.4 cm. However, there are maximum differences of up to 20 cm, which occur especially at the beginning or end of a flight strip and can be associated with the block triangulation.

The independent comparison shows a good agreement of GPS and block adjustment. Some parts of the differences originate from the block triangulation and are not introduced by GPS. These are especially uncertainties of photogrammetry at the edges of the block. The increasing accuracy of kinematic GPS also suggests to revisit the shift and drift models used in the GPS supported block adjustment and to investigate new models to account for false ambiguity resolution. Currently a research project is dealing with this topic in a cooperation of the Institut für Erdmessung (IfE) and the Institut für Photogrammetrie und Ingenieurvermessung (IPI) at the university of Hannover.

Acknowledgment. Part of the research work reported in this document has been supported by the German Ministry of Research (BMBW) and the German Research Foundation (DFG)

References

Denker, H. (1988). Hochauflösende regionale Schwerefeldbestimmung mit gravimetrischen und topographischen Daten. *Wissenschaftliche Arbeiten der Fachrichtung Vermessungswesen der Universität Hannover*, Nr. 156, Hannover.

Elsässer, L. (1995). Bündelblockausgleichung unter Nutzung von GPS-Positionen. Submitted to *Nachrichten der Niedersächsischen Vermessungs- u. Katasterverwaltungen*.

Goldan, H.-J. (1994). Tide observations with kinematic GPS on ships. *Proceedings, International Symposium on Marine Positioning*, INSMAP 94, Hannover, Germany

Goldan, H.J., G.Seeber, H. Denker, D. Behrend (1994). Precise height determination of the tide gauge Helgoland. *Proceedings International Symposium on Marine Positioning*, INSMAP 1994, September 19-23, Hannover

Jacobsen, K. (1993). Experiences in GPS Photogrammetry. *Photogrammetric Engineering & Remote Sensing*, PE&ER, Vol. 59, No. 11, November, 1651-1658.

Jacobsen, K. (1994). Combined Block Adjustment with Precise Differential GPS-Data. Presented at *International Archives of Photogrammetry and Remote Sensing* (ISPRS), Commission III, Working Group 1, Munich.

Seeber, G. (1993). *Satellite Geodesy.* Walter de Gruyter, Berlin, New York.

Wübbena, G. (1989). The GPS Adjustment Software Package GEONAP - Concepts and Models. *Proceedings 5th Intern. Geod. Symp. on Satellite Positioning.* Las Cruces, NM, 452-461.

Wübbena, G. (1991). Zur Modellierung vpn GPS-Beobachtungen für die hochgenaue Positionsbestimmung. *Wissenschaftliche Arbeiten der Fachrichtung Vermessungswesen der Universität Hannover*, Nr. 168, Hannover.

Wübbena, G., Bagge, A, G. Seeber (1995). Developments in Real-Time Precise DGPS Applications - Concepts and Status. These Proceedings.

Developments in Real-Time Precise DGPS Applications: Concepts and Status

Dr.Ing. Gerhard Wübbena
Dipl.-Ing. Andreas Bagge
Geo++ GmbH, Osteriede 8-10, D-30827 Garbsen

Prof.Dr.Ing. Günter Seeber
Institut of Geodesy — University of Hannover
Nienburger Str. 1, D-30167 Hannover

1 Current Situation and Development Tendencies

In many of today's precise GPS applications it is required to obtain the coordinate results in real-time. This is in particular true for marine and airborne applications, as well as for the control of land based vehicles and machines and for modern developments in cadaster and GIS surveying. Key factor is the rapid and reliable solution of ambiguities. Powerful algorithms are available, however, they still suffer from certain restrictions. The interstation distance is limited to about 10 km, and rapid (few epochs) solutions are not always available.

Another problem is the required data rate to transmit the neccessary carrier phase corrections from the reference to the mobile station. With the RTCM-2.1 format, a data frame of more than 4800 bits is required for all-in-view, that means for twelve satellites. Most real-time high-precision applications want an update rate of one second, thus a data rate of at least 9600 bits per second (bps) is required. Only some frequencies in the radio spectrum allow such an high data rate to be transmitted reliably.

2 Concepts of GNRT-K

GNRT is the software package from Geo++ for real-time applications of DGPS. GNRT-K is an optional module to GNRT which uses the carrier phase observations. It solves for the phase ambiguities and allows subcentimeter accuracies in real-time.

2.1 Development of RTCM++ Format

RTCM++ is an enhencement to RTCM-2.0 and RTCM-2.1. The RTCM++ extension is a compact data format for the carrier phase corrections. With RTCM++, the complete data set for all-in-view satellites requires less than 2400 bits. Thus many more frequencies are possible candidates for the broadcasting of carrier phase corrections. The RTCM++ extensions are compatible to RTCM-2.0 and RTCM-2.1 data frames, using

RTCM message type 59, which is reserved for proprietary messages.

The type 59 records enclose the complete information to reconstruct the code and carrier phase corrections on the mobile side. Thus, with a converter the mobile user has full RTCM-2.1 phase corrections available.

Normal RTCM-2.0 code corrections remain unchanged in the RTCM++ data stream.

2.2 Technical Concept of GNRT and GNRT-K

GNRT rsp. GNRT-K were designed with the following features in view:

► *PC based, multitasking OS*

The PC platform allows cheap and flexible hardware. The multitasking operating system OS/2 allows the integration of GNRT with commercial or user provided software.

► *Graphical user interface*

The graphical user interface, combined with a pen driver, is useful especially in the field. Moreover, it makes a keyboard no longer neccessary.

► *Real time capabilities*

are neccessary to produce reasonable response times. A predictable delay of one or two seconds between measurement and result, mostly due to delays in the receiver and in transmission paths, is in general tolerable.

► *Modular system*

GNRT as a modular system allows to build all components of a DGPS system from its components: reference station, mobile station, precise mobile station or other special approaches like reverse or relative DGPS (see below).

► *Receiver independent*

The receiver module is only one component in the GNRT system. For almost every receiver with RS-232 interface and programmable input/output a receiver module can be developed. Currently modules for Ashtech, Navstar, Novatel, Trimble and Zeiss are available, others, as for Leica, are under development.

► *Compatibility to standard formats*

Today a DGPS software package has to support the standard formats in GPS and DGPS. These standards are RTCM, NMEA and RINEX, and are all supported by GNRT. With these interfaces in a multitasking environment GNRT is a very flexible system.

► *Combined ambiguity search algorithms*

The state of the art in carrier phase ambiguity solving is a very sophisticated ambiguity search algorithm. GNRT-K, the phase module of GNRT, has implemented the algorithms from the well known GEONAP GPS postprocessing software. It starts multiple threads simultaneously to make all informations available for the ambiguity resolution.

► *Network of multiple reference stations*

To reduce the dependency from the distance to the reference station, GNRT is prepared to use some new features of the RTCM++ format. The concept is to build a net of reference stations and compute a more general set of correction parameters. These parameters are valid not only in the near environment of the reference station, but over the whole region covered by the reference station network. The GNRT net module will be able to generate the additional parameters, and RTCM++will transport them to the GNRT mobile station. The effect is that the mobile station has corrections available as if the reference station is very close to the mobile station.

2.3 GNRT Components

The components or modules of the GNRT DGPS software system are:
- *Receiver interface*
  It collects the raw data from a GPS sensor, managing data flow from and to GPS receiver.
- *RTCM interfaces*
  It consists actually of two modules, one for input and one for output of DGPS corrections in RTCM format. All input and output is RTCM-2.0 and RTCM-2.1 compatible, all RTCM++ enhancements are contained in message type 59 records.
- *Special purpose interfaces*
  Modules for Relative DGPS or Inverse DGPS, NMEA compatible output etc.
- *DGPS module*
  GNRT as the main module computes the GPS solution from all available informations. It can be run in base or mobile station mode. If running as base station, it takes GPS observtions and known base position as input and computes DGPS corrections. If running as mobile station, it takes GPS observations and DGPS corrections as input and calculates the mobile position.
- *Real time kinematics module*
  The GNRT-K module solves all ambiguities of carrier phase observations as far as possible. It starts multiple threads for optimum ambiguity search algorithms.
- *User interface*
  The GNRT system presents itself in various numerical and graphical status and control windows. It is able to run without keyboard, i.e. on pen driven notebooks in the field. The GNRT base station is remotely operable through modem or network connections.

2.4 GNRT Configurations

The GNRT modules may be combined to many different DGPS systems.
- A *Standard base station*
  operates as a local temporary DGPS reference station.
- A *Precise base station*
  adds carrier phase corrections to standard base station.
- A *Mobile station*

allows submeter accuracy in real time.

▸ A *Precise mobile station*
  allows subcentimeter accuracy in real time.

▸ A *Permanent reference station*
  called GNREF, adds RINEX data logging capabilities and integrity monitor function to precise base station.

▸ A *Node in refence station network*
  adds a GNREF network module to permanent reference station.

▸ An *Inverse or Reverve DGPS station pair*
  does all computation of mobile position on base station. Only a receiver and a radio link are required on the mobile station.

▸ A *Relative DGPS station pair*
  allows positioning of mobile station relative to moving base station, i.e. for an helicopter landing on a aircraft carrier.

Other approaches are possible and easy to implement due o the modular system. Even existing user or third company programs on the same platform with standard input/output channels may communicate with the GNRT system in realtime using pipes.

3. EXAMPLES AND RESULTS

The time needed to fix the initial ambiguities, often called the initialisation time, depends on the available signals and the distance from the reference station. GNRT-K is able to fix the ambiguities "On-The-Way", no static initialization mode is required.

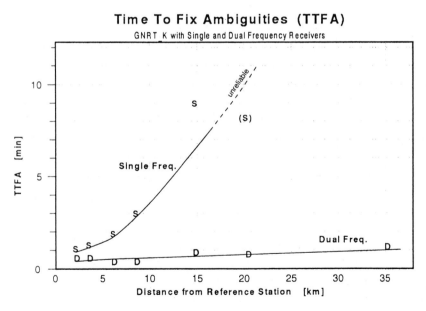

Fig.1. Required Time to Fix Ambiguities (TTFA) for single and dual frequency receivers, in relation to the distance from reference station.

215

Some first tests showed (see Fig.1), that for dual frequency receivers the time to fix ambiguities (TTFA) is nearly independent from the distance (tested up to 35 km) and normally better than one minute. For single frequency receivers the TTFA is acceptable only for short distances maybe up to 5 or 10 km. For longer distances the required time growes exponentially and, much worse, the estimated ambiguities tend to be unreliable. This will improve if reference station networks become available.

In hydrographic surveying GPS derived heights are of special importance for squat and wake determination to improve the results from echo soundings. An example for real-time applications is given in Fig.2. The precision of the height component is better than 2-3 cm. The typical wave and squat effects in the order of 10 cm are recognizable.

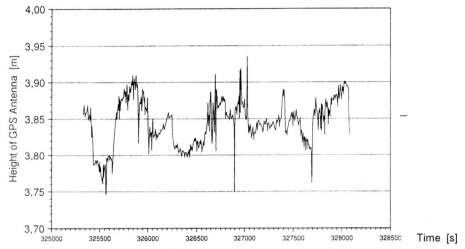

Fig.2. Height Determination with GNRT-K for hydrographic surveying.

4. CONCLUSIONS

Today it is possible to get cm accuracy within a few (ten) seconds with dual frequency receivers over 30 or more kilometers in real time. Single frequency results are still restricted to short baselines (< 10km) and require less than 1minute to get the centimeter accuracy. Future enhancements as the establishment of reference station networks will reduce the station dependent errors so that cm accuracy will be possible over more than 10 or 20 km, even with single frequency receivers.

REFERENCES

Hankemeier, P. (1995): The DGPS Service for the FRG - Concept and Status. Contributed paper to the IAG Symposium in Boulder, July 1995

Wübbena, G. and Bagge, A. (1995): GPS-bezogene Ortungssysteme. Presented at the 37th DVW-Seminar *Hydrographische Vermessungen - heute -*, 28./29. March 1995 in Hannover/Germany

A Kinematic GPS Survey at the Northern Part of Kagoshima Bay, Japan

Sugihara, M. and Komazawa, M.
Geological Survey of Japan, 1-1-3, Higashi, Tsukuba, Ibaraki, Japan

INTRODUCTION

Kagoshima Bay is a volcanologically interesting area which is divided into two parts by the active Sakurajima volcano (Fig. 1). The Aira Caldera is situated at the northern part of the bay and was formed by a large eruption about 22,000 years ago (Aramaki,1984). The existence of several craters are inferred from volcanic sediments, however, details are not clear because most parts of the caldera are undersea. We undertook a continuous kinematic GPS survey there for the purpose of detecting geoid undulations caused by the geological structures of the volcanoes.

FIELD OBSERVATIONS

Continuous kinematic GPS survey measurements were made along ten tracks in the northern part of Kagoshima Bay (Fig.1) for three days in July, 1994. We followed as straight a track as possible on each leg of our survey for two reasons: (1) to evaluate time-dependent components of the sea level, (2) to keep the boat in the same orientation. A different port was selected as a base at each day in order to cover the study area with ten tracks crossing one another. A tide gauge was set at Fukuyama port which is one of the bases. We used two sets of Trimble 4000SSTIIP GPS receiver. The reference antenna was set at an end-point of a known baseline near the port throughout the leg; the roving antenna was set at the other end-point of the baseline for survey initialization at the beginning of the leg, then loaded on a boat and the boat departed from the port. The average speed of the boat was 20 km/hour and the sampling period was three seconds.

We processed the data using the TRIMVEC-PLUS program. The result of the leg 0 is shown in Fig. 2. A prominent depression of antenna height is observed while the boat was changing course. The average amplitude of pitching during uniform motion is about 5 cm. A difference of about 50 cm was observed between the antenna heights at beginning and end of the leg. Comparing the data with the record of the tide gauge, we ascribe this difference to an ocean tide effect.

We eliminated these time-dependent components of the sea level from the GPS record to evaluate geoid undulation. Unusual parts of the data recorded during irregular motion were not used in the analysis. Usual pitching and yawing effect should be removed with averaging and gridding process. Tide correction, which is most important process, is explained in the next section.

TIDE CORRECTION

We followed the same track on each leg of our survey to evaluate time-dependent components of sea level. It is shown more evident in Fig. 3 than in Fig. 2 that the height differences between at the outgoing and returning tracks have a trend. Horizontal axis shows both longitude of the roving antenna (the top scale) and time interval between the outgoing and returning at each point (the bottom scale). High frequency components of

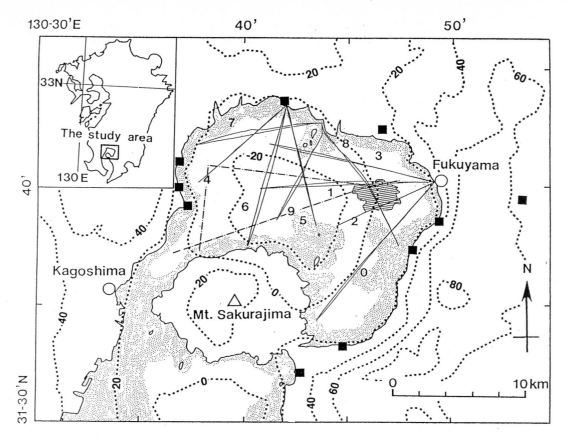

Fig. 1. Track lines (0-9) of continuous kinematic GPS survey. Solid squares show the first order bench marks where static GPS measurements were carried out to evaluate GPS/levelling geoid undulations on land around the bay. Hatched area and dotted area are the places deeper than 200 m and shallower than 100 m. Dotted lines show the contour of free-air anomaly obtained from gravity data (contour interval is 20 mgal). Broken lines show the tracks of the surface ship gravimetry (Chujo and Murakami, 1976).

the height changes were eliminated from the GPS data using a five-points smoothing filter to make clear the trend.

Sea level changes occur almost synchronously everywhere in the northern part of Kagoshima Bay (Maritime Safety Agency of Japan, *personal communication*). Our records of the tide gauge at Fukuyama port were compared with JMA's record at Kagoshima port, the differences between them were at most 5 cm in amplitude. The record of the tide gauge can therefore be regarded as a typical ocean tide for the whole study area. Subtracting it from the height changes of the antenna, we compared again the height differences between at the outgoing and returning tracks (see the lower part of Fig. 3). The residual differences are at most 5 cm and show periodic pattern with a dominant period of about 15 minutes. This may be associated with a seiche, however, such periodic pattern was hard to discern on the chart of the tide gauge, and an advanced numerical modeling will be necessary to evaluate a seiche component.

We removed the time-dependent components whose amplitude is larger than 5 cm from the height changes of the antenna at each leg using above procedure. The data were then adjusted to eliminate differences between different legs at the crossover points, and a contour map of "time-independent" sea surface topography were derived (Fig. 4). The minimum wavelength resolution of the contour map was 2 km.

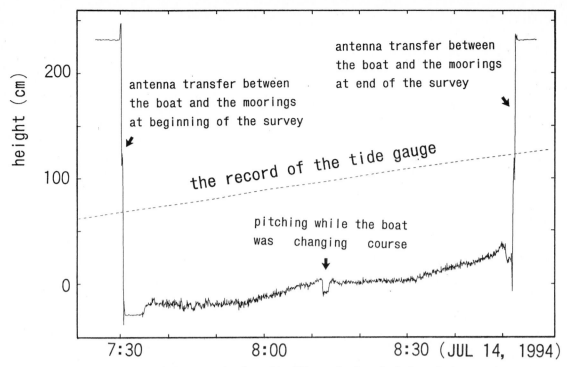

Fig. 2. An example of the result (leg 0). The relative height of the roving antenna is plotted on the vertical axis, and time on the horizontal axis. The dotted line shows the record from the tide gauge.

DISCUSSION

First, the contour map of the sea surface topography was compared with GPS/levelling geoid undulation differences determined by static GPS survey on bench marks (shown in Fig. 4). The trends of the sea surface topography is connected smoothly to the geoid undulation on land. Strictly speaking there are differences between them at most 15 cm. However, we referred the opened levelling data which had measured by GSI about 20 years before, and crustal movement during the period may cause the differences.

Next, the contour map of the sea surface was compared with the geoid map in and around Japan determined by Fukuda et al (1993) using land gravity data, surface ship gravity data and satellite altimeter data. Typical wavelengths of the geoid undulation in this map is longer than 10 km because of interval of the data. Concerning longer wavelength components than 10km, trends of our contour map agree well with their geoid map except the east region of Mt. Sakurajima, where our contour lines are parallel to the coastline. This pattern is derived from the data of leg 0 alone, however, significantly obvious: the height deference between at 130-44'E and at 130-46'E on the track is 10 cm and exceeds the amplitude of the periodic pattern like a seiche (see Fig. 3). An absence of the pattern in the geoid map by Fukuda et al (1993) may be ascribe to the wavelength resolution: there is a positive gravity anomaly whose wavelength is about 10 km at Mt. Sakurajima (see Fig.1) using denser gravity data than they determined.

The above suggests the sea surface topography (Fig. 4) shows geoid undulation in the northern part of Kagoshima Bay. To test this suggestion we derived gravity anomaly values from the sea surface topography data. The method employed for the gravity anomaly estimation was the method of Least Squares Collocation (Segawa, 1984). An outline of the procedure is: (1) fit the geoid data to an autocorrelation function and

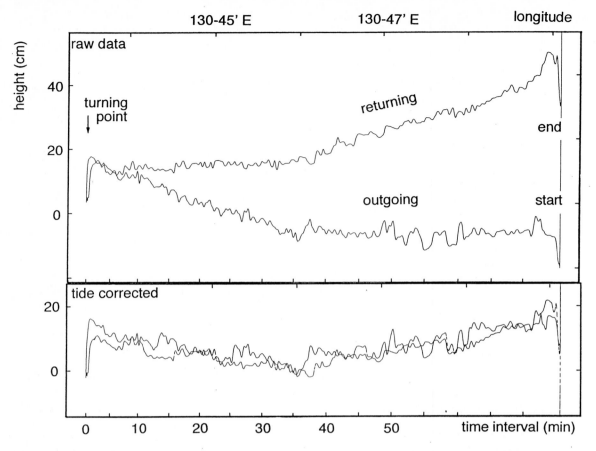

Fig. 3. The effect of tide correction (leg 0). Smoothed relative height of the roving antenna is plotted in the vertical axis and longitude in horizontal axis. The upper part shows the original data, the lower shows the result of the tide correction.

estimate the statistical parameters with which cross correlation function between gravity anomaly and geoid is expressed, (2) calculate covariance matrix and transform matrix from geoid into gravity anomaly. The anomaly at a grid point is calculated from the geoid at the grid points whose distance from the point is less than 5 km, judging from the elements of the transform matrix. The region where gravity anomalies can be estimated by this method is therefore limited to the center part of the surveyed area. The sea surface topography (Fig. 4) was transformed into the negative anomaly at the center of the bay, which is similar to the anomaly obtained from land and surface ship gravity data (Fig. 1).

It is practically efficient method of geoid mapping to take a continuous kinematic GPS survey on a bay: we carried out the survey with the small boat which displaces only 900 kg and evaluated a sea surface topography. On the other hand, it is incomplete to analyze geological structure using GPS data only: (1) geoid undulations are of longer wavelength than shallow geological structures, (2) sea surface topography inevitably contains the dynamic effect like a tidal current: the dense contour lines near the exit of the bay (Fig. 4) could be the case. However, it is promising to analyze the geological structures of the Aira Caldera based on both the GPS data and gravity data since land gravity data, surface ship gravity data, and GPS data cover the area complementally one another (see Fig. 1).

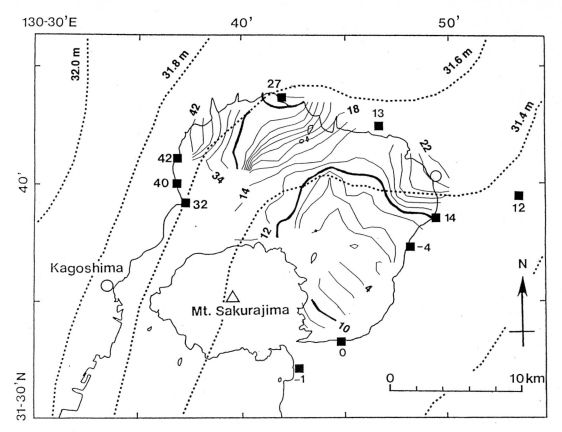

Fig. 4. Contour map of relative sea surface topography determined by the GPS survey (contour interval is 2 cm, the interval of thick lines is 20 cm). Squares are the bench marks where static GPS survey was carried out. Numbers attached to the squares indicate GPS/levelling geoid undulation differences in centimeters. Dotted lines show contour lines of the geoid map determined by Fukuda et al (1993) (contour interval is 20 cm).

Acknowledgment. Thanks are expressed to Dr. Trevor Hunt for his valuable comments. We gratefully acknowledged the use of levelling and triangulation data of Geographical Survey Institute of Japan (GSI) and tide records of Japan Meteorological Agency (JMA).

REFERENCES

Aramaki, S. (1984) Formation of the Aira caldera, southern Kyushu, ~22,000 years ago, *J. Geophys. Res.*, 89, 8485-8501.

Chujo, J. and Murakami, F. (1976) The geophysical preliminary surveys of Kagoshima bay, *Bull. Geol. Surv. Japan*, 27, 807-826.

Fukuda, Y., Shi, P. and Segawa, J. (1993) Map of geoid in and around Japan with JODC J-BIRD bathymetric chart in a scale of 1:1,000,000, *Bull. Ocean Res. Inst., Univ. of Tokyo*, no.31.

Segawa, J. (1984) Gravity anomaly and geoid height, *The Earth Monthly*, 6, 401-411, (in Japanese)

GPS KINEMATIC REAL-TIME APPLICATIONS IN RIVERS AND TRAIN

M A Campos and C P Krueger (Universidade Federal do Paraná, 81531-990 Curitiba, Brazil, e-mail: miltonac@cce.ufpr.br)

ABSTRACT

Monitoring the circulation of trains, the deformation of railway tracks, and the support of environmental studies with respect to river floods using GPS technology, is a new research task for the GPS Group at Federal University of Parana-UFPR. In cooperation with Institut fuer Erdmessung, University of Hannover - IFE HN, several tests have been performed. The first subject includes the monitoring of train's motion between Curitiba and Paranagua in the State of Parana in South of Brazil. The distance between Curitiba and Paranagua (sea level) is about 83 km, and the height difference is about 925 m. The railway line crosses a rather steep and mountaineous area in the coastal chain. The test provided very good results, and it will be described in the poster. The second subject refers to the monitoring of railway track deformations. Special equipment, software and further tools are required to start with the investigations. The research concept will be outlined. The third study started at the Parana river in Southern Brazil. Thirty benchmarks were established at both sides of the river and on some islands to control levelling lines and the installation of river cross sections for the monitoring of the changing water level. DGPS and echosounders are used to provide the necessary data input for hydrological floods models. Another area to be studied is the Iguacu river showing high flood again in January 1995.

INTRODUCTION

About 65% of State of Parana agriculture production, for exporting, is transported by the Brazilian Federal Railway Company - RFFSA - Section 5 - SR5, in South of Brazil.

The GPS Group of Federal University of Parana - UFPR in close cooperation with the Institut fuer Erdmessung - IFE of Hannover University - Federal Republic of Germany - FRG, realized studies to verify if the real-time train's motion monitoring is viable or not, in the way from CURITIBA (h=925m) to PARANAGUÁ (h=0,00m), about 85 km long.

To monitoring the water flood of Parana River, cooperative work among State University of Maringa- UEM, and Institut fuer Erdmessung - IFE, 30 banchmarks were established at both sides of the river and on some islands. with the 1994 Parana River flood some of the benchmarks was damaged or destroyed. Some of them was re-established in a more stable local. The first work with DGPS and Echosounder gave no good results. We are looking for finnancial support to acquire new digital echosounder. The same job will be done in Iguaçu River.

FIGURE 1. TEST WITH DGPS IN REAL TIME, WAY FROM CURITIBA TO PARANAGUÁ. AGE LIMIT 10 SEC

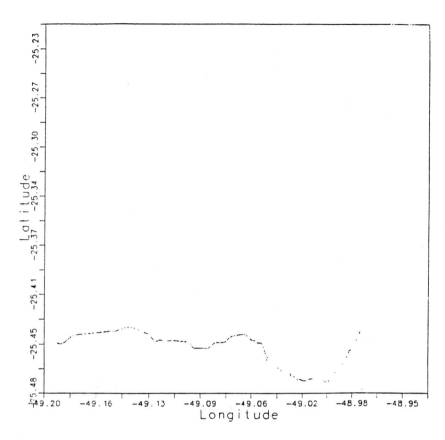

TRAIN MONITORING

To control the trains movement in real-time we used a pair of DGPS Receiver, installed one in UFPR - Federal University of Paraná and another in a special railcar. The connection between receivers was done by UHF radio system.

The results showed that the project is viable. The Figures 1 and 2 show the graphycal register got in real-time. The Figure 3 show the same trajectory, but obtainned from a pos-processing with the GEONAV software. For these tests we adopted "age limit" of 10 and 100 seconds. The results showed in Figures 1 to 3 are 10 and 100 sec.

The RFFSA company is establishing radio repetition stations in the railway to get radio communication between stations and to send the DGPS corrections.

Is it also possible to use satellite communications but it is very expensive. The option by radio will be used. The next tests will be done in august/95 with GPS sensors in more than three trains simultaneously. Sensors for temperature, pression, electric current will be used also to Mechanical researchs personal. The processed GPS data will presented to the Operational Control Center in displays over digitalized maps.

FIGURE 2. TEST WITH DGPS IN REAL TIME, WAY FROM CURITIBA TO PARANAGUÁ. AGE LIMIT 100 SEC

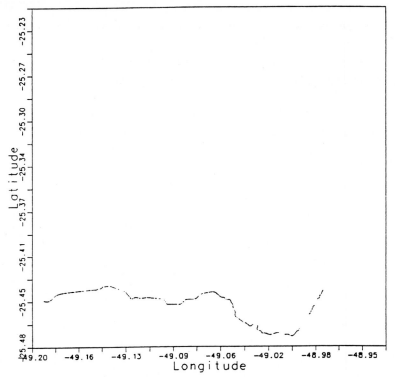

FIGURA 3 - TEST WITH DGPS. WAY FROM CURITIBA TO PARANAGUÁ. PÓS-PROCESSED.

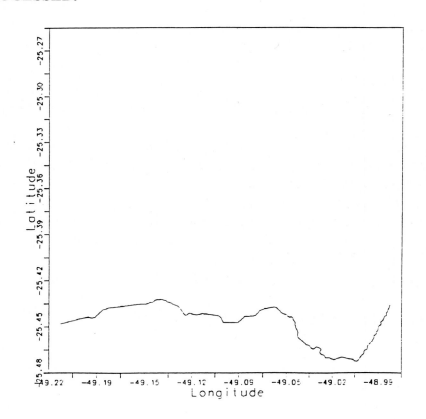

MONITORING OF PARANÁ AND IGUAÇU RIVERS.

This section of the Project will be realized in close cooperation between **UFPR/UEM/ IFE.** Figure 4 show the monumented points in the Parana River. This project wait for finnancial support to acquire a digital echosounder.

In the Iguaçu River the research works will be done simulyaneously with Parana River and will consist of cross sections DGPS surveys for flood studies.

UFPR students will participate in all these jobs.

FIGURE 4 - POINTS MONUMENTED IN PARANA RIVER.

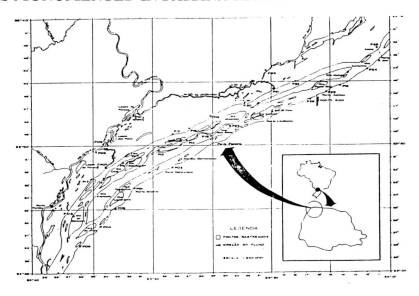

BIBLIOGRAPHY

ANDERSON, E. W., The Principle of Navigation, Hollis and Carter, London, 1966

CLAY C., MEDWIN H., Acoustical Oceanography, John Wiley and Sons INc.,1977

EBERLE, L., Satellite Navigation (GPS) in Combination with echosounding, Munich Polithechnic and CEFET, 1994

FUEM / PADCT/CIAMB, Estudos Ambientais da Planície de Inundação do Rio Paraná no Trecho Compreendido entre a Foz do Rio Parapanema e o reservatório de Itaipu. Maringá. Fund. Uni. Est. de Maringá. 3 v. 1993.

GPS WORLD - Artigos relacionados à pesquisa

HAYNES, G. Sound Underwater, David and Charles (Holdings) Ltd., London, 1974

HOFMANN-WELLENHOF B. ,Real Time GPS: an Application for Railway Engineering. Book of Abstracts. Beijing,China, 1993.

INGHAM, A. E., Sea Surveying, North East London Polithechnic, William Clowes & Sons Ltd., London, 1975

SEEBER,G. - SATELLITE GEODESY , Foundations, Methods and Applications. Walter de Gruyter, Berlin, 1993.

TESTING HIGH-ACCURACY, LONG-RANGE CARRIER PHASE DGPS IN AUSTRALASIA

Oscar L. Colombo, University of Maryland and NASA Goddard S.F.C., Code 926, Greenbelt 20771, Maryland, U.S.A. (ocolombo@geodesy2.gsfc.nasa.gov

Chris Rizos and Bernd Hirsch, School of Geomatic Engineering, University of New South Wales, Sydney 2052, Australia (C.Rizos@unsw.edu.au)

ABSTRACT

In August 12, 1994, a test was conducted to assess the precision of carrier-phase kinematic differential GPS over long baselines, using reference stations situated in both Australia and New Zealand. A boat was positioned (off-line) while going around Sydney Harbour for about 2 hours. Some stations were very far; baselines from them to the boat ranged from 736 to 2200 km. The boat was also positioned relative to nearby receivers, in Sydney, using a standard, short-baseline kinematic technique, and resolving the phase ambiguities as exact integers. Finally, the short- and the long-baseline solutions were compared. Depending on the choice of distant stations, the position estimates differed between 7 and 14 cm r.m.s. about the mean. The mean was either quite small, or quite large, depending on just how the starting and finishing positions of the boat were obtained.

INTRODUCTION

Precise navigation based on the Global Positioning System (GPS) can be of value to those engaged in photogrammetry, gravimetry, or topographic mapping, particularly when surveys are repeated to monitor geophysical changes (e.g., in terrain height due to subsidence, or in the position of markers due to the flow of ice in a glacier). True vertical or horizontal movements between epochs must be distinguished from a change in position errors, so these must be kept small

Differential GPS (DGPS) navigation with carrier phase is an interferometric technique used to locate the antenna of a GPS receiver within centimeters of its true position relative to a fixed reference receiver. Speeds range from zero to several kilometers per second.

DGPS accuracy has been tested for satellites and terrestrial vehicles by looking at misclosures in round-trip navigation, crossover differences of altimetry over smooth ice, navigation of static receivers, or simultaneous tracking with GPS and lasers (e.g., Tapley et al., 1994; Hermann et al., 1994). Accuracy appears to be at the level of a few cm for short-baseline kinematic, long baseline static, and satellite orbit determination.

LONG-RANGE DGPS NAVIGATION OF A TERRESTRIAL VEHICLE

The precise, long-range, DGPS navigation of a terrestrial vehicle requires determining an often unpredictable trajectory while filtering out certain error sources that do not cancel out entirely, as they do over short-baselines. Such errors include: uncertainties in the ephemerides of the GPS satellites (6 states per satellite); in the position of reference stations (3 states per station); and in the correction of tropospheric refraction with imperfect models (1 state per station) Observations at "fixed" stations have to be corrected for local earth-tidal motions (up to 30 cm).

Estimating additional error parameters to model those unwanted effects is common practice in long-baseline static geodetic positioning (e.g., King et al., 1985), in precise satellite navigation (Tapley et. al, ibid.), or in Wide Area Augmentation System (WAAS) DGPS (Brown, 1989; Kee and Parkinson, 1993, Pullen et al., 1994). However, the *practical* aspects of precise kinematic positioning (decimeter-level, high-rate ambiguous carrier-phase as main data, poorly known vehicle dynamics, off-line data editing and processing) are very different from the positioning of fixed stations or satellites ("vehicle" stationary, or with highly predictable and understood dynamics; much lower data rates), or of WAAS (meter-level, real-time navigation, real-time integrity monitoring, unambiguous pseudorange as primary data).

Achieving exact *ambiguity resolution* is rare because of ionospheric refraction. Instead, *biases in the phase ion-free combination* are often estimated as part of the overall solution (1 state per bias). This is sometimes called *"floating the ambiguities"* (e.g., King et al., ibid.; Sauer, 1994). With long baselines, this works better if one combines phase with good pseudorange data (Blewitt, 1989).

Without carrying atomic clocks or other special equipment, one needs 5 satellites in common view to have enough redundant observations (Loomis, 1989): 4 on account of the instantaneous vehicle coordinates and clock errors, plus 1 more for all the other error states. Satellites with elevations much lower than 20° should be excluded, to reduce the effect of refraction and multipath. More than one fixed reference receiver may be used, in case of equipment problems, and to filter out GPS orbit errors better, as these can be observed from more than one direction (Colombo, 1991).

Hundreds of thousands of observations may be collected over several hours, with two or more receivers measuring at a rate of 1 or 2 times per second (Hz). The trajectory estimator may have more than 50 error states. As a consequence, one has to try and reduce computing time, numerical instability in the filter and smoother, and size of memory needed (RAM and hard disk). To avoid these problems, one may use a two-step procedure (Colombo, 1991, 1992), partitioning the solution to exploit the fact that the biases and other error states are constant or nearly constant over considerable periods of time.

That technique has been implemented by O. Colombo in a computer program called **IT** (for "Interferometric Translocation"). To prepare the GPS data for processing, preliminary editing, clock corrections, preliminary pseudorange navigation, double-differencing, and ephemerides interpolation to all measurement epochs, a "front end" program **DOIT4_IT**, developed by C. Rizos at the Geodesy Laboratory of the University of New South Wales (UNSW). **IT** has been used in different work-stations, with running times that, depending on CPU, range from 0.5 to 2 minutes per hour of observation, per baseline, and per Hz of data rate (with 7 satellites in common view on each baseline). **IT** can also be used for long-range static surveys, and for short-range navigation and surveying, including stop-and-go kinematic and rapid-static surveys.

THE TEST

On August 12, 1944, the trajectory of a small boat cruising Sydney harbor was estimated with the long-range DGPS technique described above, using reference receivers as far as 2200 km away. Both Selected Availability (S/A) and Anti-Snooping (A/S) were on. The same trajectory was then determined relative to a nearby reference receiver at (UNSW, 6 - 14 km away), after resolving the phase ambiguities as integers by static initialization. This kind of short-range DGPS is usually very accurate, so the difference between the two navigations can be used as a measure of the long-range errors. In practice, uncertainties in short-range navigation and in reference receiver coordinates limit the smallest error that can be reliably detected to 5 - 10 cm. There were two other fixed receivers near the point of departure. Antenna swaps, in addition to static initialization periods at the beginning and end of the run, were used to verify that the ambiguities had the correct integer values. The trajectory relative to UNSW was compared to that relative to one of the harbour receivers, which was closer. Since the roving receiver started and ended at the same tripod on the place on embarkation, the misclosure in position provided another check on the short-range results.

The fixed reference stations were positioned accurately and in the same reference frame. This was done by Peter Morgan and Russell Tiesler, at the University of Canberra, employing the geodetic software GAMIT (Release 9.23, 1994).

The boat was a 6-meter, double-hulled craft with twin outboard motors, belonging to the Waterways Authority of the State of New South Wales. Very light and without a keel, it bobbed and swayed easily even in moderate seas. On 12 August, 1994, her trip began and finished at Fort Denison, on a small rocky island near the Sydney Opera House. After reaching the suburb of Manly, about 6 km to the NE of the starting point, she was turned around without stopping, and traveled back to Ft. Denison. The distance to the reference site at UNSW varied between 6 and 14 km. Speed was less than 10 knots, to reduce sway and vibration, and decrease the chance of cycle-slips or even of completely losing signal lock. At times, the craft was rocked by waves made by the very large ferry-boats of the Sydney-to-Manly run. Luckily, modern receivers are almost free of cycle-slips.

Nearby, there were two fixed receivers at Ft. Denison, and one at UNSW, atop the Mather Pillar, on the roof of the Geography and Surveying building, some 7 km south of Ft. Denison. Mean distances from the boat to the distant receivers are given with the Table of results, in next section.

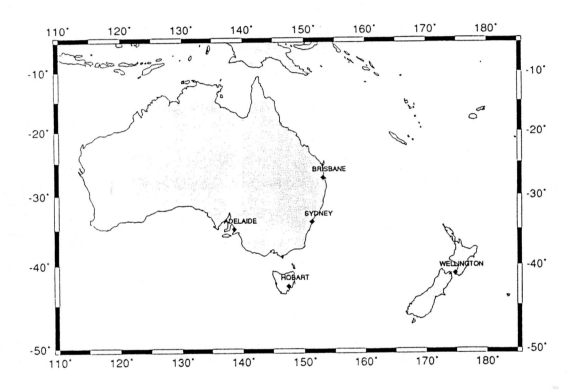

Figure 1. Sites of the reference GPS receivers during the August 12 experiment.

The observing period was chosen in the early afternoon, local time, when there were more than 6 BLOCK II satellites in view for longer than 3 hours and above 20° elevation.

RESULTS AND CONCLUSIONS

The long-range navigation was calculated in two different ways: (1) Using only data from the boat and the distant receivers, with the initial and final positions constrained to be the same during both static initialization periods. (2) Adding data from a nearby fixed site (UNSW) — with the ambiguities resolved — to the first and last part of a continuous, unconstrained navigation. This corresponds to the situation in which there is a reference receiver near the place of departure or arrival (air strip dock, etc.), making the starting and final positions well-determined. Solving for position while floating the ambiguities *and* the orbits can make the solution ill-conditioned if the survey session is a short one. The change in position may be well-determined, but not the initial position vector; this adds a constant, parallel shift to the computed trajectory. This happened when strategy (1), above, was used, but not with strategy (2).

The *a priori* uncertainties (1 sigma) were: white measurement noise of phase ion-free double differences, 2 cm; instantaneous vehicle position, 100 m per coordinate (modelled as an unpredictable "white-noise" process); satellite initial position, 10 m per coordinate; satellite initial velocity, 1 mm/s per coordinate; refraction correction error at each station, 10 cm plus a small random walk; reference station coordinates, fixed to precisely surveyed values; ion-free biases, 10 m.

R.M.S. Position Discrepancy and Offset Between Short and Long-Range Navigation

Case No.	Far Stations (*)	R.M.S. (CM)	Shift (M)
1	SUNM, ADEL, WELL	4	1.3
2	ADEL, WELL	14	3.5
3	SUNM	11	3.2
4	ADEL, WELL	17	2.0
5	ADEL, WELL (+ UNSW)	4	0.05

(*) Distances from Sydney to: SUNM, 736 km; ADEL, 1100 km; WELL, 2200 km.

Processing strategy 1 was used with all the combinations of reference stations shown in the Table, except for case 5. Processing strategy 2 was used in Case 5, with Adelaide and Wellington as the only reference stations from 3.5 to 4.5 hours (GPS time), but with UNSW added during the first hour and the last 5 minutes. The Vertical, North, and East discrepancies for Case 5 are shown in Figure 2. In Case 4, the estimation of all the ambiguities was re-started in mid-journey, as if there had been a catastrophic loss of lock in all channels.

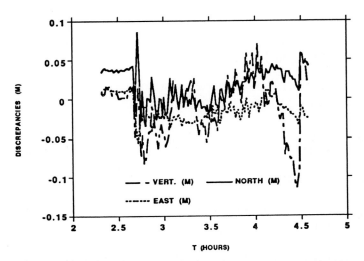

Figure 2. Discrepancies (long- vs. short-range navigation) in the position of the boat as estimated in Case 5.

The r.m.s. about the mean and the average position differences between the long-baseline trajectory and the short-baseline control were calculated over the period of motion (2.67 - 4.45 hours, GPS time). This was further restricted, in Case 5, to the interval when Adelaide and Wellington were the only reference sites (from 3.5 to 4.5 hours, GPS time). The data compression interval was always 2 minutes.

The main conclusion to be drawn from this experiment is that decimeter-level navigation is feasible in practice, with an appropriate procedure. The one we tested appears to be robust and flexible. We are planning future work with different vehicles under various operating conditions.

AKNOWLEDGEMENTS. Many colleagues in Australia and New Zealand gave us encouragement, equipment, measurements, advise, information, and other help, making their resources and talent available to us at no cost. Most funding has been provided by the School of Geomatic Engineering of the University of New South Wales. Dr. Colombo also has received financial support for related work, in the U.S.A., from NASA (grant NAG 5245), and in Denmark, from KMS (the Danish national Cadastre and Mapping Service).

REFERENCES

Bierman, G.J., 1977, Factorization Methods for Discrete Sequential Estimation; Academic Press.

Blewitt, G., 1989, Carrier Phase Ambiguity Resolution for the Global Positioning System Applied to Geodetic Baselines up to 2000 km; J. Geophys. Res. (Red), 94, B8, 10187-10203.

Brown, A., 1989, Extended Differential GPS; "NAVIGATION", Journal of the Institute of Navigation (ION), Vol. 36, No. 3, Fall 1989 Issue.

Colombo, O.L., 1991, Errors in Long-Distance Kinematic GPS; Proceedings GPS-91, ION, Satellite Division 4th International Technical Meeting, Albuquerque, New Mexico.

Colombo, O.L., 1992 Precise, Long-Range Aircraft Positioning with GPS: The Use of Data Compression; Proceedings VI International Symposium on Satellite Positioning, Columbus, Ohio.

GAMIT: Documentation for the GAMIT GPS Analysis Software; Joint publication from the Dept. of Earth, Atmospheric and Planetary Sciences, MIT, and Scrips Inst. of Oceanography, UC San Diego, Release 9.23, March 1994.

Hermann, B., Evans, A., Law, C., and B. Remondi, 1994, Kinematic On-the-Fly GPS Positioning Relative to a Moving Reference; Proceedings GPS-94, ION, Salt Lake City, September 1994, pp. 1498-1499.

Kee, Ch., and B. Parkinson, Static Test Results of Wide Area Differential GPS,' Proceedings GPS-93, ION, Salt Lake City, Utah, pp. 1233 - 1243.

King, R.W., Masters E.G., Rizos C., Stolz A., and J. Collins, 1985, Surveying with GPS; Monograph No. 9, School of Surveying, The University of New South Wales, Sydney.

Loomis, P., 1989, A Kinematic GPS Double-differencing Algorithm; Proceedings 5th International Symposium on Satellite Positioning, Las Cruces, N.M., N. Mexico State U., 611-620.

Sauer, D. B., 1994, Determination of High-Precision Trajectories without Fixing Integer Ambiguities; Proceedings 1994 IEEE PLANS Meeting, Las Vegas, Nevada.

Tapley, B., et al. (many co-authors), Precision Orbit Determination for TOPEX/POSEIDON; J. Geophys. Res. (Oceans),TOPEX/POSEIDON Special Issue, Vol. 99, C12, Dec. 1994.

Chapter 4

The GPS and its Relations to Geophysics

USING THE GLOBAL POSITIONING SYSTEM TO STUDY THE ATMOSPHERE OF THE EARTH: OVERVIEW AND PROSPECTS

James L. Davis and Mario L. Cosmo
Harvard-Smithsonian Center for Astrophysics
60 Garden Street
Cambridge, Massachusetts 02138 USA

Gunnar Elgered
Onsala Space Observatory
Chalmers University of Technology
S-439 92 Onsala Sweden

INTRODUCTION

The Global Positioning System (GPS) is rapidly finding increasingly wide application in the measurement (often in real time) of a variety of properties of both the neutral atmosphere and the ionosphere. Whereas earlier space geodetic techniques were also sensitive to the atmosphere, GPS provides a much denser spatial and temporal sampling of the atmosphere. The improved spatial sampling is due mainly to the low price of GPS receiver systems—which are $\sim$0.5% of the cost of, say, a Mark III Very Long Baseline Interferometry (VLBI) receiver/recorder system—thereby enabling a truly large number of GPS receivers to be deployed. An improved sampling is obtained at high temporal frequencies because the high signal-to-noise ratio (SNR) of the GPS signals enables very low integration periods (< 1 sec). At low temporal frequencies, the low cost of the GPS receiver systems and the simplicity of data reduction enable continuously operating GPS receivers to be employed. Thus, data may be obtained for extremely long periods of time.

REVIEW OF ATMOSPHERIC EFFECTS ON GPS OBSERVABLES

In this section we briefly review the manner in which the GPS observables are affected by the atmosphere. The two main GPS observable types are the code (either P-code or C/A code) observables and the carrier beat phase observables. In this paper, we will concentrate on the phase observables, which have a much greater potential for accuracy than the code types.

The carrier beat phase (or simply "phase") is the difference between the phase of the signal received by the GPS antenna/receiver system from a given satellite and a signal of the appropriate frequency generated by the oscillator within the GPS receiver. The phase is determined independently for each frequency channel, L1 and L2. The phase observable ϕ_i for the ith frequency channel ($i = 1$ for L1 and $i = 2$ for L2) can thus be expressed (in cycles) at some epoch t (e.g., King et al., 1985) as

$$\phi_i = \frac{\rho}{\lambda_i} + C_i^{\text{rec}} + C_i^{\text{sat}} + \phi_i^{\text{atm}} + \phi_i^{\text{ion}} + \phi^{\text{apr}} + N + \epsilon_i \tag{1}$$

where ρ is the instantaneous distance between the receiving antenna and satellite, λ_i is the wavelength associated with the ith channel, C_i^{rec} is the receiver "clock" phase error due mainly to drifting of the receiver frequency standard from the nominal frequency, C_i^{sat} is the satellite "clock" phase error due to the same problem in the transmitting satellite, ϕ_i^{atm} is the phase delay due to the neutral atmosphere, ϕ_i^{ion} is the dispersive ionospheric phase delay, ϕ^{apr} is the a priori phase difference between the initial phases of the satellite and the receiver, N is an integer cycle bias or "ambiguity," and ϵ_i is the phase measurement error, random and otherwise, due to other sources. In (1), all the phase quantities are expressed in units of cycles.

The most general form for ϕ_i^{atm} and ϕ_i^{ion} can be written by noting that the propagation delays are defined as the difference between the observed phase and that which would have been observed had there been no propagation medium. Then we can write most generally

$$\phi_i^{\text{atm}} + \phi_i^{\text{ion}} = \frac{1}{\lambda_i} \left[\int_S ds\, n_i(\mathbf{x}) - \int_V ds \right] \tag{2}$$

where S represents the path of propagation, V the *in vacuuo* path, ds the element along those paths (whichever is appropriate for the particular integral), and $n_i(\mathbf{x})$ the index of refraction at the position $\mathbf{x}$ along the path of propagation.

It is usual to break the right-hand side of (2) up into separate contributions for the ionosphere and the neutral atmosphere, since the contributions from each represent different physical phenomena. Nevertheless, doing so is an approximation, since the path along which the signals propagate is determined by the total refractive index.

The Ionospheric Delay

The ionospheric delay in radio interferometric techniques is estimated by making use of the dispersive properties of the ionosphere. The ionospheric phase delay (in cycles) is given approximately by (Herring, 1983)

$$\phi_i^{\text{ion}} \simeq -\frac{e^2}{8\pi^2 m_e \epsilon_\circ f_i} \int_S ds\, N_e(\mathbf{x}) \tag{3}$$

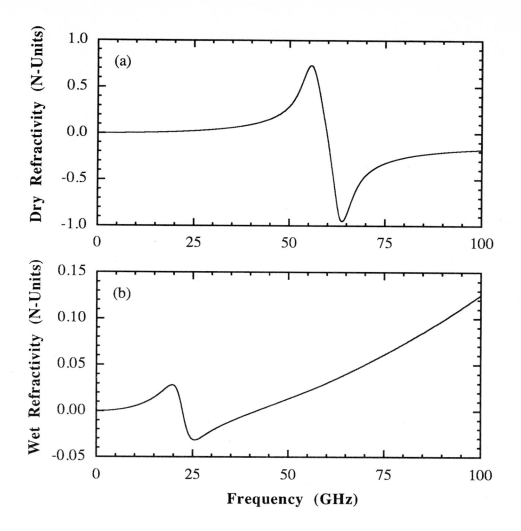

Fig. 1. The dispersive parts of the atmospheric refractivity for (a) the "dry" components and (b) water vapor. Values for the dry (not total) pressure, temperature, and relative humidity of 1013.25 mbar, 300 K, and 50% were used. The Millimeter-Wave Propagation Model (MPM) of Liebe (1989) was used.

where m_e is the electron mass, ϵ_o is the permittivity of free space, e is the electron charge, and $N_e(\mathbf{x})$ is the density of free electrons. The negative sign in (3) indicates that the effect of the ionosphere is a phase advance. The approximate sign indicates that we have neglected nonlinear polarization effects, polarization and damping terms, bending (i.e., the difference between the paths S and V), and scintillation effects. All these effects have been considered in detail for the VLBI frequencies (S- and X-band) by Herring (1983) and found to be negligible, but no similar complete study has been performed for the GPS frequencies.

Of the quantities in (3), only the integrated electron content (IEC) is not a fixed quantity. For ground-based receivers, this quantity is usually referred to as the total electron content (TEC), and can vary by orders of magnitude with variations of site location, time of year, and time of day. A number of specific methods can be used to estimate the IEC from GPS data, but all rely on the combination of observables at the two L-band frequencies (e.g., Sardon, 1993).

The Neutral Atmospheric Delay

Unlike the ionospheric delay, bending can be significant for the neutral atmospheric delay (hereafter simply the "atmospheric delay"). Thus, the expression for the atmospheric delay can be written

$$\phi_i^{\text{atm}} = \frac{1}{\lambda_i} \left[\int_S ds \, n_i^{\text{atm}}(\mathbf{x}) - \int_V ds \right]$$

$$= \frac{10^{-6}}{\lambda_i} \int_V ds \, N(\mathbf{x}, f_i) + \frac{1}{\lambda_i} \left[\int_V ds - \int_S ds \right] \tag{4}$$

where $N(\mathbf{x}, f_i) = 10^6 \times [n_i^{\text{atm}}(\mathbf{x}) - 1]$ is the refractivity.

The primary contribution to the refractivity of dry air below 1000 GHz is the effect of induced and orientation polarization of the atmospheric molecules (see, e.g., Jackson, 1975). Both these effects are nondispersive at microwave frequencies. For a gas made up of q constituents, these contributions to the refractivity can be written (Debye, 1929) as

$$N_{\text{nd}} = \sum_{k=1}^{q} \left(A_k + \frac{B_k}{T} \right) \rho_k \tag{5}$$

where N_{nd} is the nondispersive (i.e., frequency independent) part of the atmospheric refractivity. A_k and B_k are constituent-dependent constants describing the induced and orientation polarization effects which may be determined by laboratory measurements, ρ_i is the density of the ith constituent, and T is the temperature. Of the eight primary constituents of dry air—N_2, O_2, Ar, CO_2, Ne, He, Kr, and Xe together make up over 99.999% of dry air by volume (Gleuckauf, 1951)—none have permanent dipole moments, so that $B_k = 0$ for these constituents. Water vapor, on the other hand, does have a permanent dipole moment. Thus (5) can be rewritten

$$N_{\text{nd}} = \sum_{k=1}^{q_{\text{dry}}} A_k \rho_k + \left(A_v + \frac{B_v}{T} \right) \rho_v \tag{6}$$

where the summation is performed over the dry components and the subscript v indicates water vapor.

The dispersive part of the refractive index for the neutral atmosphere for frequencies below 100 GHz is quite small. The main contributions are the anomalous dispersions associated with the 22.23508 GHz rotational transitions of water vapor and the band of rotational transitions of O_2 near 60 GHz. There is also a significant dispersion from with the water vapor continuum, and a nearly negligible dispersion from the dry components (Liebe, 1989). The dispersive contributions to the refractivity are shown in Fig. 1. Since the value of N_{nd} for these values of pressure, temperature and humidity is 340 N-units, the contribution of dispersion across the frequency band of GPS can be seen to be negligible. (The frequency-squared dispersion associated with the water vapor continuum furthermore induces a negligible error in the determination of the ionospheric delay.)

ESTIMATION OF ATMOSPHERIC QUANTITIES

In the previous section, we showed that the GPS observables are sensitive to various atmospheric parameters integrated along the line of site. For the ionosphere, (3) indicated that the only unknown quantity was the integrated electron content (IEC). For the neutral atmospheric propagation delay, from (4) and (6), we have

$$\phi_\ell^{atm} = 10^{-6} \left[\sum_{k=1}^{q_{dry}} A_k \int_V ds\, \rho_k(\mathbf{x}) + A_v \int_V ds\, \rho_v(\mathbf{x}) + B_v \int_V ds\, \frac{\rho_v(\mathbf{x})}{T(\mathbf{x})} \right]$$
$$+ \left[\int_V ds - \int_S ds \right] \qquad (7)$$

where the frequency dependence has been removed from the atmospheric delay and the subscript ℓ indicates that the phase is to be expressed in units of length. Equation (7) tells us that the atmospheric delay contains information concerning the integrated constituent densities of moist air, as well as the temperature profile. Furthermore, through the ideal gas law (modified, if accuracy requires, for the nonideal behavior of real gases), one can write (7) in terms of partial pressures.

The methods used to estimate the atmospheric propagation delays are similar to those developed for VLBI (Davis, 1986; Herring et al., 1990). The atmospheric delay at a given site is modeled as

$$\phi_\ell^{atm}(\epsilon, t) = \tau_h^z(t) m_h(\epsilon) + \tau_w^z(t) m_w(\epsilon) \qquad (8)$$

where $\tau^z(t)$ is the zenith delay at epoch t and $m(\epsilon)$ is the "mapping function" evaluated at elevation angle ϵ. The subscripts h and w denote "hydrostatic" and "wet." The mapping functions are assumed known. The zenith hydrostatic delay is also assumed known. This term is proportional to the vertical integration of the total density (not the density of the "dry" constituents), which if the atmosphere is in hydrostatic equilibrium yields a term proportional to the surface pressure; hence the term "hydrostatic delay" (Davis et al., 1985). The time-dependent values for $\tau_w^z(t)$ can then be estimated from the GPS data in the combined solution for all the parameters. The preceding discussion is, of course, appropriate only for ground-based GPS receivers.

APPLICATIONS

The Ionosphere

Historically, the interest in the ionosphere was created by the need for global (or at least long distance) short-wave radio communication. In the 1930's and 40's no artificial communication satellites were available and short-wave communication relied on one (or multiple) reflections of the radio signal from the bottom-most layers of the free electron distribution.

The early communication satellites were therefore necessarily designed to operate at significantly higher frequencies in order to penetrate the ionosphere. Geostationary satellites are located at an altitude of ~36,000 km above the equator, whereas the ionosphere extends typically to several 1000's of km.

The main influences on satellite communication from the free electrons and their interaction with the magnetic field of the Earth causes phase delays (as mentioned above) and a rotation of the electric field vector (polarization). Both these effects are, however, reduced for higher frequencies and since the 1960's there has been a continuous trend towards operating satellite links at ever higher frequencies. This trend has not been driven by the ionospheric difficulties but by the need for greater bandwidth. Thus the ionosphere is not a major problem for today's satellite communication systems.

Today, applications of ionospheric studies range from basic research (e.g., magnetosphere-ionosphere research) to terrestrial hazards (e.g., space weather and geomagnetic storms). There is also a number of applications for microwave techniques which are affected by the ionosphere but which do not employ dual-frequency systems, such as radio astronomy (including single-frequency VLBI using the Very Long Baseline Array) and navigation using single-frequency GPS receivers.

The Neutral Atmosphere

A number of applications exist for atmospheric studies. Partly this situation exists because the atmospheric propagation delay is, per (7), sensitive to a number of different quantities which may be of interest. But mainly it exists because changes in the atmosphere, especially the troposphere, has such a fundamental impact on our lives. On short timescales, changes in the atmosphere can indicate the passages of fronts or more violent systems. On long timescales, secular changes in the atmosphere can reflect long-term climate change. Similarly, variations in the atmosphere on short spatial scales (< 50 km) can yield information on the interaction of the atmosphere with land or ocean elements. On longer spatial scales, synoptic information is yielded.

Furthermore, applications are not limited to the troposphere (or any particular part of the atmosphere), since the GPS phase observables are influenced by the total integrated line-of-sight quantities. For ground-based receivers, this line-of-sight is a path from the ground to the GPS satellite (or vice versa), and so all atmospheric "layers" are sampled. Occultations of the GPS satellites by the Earth as viewed from a satellite with a GPS receiver in low-Earth orbit (LEO) could be used to sample specific altitudes within the atmosphere (e.g., Bevis et al., 1992).

Clearly, the number of possible applications of GPS to atmospheric problems is immense. Below, we review two such applications, one relying on ground-based observations and one space-based, to provide examples.

Global change studies. Studies indicate that concentrations of atmospheric carbon dioxide and methane ("greenhouse" gases) will double by the end of the 21st century. These increases will, through the greenhouse effect, increase temperatures in the troposphere by 1–5 K (e.g., Roble, 1993). In the stratosphere, temperatures are predicted to decrease by 10–20 K (Brasseur and Hitchman, 1988). Recent studies by Roble and Dickinson (1989) indicate that this cooling might extend into the mesosphere and thermosphere/ionosphere as well. The altitude of the ionization layers of the ionosphere could lower by 20 km (Rishbeth and Roble, 1992).

The uncertainties in the model calculations could be significantly reduced by measurements. Two methods could be used, both of which employ GPS receivers. One of the methods, mentioned above, involves occultations by LEO satellites. According to Bevis et al. (1992), a GPS receiver in low Earth orbit could observe ∼600 occultations per day. Resolution of sampling would be ∼200 km horizontal and ∼1 km vertical, with a temperature resolution of 1 K down to the upper troposphere.

A method which could yield even greater spatial resolution, albeit over a more limited region, is direct *in situ* sampling of the atmosphere by a LEO satellite. In order to be useful, such a satellite would have to have an altitude of only ∼100 km, which would mean that the satellite would be subjected to significant atmospheric drag, decreasing the lifetime of the satellite and making position determination difficult. One possible solution would involve a tethered satellite. Developed at the Smithsonian Astrophysical Observatory (Colombo et al., 1974), thethered satellites offer the capability of trawling instrumented probes deep into the upper atmosphere at altitudes of 110–130 km. The "multiprobe" technique (Lorenzini et al., 1989) could furthermore be used to sample at several levels. A GPS antenna included in each of the probe packages would enable the positions of the probes to be accurately determined.

Meteorology. It has been known for many years that the "wet" atmospheric propagation delay is nearly proportional to the integrated precipitable water vapor (IPWV). This is because the temperature structure within the troposphere, wherein most of

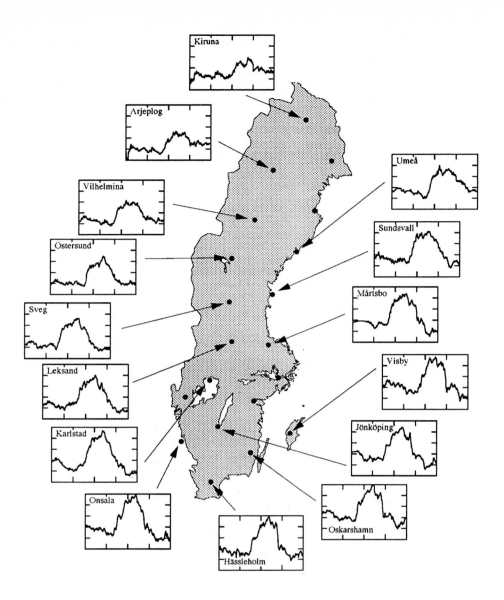

Fig. 2. Determinations of IPWV from stations of the Swedish permanent GPS network, from Elgered et al. (in press). Horizontally, each plot spans a total of four days (beginning 17 December 1993 0 UT); vertically, each plot spans 0–25 mm of IPWV. At the time of these measurements, not all the sites of the network were operational yet, accounting for the lack of data for some sites. The Onsala Space Observatory group is studying the application of ground-based GPS to meteorology.

the water vapor resides, is very regular. An approximate relationship between the wet delay τ_w and the IPWV is $\tau_w \simeq 6.5 \times \text{IPWV}$ (e.g., Resch, 1984).

As Bevis et al. (1992) have pointed out, ground-based monitoring of IPWV using GPS receivers could improve the temporal and spatial resolution of water-vapor determinations. In order to be effective, a global coverage of densely-spaced (intersite distances on the order of 100 km) are required. As an example of the information which may be obtained, Fig. 2 shows estimates of IPWV (Elgered et al., in press) from sites of the Swedish permanent GPS network. Each plot shows the IPWV for the site indicated between 17–21 (0 UT) December 1993. During this period a warm front (represented by an increase in the IPWV) clearly passes through the area.

FINAL REMARKS

The the estimation of atmospheric parameters from GPS data is far from being a mature technique. The most advanced methods used to estimate the time-dependent neutral zenith propagation delay from GPS data are essentially identical to those employed for VLBI analysis in the mid-1980's. The model (8) allows for no azimuthal variations in the "mapping function," which can be significant (Davis et al, 1989). Thus, there is much room for improvement in the atmospheric models used to retrieve relevant parameters from GPS data.

Furthermore, GPS is a relatively young technique, and as of this writing the understanding of its error sources is continuously evolving. Recently, for example, Elósegui et al. (1995) demonstrated that signal scattering can cause significant errors in the estimation of zenith-delay parameters from permanent GPS systems mounted on pillars. Moreover, no investigation of the (temporal) error spectrum of atmospheric parameters determined from GPS data has yet been undertaken.

Thus, a great deal of work remains to be done both on the technique of atmosphere parameter retrieval from GPS data and on specific applications of such retrieval (e.g., How can the GPS determinations of IPWV best be utilized for weather prediction?) What is clear already is that the spatial and temporal sampling achievable with GPS make it a unique tool for studying an energetic global system such as the atmosphere.

Acknowledgments. This work was supported by NASA grant NAG5-538 and the Smithsonian Institution.

REFERENCES

Bevis, M., Businger, S., Herring, T. A., Rocken, C., Anthes, R. A., and Ware, R. H. (1992) GPS meteorology: Remote sensing of atmospheric water vapor using the Global Positioning System, *J. Geophys. Res.* **97**, 15,787–15,801.

Brasseur, G., and Hitchman, M. H. (1988) Stratospheric response to trace gas perturbations: Changes in ozone and temperature distributions, *Science* **240**, 634–637.

Colombo, G., Gaposhkin, E. M., Grossi, M. D., and Weiffenbach, G. L. (1974) Shuttle-borne "Skyhook:" A new tool for low-altitude research, *Rep. Radio and Geoast. 1*, Smithsonian Astrophysical Observatory, Cambridge, Massachusetts.

Davis, J. L., Herring, T. A., Shapiro, I. I., Rogers, A. E. E., and Elgered, G. (1985) Geodesy by radio interferometry: Effects of atmospheric modeling errors on estimates of baseline length, *Radio Science* **20**, 1593–1607.

Davis, J. L., Elgered, G., Niell, A. E., and Keuhn, C. E. (1989) Ground-based measurement of gradients in the "wet" radio refractivity of air, *Radio Science* **28**, 1003–1018.

Davis, J. L., (1986) Atmospheric propagation effects on radio interferometry, *USAF Rep. AFGL-TR-86-0243*, Air Force Geophysics Laboratory, Hanscom AFB, Mass.

Debye, P. (1929) *Polar Molecules*, Dover, New York.

Elósegui, P., Davis, J. L., Jaldehag, R. T. K., Johansson, J. M., Niell, A. E., and Shapiro, I. I. (1995) Geodesy using the Global Positioning System: The effects of signal scattering on estimates of site position, *J. Geophys. Res.* **100**, 9921–9934.

Gleuckauf, E. (1951) The composition of the atmsophere, *Compedium of Meterology*, T. F. Malone (ed.), American Meteorological Society, Boston.

Herring, T. A. (1983) Precision and accuracy of intercontinental distance determinations using radio interferometry, *USAF Rep. AFGL-TR-84-0182*, Air Force Geophysics Laboratory, Hanscom AFB, Mass.

Herring, T. A., Davis, J. L., and Shapiro, I. I. (1990) Geodesy by radio interferometry: The application of Kalman filtering to the analysis of very long baseline interferometry data, *J. Geophys. Res.* **95**, 12,561–12,581.

Jackson, J. D. (1975) *Classical Electrodynamics*, Wiley, New York.

King, R. W., Masters, E. G., Rizos, C., Stolz, A., and Collins, J. (1985) *Surveying With Global Positioning System*, Dummler, Bonn.

Liebe, H. J. (1989) MPM—An atmospheric millimeter-wave propagation model, *Int. J. Infrared Millimeter Waves* **10**, 631–650.

Lorenzini, E. C. (1989) Multiple Thethered Probes, *Thethers in Space Handbook, 2nd Edition*, P. Penzo and P. Ammann (eds.), NASA, Washington, D.C.

Resch, G. M. (1984) Water vapor radiometry in geodetic applications, *Geodetic Refraction*, F. K. Brunner (ed.), Springer-Verlag, New York.

Rishbeth, H., and Roble, R. G. (1992) Cooling of the upper atmosphere by enhanced greenhouse gases—Modeling of the thermospheric and ionospheric effects, *Planet. Space Sci.* **40**, 1011–1026.

Roble, R. G. (1993) "Greenhouse cooling" of the upper atmosphere, *Eos Trans. AGU* **74**, 92–93.

Roble, R. G., and Dickinson, R. E. (1989) How will changes in carbon dioxide and methane modify te mean structure of the mesosphere and thermosphere? *Geophy. Res. Lett.* **16**, 1441–1444.

Sardon, E., (1993) Calibraciones ionosfericas en geodesia espacial mediante el uso de datos GPS, Ph.D. Thesis, Universidad Complutense de Madrid.

ON ATMOSPHERIC EFFECTS ON GPS SURVEYING

Jikun Ou

Institute of Geodesy & Geophysics (IGG), Chinese Academy of Sciences
54 Xu Dong Road, Wuchang, Hubei, 430077, P.R. of China

INTRODUCTION

The aim of this paper is to study the atmospheric effects and the mechanics of delay of radio signals, and to search for a better approach for reducing the effects of the atmosphere on GPS surveying. The stresses of our investigations are put on cases with low elevation angles and single-frequency receivers. Several important models, widely used , are tested by direct comparison of their effects on positioning with the real GPS measurement data, instead of comparison of compensation or characteristic quantities, such as TEC , etc.

The results of the investigations demonstrate that

1). more attention should be payed to the ionospheric irregularities. The only practical way to significantly raise the accuracy of an ionospheric model is to incorporate near real-time measurement or estimation of the ionospheric parameters into the model.

2). after the comparison of seventeen neutral atmospheric models, Herring's model is recommended for application since not only its accuracy is relatively higher, but also its algorithm is convenient.

3). in our experiments an interesting phenomenon has been found, which indicate that the correction only with a neutral atmospheric model is not enough (underestimate); quite the contrary, the correction with an ionospheric model is excessive (overestimate). A question is presented: "Should the ionospheric mapping function be improved ?" The primary result of a simple test shows that it is necessary to reduce the correction at some lower elevation angles. Therefore a new mapping function is proposed.

In what follows, only some parts of the results are shown, for further details, please read [Ou,1994].

EXPERIMENTS AND RESULTS OF CALCULATION

The comparisons of the atmospheric models, special of the ionospheric models have been performed by many scholars and researches. Their conclusions are interesting and useful.

We compare various atmospheric models through evaluation of the effects of these models on the position of the tracking station instead of comparison of compensation amounts from these models. This is a direct approach to know which model is more suitable for geodesy, because the interest of the users for geodesy and navigation is the precise position of the tracking station.

The conditions in the calculations for the comparison are not changed except various atmospheric compensations because of the different models used.

The observation equations for single point positioning with GPS pseudorange measurements are read

$$ob^s = \rho^s + t_{cr} + I_o^s + T_r^s \qquad s \geq 4 \qquad (1)$$

where ob is the pseudorange observation from the tracking station to the sth satellite, ρ is the geometric distance between the tracking station and the sth satellite, t_{cr} is the offset of the receiver clock. I_o^s and T_r^s are the compensations from the ionospheric and tropospheric models, respectively. If no atmospheric model is used, these two terms are equal to zero.

The orbit error and the clock bias of the observed satellite are not taken into account.

In our experiments, only data for which the broadcast ionospheric coefficients, the values of the solar flux and sunspot number for a 12-month running average (full month) and the daily value were available, were calculated.

The I_o^s and T_r^s terms in eq.(1) get different values from different atmospheric models, which results in the different adjustments.

From the comparison of the tropospheric models in [Ou,1994], the DGS model which is combined by Davis' dry mapping function, Goad's wet mapping function and Saastamoinen's zenith delay model proposed by Janes et al(1991), and Herring's model (HRG) are considered as the more useful models than the others, therefore these two models are used in our experiments.

The ionospheric models used in our experiments are Klobuchar's model (KLB) and Bent's model (BENT).

If none of the atmospheric models are used, this is indicated as "ATM".

The coordinates of the tracking station Delft, in the Netherlands, are known from other campaigns, which the accuracy of the coordinates of Delft is precise enough for comparison of the atmospheric models.. These coordinates are considered as the truth reference values. The magnitude of the differences between the calculated values and these truth values can be considered as a measure of accuracy for the different models, since the only change in the different adjustments is the atmospheric model used.

Experiment at date 14/03/1991
Observation time : 14^h 27^m 29.1^s ~16^h 24^m 14.9^s . Epochs : 475 * 15 seconds.Number of the observed satellites: 4 ~ 6.

These data were processed in seven ways. ATM represents a calculation of without any compensation from atmospheric models. DGS and HRG are used to correct the tropospheric delay, respectively. KLB and BENT are used to compensate the ionospheric effects,respectively.

The final two calculations make use of the combination of Herring's model with Klobuchar's or Bent's model , so that the tropospheric and ionospheric effects are compensated simultaneously.

The accuracies of the differences between the calculated and the truth values of coordinates at the tracking station corresponding to the various models can be found in table 1.

Experiment at date 31/10/1991
Observation time :10^h 58^m 35^s ~ 11^h 43^m 29.9^s . Epochs : 539 * 5 seconds.
Number of the observed satellites : 5.
We used the same seven models as in above experiment. The results calculated are shown in table 2.

Summary   From the above results, we can draw the following conclusions:
1. The influence of the compensation from the atmospheric models on the height component is more exhilarative than that on the horizontal component.
2. The accuracies of the height and the north (or x and z) components can be improved more or less by making use of either the tropospheric models DGS and HRG, or the ionospheric models KLB and BENT, or the combined models HRGKLB and HRGBENT, but the quality of the east (or y) component becomes worse than without compensations.

Table 1. The result of the test at 14/03/1991.

No.	Name of model	σ_x (m)	σ_y (m)	σ_z (m)	CPU(s)
1	ATM	11.528	9.263	17.106	14.
2	DGS	7.898	9.823	12.902	39.20
3	HRG	7.899	9.823	12.902	37.87
4	KLB	8.063	9.566	7.373	99.11
5	BENT	7.658	11.516	7.498	227.75
6	HRGKLB	8.626	10.197	7.913	135.92
7	HRGBENT	9.518	12.174	8.251	321.42

Table 2. The results of the test at 31/10/1991 (σ:in meters).

No	Name of model	σ(Height)	σ(North)	σ(East)	σ(Horizontal)	CPU (s)
1	ATM	15.119	9.083	4.583	10.174	17.15
2	DGS	8.534	8.204	4.952	9.574	23.95
3	HRG	8.533	8.205	4.952	9.575	22.83
4	KLB	7.127	7.601	5.033	9.108	121.52
5	BENT	8.464	8.604	5.089	9.987	276.43
6	HRGKLB	13.401	6.805	5.404	8.689	162.06
7	HRGBENT	14.731	7.871	5.460	9.580	388.70

3. The corrections from the ionospheric model KLB are less than those from model BENT in generally, which results in the fact that influences of the previous on the coordinate components are more effective than those of the latter.
4. The efficiencies of the models HRGKLB and HRGBENT are less than that only using model KLB.

SHOULD THE IONOSPHERIC MAPPING FUNCTION BE IMPROVED ?

We have found a very interesting phenomenon in the experiment at date (31/10/1991.), which indicates that an underestimate of the height component would be resulted if only a tropospheric model is used to correct the atmospheric delay, and on the contrary, an overestimate would be obtained if an ionospheric model is used to compensate the effects of the atmospheric delay.

We believe that there exists a set of middle-values between the positive and the negative values which are the differences between the computed and truth coordinates . This can be thought of the fact that is resulted from the effects of a turbulence part on the regular part of the ionospheric delay.

We supposed that the ionospheric mapping function needs to be adjusted in this situation because of too large compensations resulting from old mapping functions.

Some tests were performed in which the model HRGBENT were used but instead of the original mapping function, a simple truncated function was used. For example, a new correction of the delay may be

$$Io = \begin{cases} 18 \quad metres & for \ el \leq 40^{\circ} \\ using \ BENT's \ model & for \ others \end{cases} \tag{2}$$

where   Io indicates the ionospheric delay.

The calculated results are listed at the row of No.1 HRGBENT18 in table(3),in which much improvement in the height component is achieved.

Table 3. Comparison of the effects of the different ionospheric
mapping functions on positioning , σ (in meters).

No.	Name of models	Height	North	East
1	HRGBENT 18	3.872	9.013	4.160
2	HRGKLB	15.178	9.530	4.655
3	HRGBENT	17.122	11.264	4.899
4	HRGBENT 25	14.541	9.300	5.539
5	ATM	14.414	8.623	4.561

Note: Observation time : $10^h \ 58^m \ 35.0^s \sim 12^h \ 27^m \ 59.9^s$, Epochs: 1071 * 5 seconds, Numbers of the observed satellites: 4 ~ 6.

Similar to the simple Hopfield model, a new ionospheric mapping function is suggested, which may improve the compensation at low elevation angles. The new ionospheric mapping function reads

$$MF = MF_{old} * \begin{cases} \dfrac{\sin(5° + 50°)}{\sin(el + 50°)} & (el < 5°) \\ & (5° \leq el < 40°) \\ 1 & (el \geq 40°) \end{cases}$$

$$where: \quad MF_{old} = \frac{1}{\sin El'} = \frac{1}{\sqrt{1-A^2}}$$

$$A = R_s \cos el \,/\, (R_s + H_{io})$$

$$R_s = 6365 \quad km$$

$$H_{io} = 350 \quad km$$

(3)

Acknowledgment. I would like to thank prof. P.J.G.Teunissen and the colleagues at Delft Geodetic Computing Centre for their discussions and remarks during this research, and also to acknowledge the laboratory of dynamical geodesy, IGG, Chinese Academy of Sciences for her special support.

REFERENCES

Beutler, G. and I. Bauersima et al (1989). Accuracy and biases in the geodetic application of the Global Positioning System, Manuscripta Geodaetica 14:28-35.

Brown, L.D. et al (1991). Evaluation of six ionospheric models as prediction of total electron content, Radio Sci., Vol.26, No.4, pp.1007-1015.

Herring,T.A.(1992). Modelling atmospheric delay in the analysis of space geodetic data, in De Munck,J.C. and T.Spoelstra (ed):Refraction of transat mospheric signals in geodesy,N.G.C.,new series No.36.

Klobuchar,J.A.(1987). Ionospheric time-delay algorithm for single-frequency GPS user, IEEE Transactions on Aerospace and Electronic Systems, Vol.AES-23,No.3.

Langley,R.B.(1992). The effect of the ionosphere and troposphere on satellite positioning system, in De Munck,J.C. and T.Spoelstra (ed):Refraction of transatmospheric signals in geodesy,N.G.C.,new series No.36.

Newby,S.P., R.B.Langley and H.W.Janes(1990). Ionospheric modelling for single frequency users of the Global Positioning System: A status report, GPS'90, Ottawa,Canada, Sept.3-7,1990.

Ou Jikun (1993). Atmosphere and Its Effects on GPS Surveying , Faculty of Geodesy, TU Delft.

Ou Jikun (1994). Research on atmospheric effects on GPS surveying, Institute of Geodesy and Geophysics,Chinese Academy of Sciences.

Saastamoinen,J.(1972-1973). Contributions to the theory of atmospheric refraction, Bull. Geode. Nos.105(pp.279-298),106(pp.383-397),107(pp.13-34).

ABOUT THE USE OF GPS MEASUREMENTS FOR IONOSPHERIC STUDIES

N. Jakowski, E. Sardon, E. Engler, A. Jungstand, and D. Klähn
DLR, Fernerkundungsstation Neustrelitz,
Kalkhorstweg 53, D–17235 Neustrelitz

INTRODUCTION

Radio beacon measurements have effectively been used in exploring the temporal and spatial structure of the ionosphere since nearly 3 decades.

The Global Positioning System (GPS) provides new possibilities to monitor the total electron content (TEC) of the ionosphere/plasmasphere on global scales. In particular a well established network of GPS–receivers, as provided by the International GPS Service for Geodynamics (IGS) (e.g. Zumberge at al., 1994) may be used for large scale TEC analysis and monitoring reaching absolute accuracies of about 2...5 TECU ($1\,TECU=1*10^{16}m^{-2}$). The accuracy of TEC determination is constrained by different factors such as satellite and receiver instrumental biases, multipath, mapping function and plasmaspheric contribution (e.g. Klobuchar et al., 1994).

LIMITATIONS IN TEC DETERMINATION

The basic problem to derive reliable TEC data from dual–frequency GPS measurements is the determination of satellite and receiver instrumental biases.

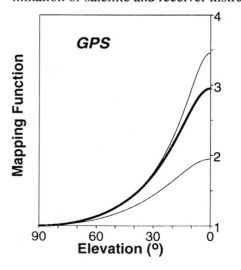

Fig. 1
Variability range of the mapping function for the conversion of vertical to link related slant TEC values. The variability range has been derived on the basis of simulation calculations taking into account the large variability of the ionosphere/plasmasphere electron distribution.

——— single layer model
$h_{sp}=400$ km

Assuming a second order polynomial approximation for TEC over each station, the instrumental biases for each satellite–receiver combination are estimated for each 24–hour period using a Kalman filter (Sardon et al., 1994). Over this period the instrumental biases are assumed to be constant. When using this technique, major accuracy limiting factors are introduced by the simplified mapping function which is based on a single layer approximation. Under day–time conditions this assumption is sufficiently fulfilled but during night–time, if the plasmaspheric electron content N_p is comparable with the ionospheric electron content N_I the mapping function is not well defined by a single layer approximation especially at elevation angles less than 30°. As simulation calculations for a spherically layered ionosphere have shown, the corresponding error in TEC may reach up to 30% (Fig. 1).

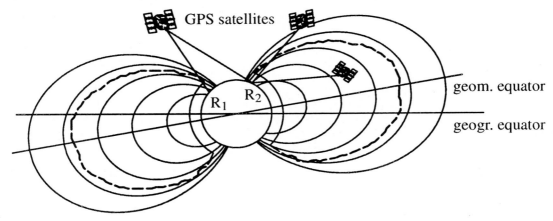

Fig. 2
Illustration of different ray path geometries to GPS–satellites indicating that radio links at mid– and high–latitude ground stations may be outside the plasmasphere. The outside border of the plasmasphere is defined by the plasmapause (dashed line) whose position is sensitive to geomagnetic activity.

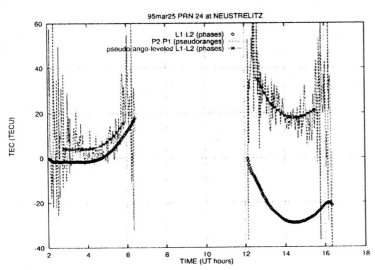

Fig. 3
Comparison of P–code pseudorange difference (dotted line) with differential carrier phases (diamonds) to illustrate the large noise and multipath in deriving TEC from P–code data for a selected data sample. The problem is solved by smoothing the noisy P–code data with the much more accurate carrier phase data (crosses).

When analyzing GPS data received at mid–European stations the field aligned structure of the plasmasphere has to be taken into account (Fig. 2). Instrumental bias estimation algorithms do not distinguish between radio links crossing the plasmasphere and those which ly outside. So errors are introduced in the bias estimation.

Multipath effects reduce the absolute accuracy of P–code measurements considerably, especially at low elevation angles. In order to reduce noise and multipath in the pseudo-range measurements, the more accurate carrier phase observations are fitted to the pseudoranges at elevation angles higher than 20°(Fig.3).

MONITORING THE IONOSPHERIC TEC OVER EUROPE

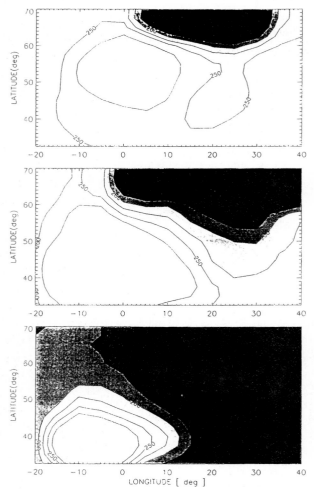

The ionosphere in view covers the geographic ranges $32.5°N \leq \phi \leq 70°N$ and $-20°E \leq \lambda \leq 60°E$. For monitoring tasks this area is presented by a grid with pixel sizes of $2.5° \times 5°$ in latitude ϕ and longitude λ, respectively. Corresponding TEC values are generated in a first order approximation by an empirical model (NTCM1) based on former long–term studies of TEC by means of the Faraday rotation technique (e.g. Jakowski and Paasch, 1984).

Taking into account relationships of TEC with solar activity, season and day–time, the model provides a first order estimation of TEC which is modified by the GPS measured data (Jakowski and Jungstand, 1994). The TEC model is constructed for any time and location within the defined area by the following formula:

Fig. 4

Sequence of three contour plots of TEC over Europe derived from GPS/IGS station measurements during the ionospheric perturbation on November 26, 1994 at hours 12, 13 and 14 UT (downward). TEC is measured in units of $10^{15}m^{-2}$. It can be seen that a zone of reduced ionization is propagating equatorward.

$$TEC(h, d, \lambda, \phi, F10.7) = \sum_{i=1}^{5} \sum_{j=1}^{3} \sum_{k=1}^{2} \sum_{l=1}^{2} H_i (h) * Y_j(d) * L_k(\phi, \lambda, h, d) * S_l (F10.7)$$

where H describes the diurnal and semidiurnal variation, Y describes the annual and semiannual variation and L and S describe the latitude dependence and the relation to solar activity, respectively. This model describes the mean behaviour of the mid–European ionosphere with an r.m.s. error of about $6*10^{16}m^{-2}$ for any time and solar activity level. Since the actual ionosphere may substantially deviate from the average behaviour, especially under perturbed conditions, measured GPS data are analyzed and combined with the NTCM1 model to construct TEC maps. Using the European IGS stations as GPS data sources, a total number of about 40...60 measuring points may be used for the construction of the horizontal TEC distribution over Europe.

OBSERVATIONS AND COMPARISON WITH DIFFERENT TECHNIQUES

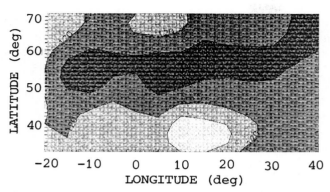

Fig. 5

Illustration of the geographic extent of the mid-latitude trough during the ionospheric storm on April 5, 1993 at 00:00 UT. TEC contour lines are given in units of $10^{15}m^{-2}$.

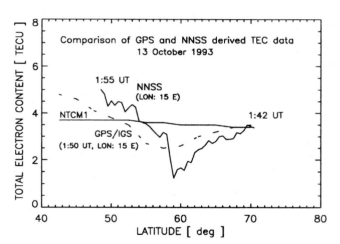

Fig. 6

Comparison of TEC data derived from GPS/IGS station measurements with corresponding NNSS data deduced from NNSS observations in Neustrelitz on October 13, 1993 around 1:50 UT. Reference is also made to the smooth NTCM1 model values.

Taking into account the still existing limitations in TEC accuracy in this study, we confirm our attention to significant large scale phenomena such as ionospheric storms or the mid-latitude electron density trough.

Although ionospheric storms are studied since more than 50 years, our knowledge about the nature of this complex phenomenon is still imperfect. Major ionospheric perturbations are generated by particle and energy input in the auroral zone followed by a horizontal mass and energy transport towards lower latitudes. The GPS technique provides a unique opportunity to study ionospheric storms continuously on a large scale. So the equatorward propagation of the so-called "negative phase" in the course of the ionospheric storm on November 26, 1994 is well illustrated in Fig. 4. In combination with vertical sounding techniques additional information may be derived about the shape of the vertical electron density profile characterized by the equivalent slab thickness τ=TEC/NmF2.

The behaviour of τ provides information about dynamic forces such as thermospheric winds and electromagnetic drifts which play an important role during ionospheric storms.

Closely related to ionospheric storms is the development of a pronounced electron density trough in subauroral latitudes which is well documented in relation to the ionospheric storm on April 4/5, 1995 (Fig. 5). To study the trough in more detail, a comparison with NNSS observations is quite useful (Fig.6).

GPS measurements provide also a good chance to check and/or improve ionospheric models. A comparison of actual measured TEC data and those derived from IRI90 is presented in Fig. 7.

It is shown that IRI90 modelling agrees rather well with the GPS data if actual vertical sounding data (IS) are used as input parameters. It is evident that IRI90 based on CCIR tables reflects only an

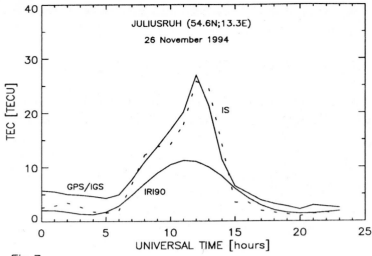

Fig. 7

Comparison of vertical TEC over Juliusruh derived from GPS measurements at European IGS stations with TEC data deduced from the integration of the IRI 90 electron density profile up to 1000 km height based on CCIR maps (IRI90) and actual ionosonde parameters (IS) obtained in Juliusruh.

average behaviour of the ionosphere which may be quite different from the actual values especially under perturbed conditions.

The accuracy of GPS–derived TEC data may be checked by different techniques as vertical sounding (Fig. 7) or NNSS differential Doppler measurements (Fig. 6).

CONCLUSIONS

The dual frequency L–band signals of GPS satellites provide a good opportunity for studying the spatial and temporal behaviour of the ionosphere on a large scale.

Although there still exist some accuracy limiting problems in TEC estimation and further improvements are necessary, the GPS technique is already effective in exploring well–pronounced large scale effects such as ionospheric storms and related phenomena. The derived TEC data are also useful for developing and testing ionospheric models. In particular the well established global distribution of GPS receivers as f.i. the IGS–network provide good conditions for comprehensive studies of the ionosphere.

Acknowledgements
The authors express their deep thanks to all the colleagues from the IGS community who made available the high quality GPS data sets. We thank J. Steger and F. Porsch for their help in data processing and simulation calculations. We are also grateful to Dr. Werner Singer for providing ionosonde data.
The study was supported under contract 50YI9202 with the Deutsche Agentur für Raumfahrtangelegenheiten (DARA).

References
Jakowski N. and E. Paasch, Report on the observations of the total electron content of the ionosphere in Neustrelitz/GDR from 1976 to 1980, Ann. Geophys. **2**, 501–504, 1984.
Jakowski N. and A. Jungstand, Modelling the Regional Ionosphere by Using GPS Observations, Proc. Int. Beacon Sat. Symp. (Ed.: L. Kersley), University of Wales, Aberystwyth, 11–15 July 1994, pp 366–369.
Klobuchar, J.A., P.H. Doherty, G.J. Bailey, and K. Davies, Limitations in Determining Absolute Total Electron Content from Dual–Frequency GPS Group Delay Measurements, Proc. Int. Beacon Sat. Symp. (Ed.: L. Kersley), University of Wales, Aberystwyth, 11–15 July 1994, pp 1–4.
Sardon. E., A. Rius, and N. Zarraoa, Estimation of the receiver differential biases and the ionospheric total electron content from Global Positioning System observations, Radio Science, **29**, 577–586,1994.
Zumberge, J., R. Neilan, G. Beutler, and W. Gurtner, The International GPS–Service for Geodynamics–Benefits to Users., Proc. ION GPS–94, Salt Lake City, September 20–23, 1994.

ASSESSMENT OF TWO METHODS TO PROVIDE IONOSPHERIC RANGE ERROR CORRECTIONS FOR SINGLE-FREQUENCY GPS USERS

Attila Komjathy and Richard B. Langley
Geodetic Research Laboratory
Department of Geodesy and Geomatics Engineering
University of New Brunswick
Fredericton, N.B. E3B 5A3 Canada

Frantisek Vejražka
Department of Radio Engineering
Czech Technical University
Prague, 166 27 Czech Republic

INTRODUCTION

One of the major error sources in GPS positioning is ionospheric refraction which causes signal propagation delays. The disturbing influences of the temporally and spatially varying ionization of the ionosphere have great impact on satellite geodesy, especially on GPS. To correct data from a single-frequency GPS receiver for the ionospheric effect, it is possible to use empirical models. In this research, we investigated the GPS single-frequency Broadcast [Klobuchar, 1986] and the International Reference Ionosphere (IRI90) [Bilitza, 1990] models. The GPS single-frequency Broadcast model is available to GPS users as part of the navigation message. The IRI90 model is a standard ionospheric model developed by the International Union of Radio Science (URSI) and the Committee on Space Research (COSPAR). After Newby [1992] investigated the IRI86 model's performance, we decided to include the new IRI90 model in our ionospheric research. We have used Faraday rotation data as 'ground-truth' with which we compared the vertical ionospheric range error corrections predicted by the Broadcast and IRI90 models. Some of the results shown here have been presented earlier [Komjathy et al., 1995].

BEHAVIOUR OF THE GPS SINGLE-FREQUENCY BROADCAST MODEL

We have examined 53 sets of Broadcast model coefficients archived by the Department of Radio Engineering of the Czech Technical University in Prague which started archiving the coefficients in April 1994. The data set we used covers a period of 11 months, from April 1994 to March 1995. There have been some earlier studies investigating the ionospheric range error correction accuracies predicted by the Broadcast model based on limited sets of coefficients covering different solar activity levels [e.g., Newby, 1992; Coco et al., 1990]. In this research, we investigated the long-term stability of the model in describing the diurnal and seasonal variations of the mid-latitude ionosphere at a level of low solar activity.

We computed hourly vertical ionospheric range error corrections from the Broadcast model for those days when a new set of model coefficients were transmitted and received. This represents a discontinuous time series covering 53 days. The model was evaluated for the Boulder, CO region from where we also have the Faraday rotation data for comparison purposes. In Figure 1, the predicted diurnal variation of the vertical ionospheric range error corrections are plotted. Each successive curve represents a new set of Broadcast model coefficients describing the diurnal variation of the ionosphere. A distinct seasonal dependence with peaks around February-March and September-October is evident.

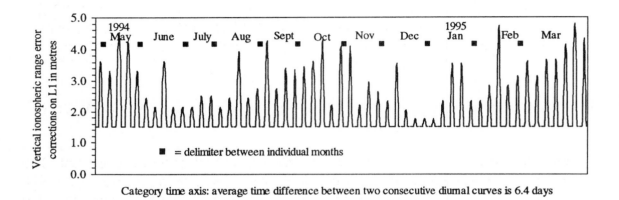

Category time axis: average time difference between two consecutive diurnal curves is 6.4 days

Fig. 1. Vertical ionospheric range error corrections on L1 predicted by GPS single-frequency Broadcast model.

The selection of a particular set of Broadcast model coefficients to be sent in the navigation message is based on the day of year and the running average of the observed solar flux numbers for the previous five days. One year is divided into 37 ten-day periods. Each period is represented by ten different solar activity levels. Based on our sample data set, the average update rate of the coefficients turned out to be 6.4 days with a range of 1 to 10 days. The diurnal curves with larger amplitudes correspond to solar activity levels when the five-day running average of the observed solar flux numbers exceeded a predetermined flux level: for the period covered by our time series, the 75 solar flux unit level.

BEHAVIOUR OF THE INTERNATIONAL REFERENCE IONOSPHERE MODEL

The second model that our paper focuses on is IRI90. We have modified the original version of the model to suit our needs for computing total electron content (TEC) profiles for multiple epochs. Using the model, we computed the diurnal variation of the vertical ionospheric range error correction for those days for which we computed the corrections using the Broadcast model. In Figure 2, the predicted corrections are plotted.

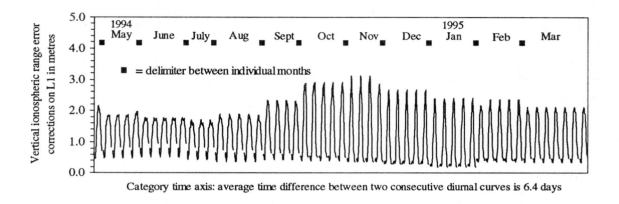

Fig. 2. Vertical ionospheric range error corrections on L1 predicted by the IRI90 model.

A distinct seasonal dependence with peaks around October-November is evident. The shape of the diurnal curves is quite different from season to season. It is interesting to note that the shape of the diurnal curves exhibits a double-peak from May to August. Not only do the day-time curves show a strong seasonal variation but also the night-time values are different from season to season. The IRI90 model uses F2 layer critical frequencies (Fof2) as well as F2 layer critical heights (hmF2) to compute vertical TEC profiles. These coefficients are stored as monthly median sets; this explains the monthly variations seen in Figure 2. The input parameters of the IRI90 model are the day of the year and the 12-month smoothed sunspot numbers. In the case of the IRI90 model predictions, there are no large day to day variations in diurnal amplitudes indicating that short-term changes in the solar activity level are not represented. Clearly, this is a consequence of using 12-month smoothed sunspot numbers that average out the short-term variations in solar activity.

COMPARISON OF PREDICTED AND MEASURED IONOSPHERIC RANGE ERROR CORRECTIONS

We decided to compare the predictions of the Broadcast and IRI90 models with Faraday rotation data in an effort to see which model is more accurate. The Faraday rotation data was provided by the Solar Terrestrial Physics Division of the U.S. National Oceanic and

Atmospheric Administration's National Geophysical Data Center. Due to instrumental malfunction, we were able to use only 27 day's worth of Faraday rotation data to assess the accuracy of the two models.

In Figure 3, a sample of predicted and measured vertical ionospheric range error corrections can be seen for 5 days in April/May 1994.

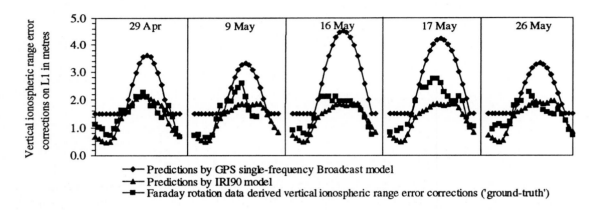

Fig. 3. Predicted versus measured vertical ionospheric range error corrections on L1.

We calculated the r.m.s. of the day-time and night-time differences of the vertical ionospheric range error corrections computed from the Faraday rotation data and IRI90/Broadcast models. Figure 4 shows that the *day-time* IRI90 model predictions are more accurate than those of the Broadcast model in 20 cases out of 27. Figure 5 reveals that the *night-time* IRI90 model predictions are more accurate than those of the Broadcast model in 21 cases out of 27.

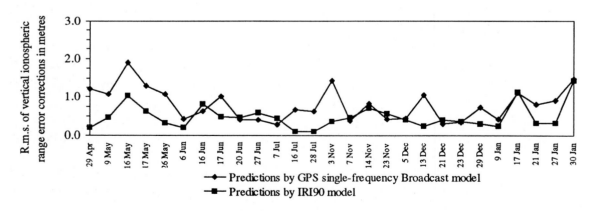

Fig. 4. R.m.s. of day-time differences between vertical ionospheric range error corrections computed from Faraday rotation data and the two empirical models.

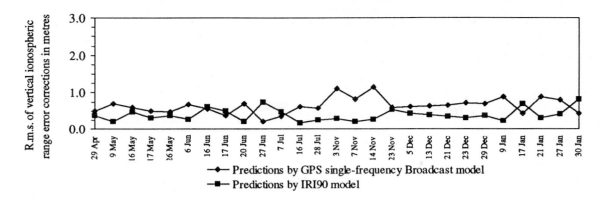

Fig. 5. R.m.s. of night-time differences between vertical ionospheric range error corrections computed from Faraday rotation data and the two empirical models.

CONCLUSIONS

Based on the comparison between the Broadcast and IRI90 models, we can conclude that both for day-time and night-time periods the IRI90 model appears to be more accurate than the Broadcast model. This conclusion is specific to low solar activity, and mid-latitude conditions. This investigation needs to be extended for medium and high solar activity periods and for low and high latitude regions of the earth. We plan on including other ionospheric models in our future investigations. We also wish to test model accuracies and efficiencies in processing static and kinematic GPS data.

REFERENCES

Bilitza, D. (ed.) (1990). *International Reference Ionosphere 1990*. National Space Science Center/World Data Center A for Rockets and Satellites, Lanham, MD. Report Number NSSDC/WDC-A-R&S 90-22.

Coco, D.S., C. Coker, and J.R. Clynch (1990). Mitigation of ionospheric effects for single-frequency GPS users. *ION-GPS-90, Proceedings of the 3rd International Technical Meeting of the Satellite Division*, The Institute of Navigation, Colorado Springs, CO, 19-21 Sept., pp. 169-174.

Klobuchar, J.A. (1986). Design and characteristics of the GPS ionospheric time delay algorithm for single-frequency users. *Proceedings of the PLANS-86 conference*, Las Vegas, NV, 4-7 Nov., pp. 280-286.

Komjathy, A., R.B. Langley, and F. Vejražka (1995). A Comparison of Predicted and Measured Ionospheric Range Error Corrections. *EOS Transactions of the American Geophysical Union*, 76(17), Spring Meeting Supplement, S87.

Newby, S.P. (1992). *An Assessment of Empirical Models for the Prediction of the Transionospheric Propagation Delay of Radio Signals*. M.Sc.E. thesis, Department of Surveying Engineering Technical Report No. 160, University of New Brunswick, Fredericton, NB, Canada, 212 pp.

Deriving Ionospheric TEC from GPS Observations

Steven Musman
NOAA Geosciences Laboratory and National Geophysical Data Center
Boulder, Colorado 80303 USA

INTRODUCTION

Global Positioning Satellites (GPS) broadcast at two different frequencies. These are usually used to correct positional measurements for ionospheric effects. Alternately information can be obtained about the ionosphere. A method for deriving total electron content (TEC) will be described. Observations taken in Boulder during January 1995 will be utilized and compared with Faraday rotation Both carrier phase and group delay will be used

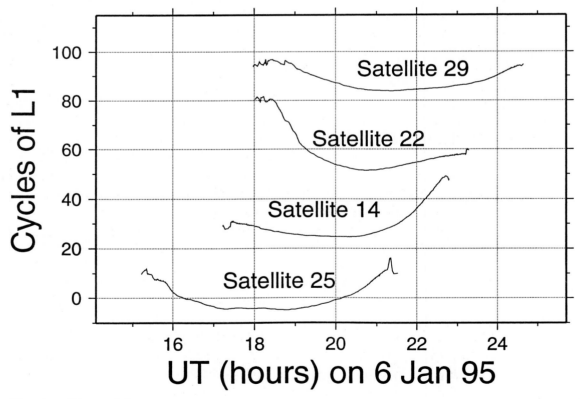

Figure 1 Phase delay

CARRIER PHASE

The phase delay observed by four different satellites near local midday on January 6, 1995 is presented in Figure 1. This was observed at the Table Mountain Gravity Observatory north of Boulder operated by the NOAA Geosciences Laboratory. An empirical model ionosphere, similar to that described in Musman et al. (1990), was constructed to fit these phase delays. This contained twelve parameters. Eight of the twelve parameters represent the time change of the zenith delay of the ionosphere: one is a constant, six represents three fourier components (twenty four, twelve, and eight hour periods) and one describes the north-south variation of the ionosphere. Note how satellites 14 and 22 are asymmetrical. Satellite 14 has more delay late in the pass, while satellite 22 has more delay early. This is a consequence of the fact that these two satellites are moving nearly north-south, in opposite directions, and the content of the ionosphere at mid latitudes generally increases toward the equator. The four other parameters are offsets for each of the four satellite passes. The parameters are determined by a weighted least squares solution which minimizes the difference between the observed and predicted delays. The weighting function is the cosine of the zenith angle. The modeled delays are

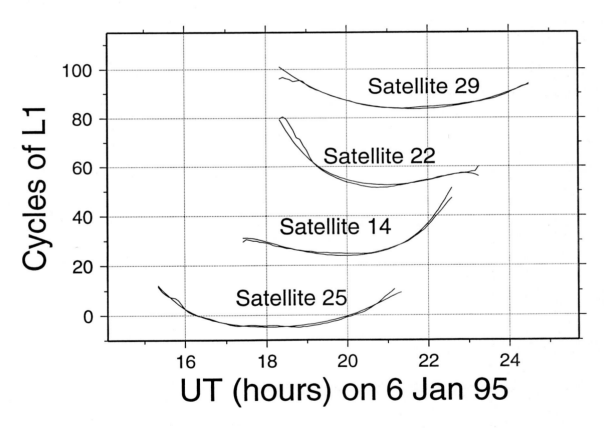

Figure 2 Modeled and observed delay

259

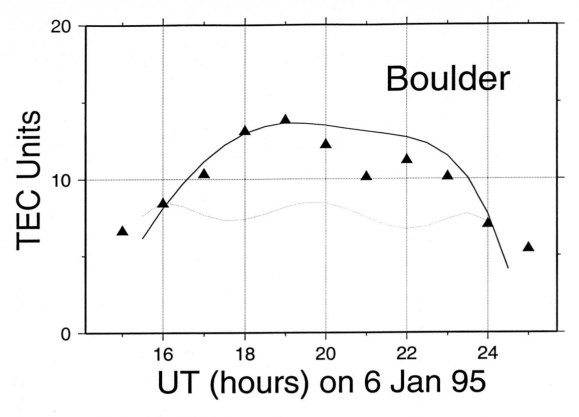

Figure 3 Modeled TEC and Faraday rotation

shown in Figure 2. The smooth curve is the modeled delay. For convenience only one observation per five minutes was used rather than all of the thirty second sampeled data.

Figure 3 presents the zenith delay. TEC units of 10^{16} electrons per meter^2 are used. The scaling factor is one cycle of L1 equals 1.17 TEC unit..The heavy line is the model. Faraday rotation measurements are indicated by the triangles. These values contain corrections by Kenneth Davies and were reported earlier by Conkright et al.(1995).

GROUP DELAY

The group delay is displayed in Figure 4 for satellite 25. This was obtained by taking the difference between the two pseudoranges and dividing by the wave length of L1 Here the sign convention is chosen so that group and phase delay can be compared directly. The smoothed group delay is obtained by using a 35 point 20% trim filter. Thirty five consecutive observations are arranged in order of amplitude. The smallest and largest 20% are eliminated and an average is taken of the remaining values which is assigned to the average time of the thirty five values. This is done for all possible sets of thirty five consecutive values. Note that Lanyi and Roth (1989) used a directional antenna in their measurements of group delay. The omnidirectional antennas used now have the advantage of not needing to track satellites, but measurements of group delay are far noisier. Another model ionosphere was constructed with the same four satellite passes shown in Figure 1 but using group delays instead of phase delays. The results are shown as the narrow line in Figure 3

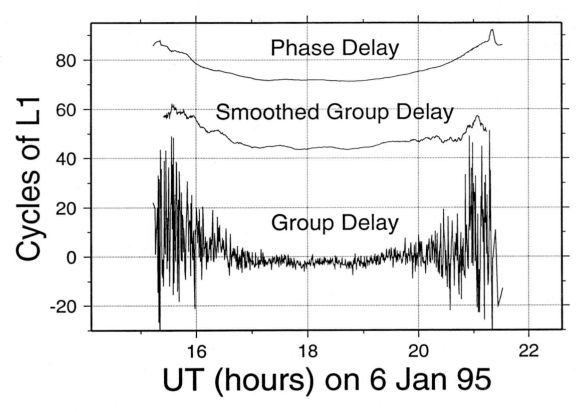

Figure 4 Comparison of phase and group delay for satellite 25

DISCUSSION AND SUMMARY

Similar models were constructed for phase delay for 7, 8,and 9 June, 1995, as well as a group delay model for the June 7. Also a model was obtained for June 6th using GPS observations taken at Platteville, Colorado about forty kilometers east of Table Mountain. The results were similar to those shown in figure 3. For this set of data the phase delay models seem better than the group delay models. Group delay does not appear to be nearly identical with phase delay with added high frequency noise. There are systematic differences when filtered group delay is compared with phase delay. Variation in group delay is generally shallower, leading to models with lower values of TEC. In addition the offsets in group delay models should be properties of the satellite-receiver system alone and independent of the ionosphere. Differences in the offsets between the two group delay models on consecutive days varied from two to four cycles of L1 and are all of the same sign.

REFERENCES

Conkright, R. O., K. Davies, and S. Musman, 1995, Comparison of ionspheric total electron contents made at Boulder, CO from GPS and GOES 2, EOS, 76, S86.

Lanyi, G. E. and T. Roth, A comparison of mapped and measured total electron content using global positioning system and beacon satellite observations, 1988, Radio Science, 23,483-492.

Musman S., A. Drew, and B. Douglas, Ionospheric effects on Geosat altimeter observations, 1990, J. of Geophys. Res. ,95, 2965-2967.

MITIGATION OF TROPOSPHERIC EFFECTS IN LOCAL AND REGIONAL GPS NETWORKS

A. Geiger, H. Hirter, M. Cocard, B. Bürki
Institute of Geodesy and Photogrammetry, ETH Zurich, Switzerland
A. Wiget, U. Wild, D. Schneider
Federal Office of Topography, Berne, Switzerland
M. Rothacher, S. Schaer, G. Beutler
Astronomical Institute, University of Berne, Switzerland

ABSTRACT

Although modern technologies, especially space borne techniques, provoked a revolutionary advance in geodetic sciences, atmospheric refraction still remains a persistent problem. Modelling the atmospheric effects and estimation of tropospheric parameters are two methods to reduce the tropospheric biases. In this paper the development and testing of both methods, separately and in combination, are presented for applications in local and regional GPS-networks. The atmospheric state variables are modelled in four dimensions (space and time) based on operational ground measurements. Parameter estimation and collocation techniques are applied for modelling. By integration along the actual microwave paths the corresponding correction values are obtained. Compared to standard corrections the modelling method considerably reduces the rms of the GPS solution. An additional estimation of zenith path delays further reduces the rms. In a local network with considerable height differences (Swiss testnet ´Turtmann´) the repeatability in height is clearly enhanced (up to a factor of 2) by introducing modelled path delays. For high precision applications the combination of modelled path delays and the estimation of tropospheric parameters is very promising.

INTRODUCTION

In many sciences the atmospheric, or more precisely the tropospheric behaviour plays an important role. In meteorology it may be one of the primary research areas. In geodesy, however, it causes major inconveniences for precise positioning. According to the wide and divergent interests in the troposphere, a great number of different approaches for modelling and mitigation of tropospheric effects on measurements have been developed. For microwave propagation the primary focus of interest is the refractivity field which can be calculated from three tropospheric state variables only: temperature T, pressure p, and humidity e. An introduction to refraction problems in satellite geodesy and a quite comprehensive bibliography on this subject is given by Langley (1995a,b). Many papers documenting advances in tropospheric corrections are found in De Munck and Spoelstra (1992).

MODELLING

The aim of this modelling is to construct a four dimensional refractivity field allowing the integration of the delay along any ray path at any epoch desired. The method used to reach this aim is based on the estimation of a functional part of the three 4D-fields T, p, and e, respectively, and a stochastic part describing the departure from the functional approach. The stochastic part is interpolated by optimizing the white noise residuals and the correlated signals simultaneously by least square conditions. This interpolation technique is mostly referred to as collocation or kriging. By introducing a covariance function which is not only dependent on spatial coordinate differences but also on temporal differences a correlation of the signal in space as well as in time is obtained, thus allowing for a four-dimensional interpolation. The functional part is represented by slowly varying standard tropospheric functions. The few necessary deterministic parameters of the functional model are estimated along with the signal optimization. The software COMEDIE (**CO**llocation of **ME**teorological **D**ata for **I**nterpretation and **E**stimation of tropospheric path delays) (Eckert et al., 1992, Cocard et al. 1992) based on the described modelling technique may deal with any kind of meteorological measurements such as balloon soundings or airborne measurements. There is no need of synoptic measurements, unevenly distributed data in space as well as in time are allowed.

For this work, however, ground measurements and one balloon sounding site, only, have been taken into account. The 72 stations used are distributed over the whole area of Switzerland and belong to the automatic meteo network (ANETZ) operated by the Swiss Meteorological Institute (SMI) at a 10 or 60 minutes sampling rate. The balloon soundings at Payerne are launched every 12 hours.

An example of the 4-D interpolation is given in Fig. 1. The 'drying and high temperature' effect to the north of the Alps due to the prevailing 'Föhn' situation can clearly be recognized.

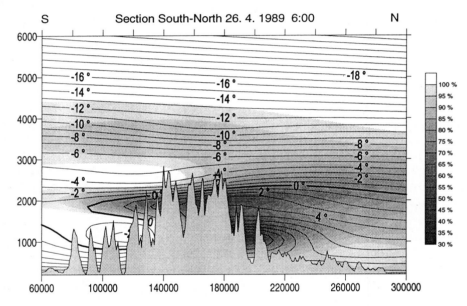

Fig. 1: Section from south to north with exagerated topography. Horizontal distance and height in meters. Temperature (isolines) and humidity (hachure) are depicted.

ESTIMATION

In the previous section estimation was understood as dealing with meteorological data only. In this section estimation of tropospheric biases means the estimation of significant parameters reflecting the tropospheric effect by using the GPS measurements only. More specifically said in most cases this refers to the estimation of a (differential) zenith path delay. Normally this zenith path delay is a departure amount from a deterministic value e.g. given by the Saastamoinen model or others. Alternatively, for the elimination of an a priori zenith path delay the modelling method described in the previous section can be used as well as a combination of hydrostatic component reduction together with water vapour radiometer (WVR) measurements. All these different possibilities of refraction correction have been implemented in the Bernese software.

It has to be pointed out, that the estimation of zenith path delays suffers from two critical elements, one being the mapping function the other being the cut off angle. The effect of a wide varity of mapping functions has been comparetively tested by Santerre et al. (1995). Down to 15 to 20 degrees cut off angle the different functions seem to perform adequately. Major discrepancies will appear below these values. This fact could lead to the conclusion that observations made at lower elevation angles should be discarded. This, however, collides with the second critical point for estimating zenith path delays. For highest possible decorrelation of height components and tropospheric parameter measurements in the low elevation range should be taken into account. The antenna phase center variation could also correlate with the height and the tropospheric parameter.

It has been pointed out (Rothacher et al.) that even in small scale networks the estimation of an exzessive zenith path delay is not sufficient for very high precision height determination if considerable height differences between stations are encountered. For these cases an estimation of a height dependent zenith path delay profile has been proposed (Rothacher et al.) by introducing a polynomial profile for the height range covered by the network. It is easy to show that the linear term corresponds to an elimination of a meteorologically induced false scale of station heights (Geiger, 1987). Obviously this method holds only for networks with well distributed stations in height and for network dimensions where the meteorological regime may be considered identical (no horizontal gradients).

RESULTS

A joint 3-days campaign of the authors institutions has been carried out in the test network 'Turtmann' in the canton of Valais, Switzerland. During the 24-hours GPS measurements two WVR equipments of the GGL-ETHZ (Bürki et al. 1993) have been operated in order to test the possibilities of radiometric measurements in small scale networks. Special emphasis has been put on the baseline with the largest (well controlled) height difference: Susten-Jeizinen, approx. 900 m, baseline length approx. 6.8 km. All GPS data treatment has been performed with the Bernese GPS Software V.3.5, where the different correction modi have been implemented. In a first step the effects of different a priori models in combination with and without parameter estimation are investigated. COMEDIE refers to the zenith path delay modelled from data of the automatic meteo network (ANETZ) and balloon soundings. At the same time two different mapping functions are evaluated.

Table 1: Comparison of rms (mm) of single differences on a baseline (Susten-Jeizinen) in the test network 'Turtmann'

A priori model	Estimation	Mapping		
		implicit	1 / cos z	Herring
Saastamoinen	n/a	7.4	-	-
	1 par	7.2	-	-
	1 par/6 h	6.8	-	-
	1 par/1 h	6.6	-	-
WVR	n/a	-	7.5	7.2
	1 par	-	7.0	7.0
	1 par/6 h	-	6.9	6.9
COMEDIE	n/a	-	7.4	7.2
	1 par	-	7.2	7.1
	1 par/6 h	-	6.9	6.8

From the table above it is interesting to see that at the level of single difference rms no significant variation appears. However, the height component may be affected by e.g. the choice of the mapping function and the estimation modus. In table 2 this effect is shown for the a priori model given by the WVR and hydrostatic correction. The differences in height are in the order of several mm.

Table 2: Impact of the mapping function on the height component (mm) on a baseline (Susten-Jeizinen, ca. 900 m height diff.) in the test network 'Turtmann'. Differences in mm. between heights calculated with 1 / cos z-mapping versus Herring-formula.

A priori model	Estimation model		
	n/a	1 par	1 par/6 h
WVR	8.2	8.2	6.6

The second step considers the effect of the cut off angle. Again, from the rms at single difference level (table 3) it is nearly impossible to conclude on the performance of the different models.

Table 3: Impact of the cut off angle in combination with different tropospheric models. Comparison of rms (mm) of single differences of 3 days measurements in the test network 'Turtmann' (all baselines)

Tropospheric correction model	15 ° Elev.	20° Elev.
Saastamoinen	7.5	6.4
Station specific estim. 1 par/6h	6.4	5.7
Station specific estim. 1 par/1h	6.1	5.5
Height dependent diff. pathdelay estim 2. polynomial	6.5	5.8
COMEDIE (no additional estim)	7.1	6.0
COMEDIE with par estim. 1 par/station	6.4	5.7

The rms of a comparison of the height component itself with an optimal GPS 4-days solution of the year before reveals considerable differences in performance (table 4).

Table 4: Impact of the cut off angle in combination with different tropospheric models. Comparison of rms (mm) of height component after a Helmert transformation onto the optimal GPS solution 1993 of the test network 'Turtmann'

Tropospheric correction model	15 ° Elev.	20° Elev.
Saastamoinen	29.9	26.3
Station specific estim. 1 par/6h	3.5	4.9
Station specific estim. 1 par/1h	3.9	5.3
Height dependent diff. pathdelay estim 2. polynomial	2.8	3.6
COMEDIE (no additional estim)	8.2	6.1
COMEDIE with par estim. 1 par/station	6.1	5.6

In the special case of the very small scale network of 'Turmann' with its favorable height distribution of stations the estimation of a 2. order polynomial per session for the zenith path delay seems to perform best. The fact that in three cases of parameter estimation the rms drops by using 15 degrees cut off instead of 20 degrees supports strongly the statement of the correlation of the height component with the tropospheric parameter.

CONCLUSIONS

It is clearly seen that the proposed modelling delivers results superior to the standard corrections. However, there is no significant amelioration compared to the estimating procedure. This fact may be due to the time span of observation which plays a considerable role in estimating tropospheric parameters. Almost permanent measurements have been treated in this example. Introducing 4-D-modelled path delays should improve short time span height determination. In this special case of 'Turmann'-network with its favorable height distribution of stations the estimation of 2. order polynomials for the zenith path delay seems to perform best. The partial drop of the rms by using 15 degrees cut off instead of 20 degrees strongly supports the statement of the correlation of the height component with the tropospheric parameter. In small networks with considerable height differences a special treatment of the tropospheric refraction is mandatory. For regional size networks this might be even more important, but has to be further investigated.

Acknowledgement. The data of the Swiss automatic meteorological network (ANETZ) needed for this study was provided by the Swiss Meteorological Institute (SMI). Research on modelling of tropospheric path delays at ETH Zurich is sponsored by the Federal Office of Topography.

REFERENCES

Beutler, G., I. Bauersima, W. Gurtner, M. Rothacher, T. Schildknecht, A. Geiger (1988): Atmospheric refraction and other important biases in GPS carrier phase observations. In: Atmospheric Effects on Geodetic Space Measurements, Monograph 12, School of Surveying, University of New South Wales, Kensington, N.S.W., Australia, pp. 15-43.

Bürki, B., Geiger, B., Kahle, H.-G., Elgered, G., Gyger, R., R. Peter (1993): ETH- and Onsala water vapor radiometers: Co-Location and comparative results. Annales Geophysicae. Suppl. 1 to Vol. 11. Abstract, p. C112.

Cocard, M., Eckert, V., Geiger, A., Bürki, B., Neininger, B. (1992): 3D modelling of atmospheric parameters for automatic path delay corrections. MUNCK, J. and T. SPOELSTRA (Eds.). Proceedings Symposium Refraction of transatmospheric signals in Geodesy Netherlands Geodetic Commiss. New Series, No. 36: 175-178.

De Munck, J.C., T.A.Th. Spoelstra (eds.) (1992): Proceedings of the Symposium on Refraction of Transatmospheric Signals in Geodesy, The Hague, The Netherlands, 19-22 May, Netherlands. Geodetic Commission, Publications on Geodesy, Delft, The Netherlands, No. 36, New Series.

Eckert, V., Cocard, M., Geiger, A. (1992): COMEDIE (Collocation of Meteorological Data for Interpretation and Estimation of Tropospheric Path delays). Teil I: Konzept und Teil II: Resultate, IGP-Bericht Nr. 194, April 1992.

Geiger, A. (1987): Einfluss richtungsabhängiger Fehler bei Satellitenmessungen. Institut für Geodäsie und Photogrammetrie, Bericht Nr. 125. p 50.

Geiger, A., M. Cocard, H. Hirter (1995): Dreidimensionale Modellierung des Refraktivitätsfeldes in der Atmosphäre. Vermessung, Photogrammetrie, Kulturtechnik, VPK 4/95, pp. 254-260.

Langley, R.B., Wells, W., V.B. Mendes (1995a): Tropospheric Propagation Delay: A Bibliography. 2nd edition. March (unpublished).

Langley, R.B. (1995b): Signals and signal propagation. In: International School: GPS for Geodesy, Delft, The Netherlands, March 26-April 1, 1995.

Rothacher, M., Beutler, G., Wild, U., Schneider, D., Wiget, A., Geiger, A., Kahle, H.-G. (1991): The role of the atmosphere in small GPS networks. Proceedings 2nd Int. Symp. Precise Positioning with GPS. Ottawa, Canada: 581-598, Springer Vlg., N.Y.

Santerre, R., I. Forgues, V.B. Mendes, R.B. Langley (1995): Comparison of Tropospheric Mapping Functions: Their Effects on Station Coordinates. IUGG XXI General Assembly, Boulder, Colorado.

GPS AS A LOCATION TOOL FOR ELECTROMAGNETIC SURVEYS

Arlie C. Huffman III
Radian Corporation
2990 Center Green Ct. S., Boulder, Colorado 80301 USA

James A. Kalinec
Radian Corporation
8501 N. Mopac Blvd., Austin, Texas 78759 USA

J. Douglas Maiden
Radian Corporation
8550 United Plaza Blvd., Baton Rouge, Louisiana 70809 USA

INTRODUCTION

Under the best conditions, running and locating an electromagnetic (EM) survey using a Geonics EM-31™ is a straightforward task. With only a few actual surveyed points, the survey grid is located precisely enough for nearly any purpose. With the complications of a Louisiana forest with dense undergrowth, the survey becomes a challenge. The cost of the project becomes prohibitive when accurate instrument location along crooked survey lines is required, if traditional surveying techniques are used. However, the use of a global positioning system (GPS) with differential capabilities nearly automates the task of acquiring highly accurate location data and EM readings. Using a combination of GPS and EM-31, over ten miles of data were collected during a 10 day period in August 1994. One GPS receiver was kept with the electromagnetic survey team, while a second was stationary, and used as a base station. The small, light GPS receiver unit allowed the operator to move through the forest relatively easily, facilitating the collection of large amounts of data in a relatively short time. As sub-meter position data were available daily, contour maps of the day's field work were available for instant analysis.

PURPOSE OF THE SURVEY

An electromagnetic survey was required to locate the edges of a landfill that had been grown over by a forest. Vegetation on the site consisted of clusters of large trees, thick undergrowth of bushes and briars, and tree roots and fallen trees on the forest floor. Since keeping the survey lines straight was nearly impossible due to these surface conditions, a differential GPS system was used along with the EM-31.

FIELD SETUP OF THE GPS

A permanent base station was located at a building about 5 kilometers from the survey site. The proximity of the base station to the area of investigation allowed for location errors well within the 3.0 - 4.0 meter uncertainty associated with the EM-31. A GPS rover unit stayed with the survey team at all times. The EM-31 and GPS operators were able to collect data as quickly as they could move through the forest, with the assistance of someone to clear the undergrowth with a machete.

Each night, the EM and location data are downloaded into a computer. A short algorithm was written to associate each EM data point with a measured position. The data are then gridded and contour plots are made. Instead of having a plot with estimated data point locations, as are created with traditional survey techniques, the contour plots show actual locations of the data. This is very useful in determining what areas should be investigated more closely, and in showing places with insufficient data point density.

RESULTS OF THE SURVEY

Figures 1 and 2 show the colored contour plots that were made from the collected data. In Figure 1, the 150 meter by 350 meter area was surveyed in approximately 4 hours without the need to set up any grid system. The survey shows two areas of disturbed soil which correlate to major slope breaks at the site. This helped to confirm the existence of landfill materials at this site.

Figure 2 is a 300 meter by 100 meter area that also was suspected of being an old landfill. The EM results show a large area of probable fill, which corresponds to other historical information available about the site.

CONCLUSIONS

By using a GPS receiver instead of the usual traditional surveying techniques, the time required to perform the surveys was cut by more than half, and the associated surveying costs were eliminated. This allows for the data to be plotted and interpreted the same day as it is taken. An area that was missed or requires further investigation is discovered while the geophysical crew is still in the field, not one or two weeks later after the EM grid has been surveyed.

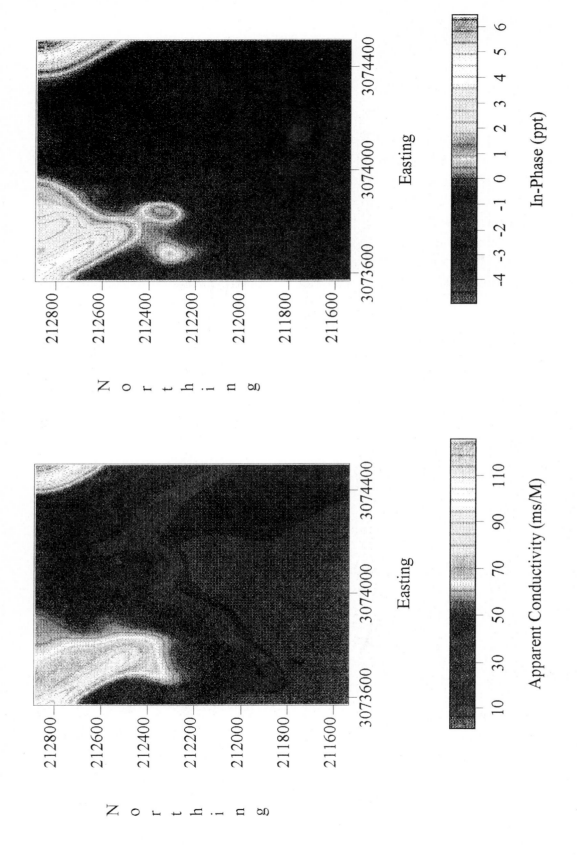

Figure 1. (a) Contour plot of EM-31 conductivity data collected with GPS. (b) Contour plot of EM-31 inphase data. Bright colors correspond to construction landfill debris filled areas.

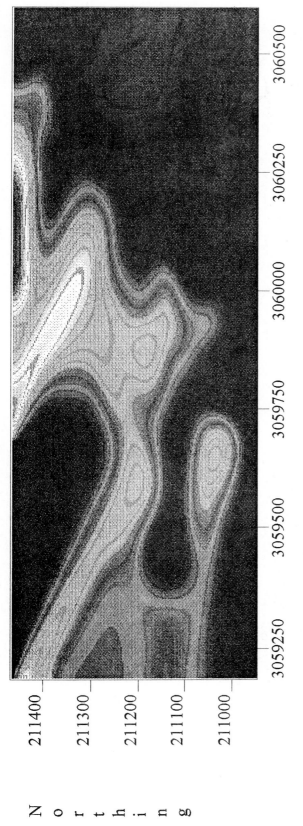

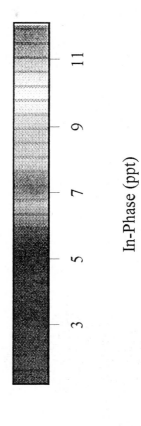

Figure 2. Contour plot of EM-31 inphase data. The bright area defines the probable extent of the landfill.

Chapter 5

Theory and Methodology

SIZE AND SHAPE OF L₁/L₂ AMBIGUITY SEARCH SPACE

P.J.G. Teunissen
Delft Geodetic Computing Centre (LGR)
Department of Geodetic Engineering
Delft University of Technology
Tel: (0)15-783546 Fax: (0)15-783711
E-mail: lgr@geo.tudelft.nl
The Netherlands

ABSTRACT

The size and shape of the ambiguity search space are instrumental for estimation and validation of the integer GPS double difference ambiguities. A qualitative analysis of the geometry of the ambiguity search space is given for the case the receiver-satellite geometry is dispensed with.

1. INTRODUCTION

In view of the ongoing technical developments for improving the precision of the code measurements, it is still of interest to consider the case of integer ambiguity estimation where the code measurements are used directly in combination with the phase measurements. In this contribution, we will therefore consider the integer ambiguity estimation problem for the case the relative receiver-satellite geometry is dispensed with. For this particular case an analytical derivation of the results can be given. This has the advantage, that a deeper insight into the integer ambiguity estimation process can be obtained. Some of the analytical results that can be derived, will be presented in this contribution. First it will be shown how the precision of the ambiguities depends on the area and elongation of the ambiguity search space, and on the ambiguity correlation coefficient. Then the size and shape of the ambiguity search space will be characterized. This will be done for the case that the ionosphere is assumed to be absent, as well as for the case that the ionosphere is assumed to be present. The size and shape of the ambiguity search space will be described analytically through its area and elongation, and through the correlation coefficient. Finally the statistics of the decorrelated ambiguities are given. The decorrelated ambiguities are obtained automatically by means of the LAMBDA-method. Examples of the decorrelating ambiguity transformations are also given.

2. AMBIGUITY SEARCH SPACE

The geometry of the ambiguity search space is instrumental for both estimation and validation. The ambiguity search space in R^2,

$$(N - \hat{N})^* Q_N^{-1}(N - \hat{N}) \leq \chi^2 \tag{1}$$

is an elliptic region, centred at $\hat{N}$, with its orientation governed by Q_N and its size controlled by χ^2. The real valued least-squares estimate $\hat{N}$ and variance-covariance matrix Q_N follow from adjusting the quartet of double differenced time series

$$
\begin{aligned}
\Phi_1(i) &= \rho(i) - (\lambda_1/\lambda_2)\, I(i) + \lambda_1 N_1 \\
\Phi_2(i) &= \rho(i) - (\lambda_2/\lambda_1)\, I(i) + \lambda_2 N_2 \\
P_1(i) &= \rho(i) + (\lambda_1/\lambda_2)\, I(i) \\
P_2(i) &= \rho(i) + (\lambda_2/\lambda_1)\, I(i)
\end{aligned}
\tag{2}
$$

The double-difference phase measurements on L_1 and L_2 at time epoch i, are denoted as $\Phi_1(i)$ and $\Phi_2(i)$, and their corresponding code measurements as $P_1(i)$ and $P_2(i)$. The wave-lengths are denoted as λ_1 ($\cong 19$ cm), λ_2 ($\cong 24$ cm), and the corresponding integer ambiguities as N_1 and N_2. The ambiguities are assumed to be constant in time. The ionospheric delay term at epoch i, is denoted as $I(i)$. The arithmetic and geometric mean of the ambiguity variances, σ_1^2 and σ_2^2, can be expressed as

$$
\begin{aligned}
(\sigma_1^2 + \sigma_2^2)/2 &= \frac{A}{2\pi\chi^2}(e + e^{-1}) \\[2ex]
(\sigma_1^2 \sigma_2^2)^{1/2} &= \frac{A}{\pi\chi^2}(1 - \rho^2)^{-1/2}
\end{aligned}
\tag{3}
$$

where A and e are respectively the area and elongation of the ambiguity search space, and ρ is the ambiguity correlation coefficient. The area A can be used as an indicative measure for the number of grid points that are located in the ambiguity search space. This measure and its higher dimensional version, the volume of the ambiguity search space, was introduced and discussed in [Teunissen, 1993, 1994a]. Graphical results, that confirm the relevance of the volume (area) as an a priori reliability measure, are given in [de Jonge and Tiberius, 1995]. The elongation, being defined as the square root of the condition number of Q_N, is a measure of shape, and the correlation coefficient is a measure for the statistical dependency of the two ambiguities. The above two equations show how the precision of the ambiguities depends on the area A, the elongation e and correlation coefficient ρ. Since admissible ambiguity transformations leave the area A invariant [Teunissen 1994b], changes in only e and ρ will allow us to change the precision of the ambiguities.

3. SIZE AND SHAPE OF SEARCH SPACE

Based on the assumption that the variance ratio $(\sigma_\Phi/\sigma_P)^2$ is small (σ_Φ, σ_P: standard deviations of undifferenced phase and code measurements), it can be shown that

$$e(I \neq 0) \cong 65 \quad ; \quad e(I = 0) \cong \tfrac{1}{2}(\lambda_1 / \lambda_2 + \lambda_2 / \lambda_1) \frac{\sigma_p}{\sigma_\Phi}$$

$$\rho(I \neq 0) \cong 1 \quad ; \quad \rho(I = 0) \cong 1 \tag{4}$$

"$I \neq 0$" refers to ionosphere present, "$I = 0$" refers to ionosphere absent. This shows that in both cases the ambiguity search space is very elongated and the ambiguities highly correlated. Only in case the ionosphere is absent, the elongation can be improved by reducing σ_p / σ_Φ. For the area of the ambiguity search space, it follows that

$$A(I \neq 0) \cong \frac{4\pi\chi^2}{\lambda_1 \lambda_2 k} \sigma_P^2 \quad ; \quad A(I = 0) \cong \frac{4\pi\chi^2}{\lambda_1 \lambda_2 k} \sigma_P \sigma_\Phi \tag{5}$$

where k denotes the number of epochs used. This result clearly shows the beneficial role played by the phase measurements in case the ionosphere is assumed absent. Typical values for $A(I \neq 0)$ and $A(I = 0)$ are given in the two tables below ($\chi^2 = 10$, $\sigma_\Phi = 0.3$ cm).

Table 1. Area of ambiguity search space in case the ionospheric delay is assumed present.

$A(I \neq 0)$	$\sigma_P = 60\,\mathrm{cm}$	$\sigma_P = 30\,\mathrm{cm}$	$\sigma_P = 10\,\mathrm{cm}$
$k = 1$	973.8	243.4	27
$k = 100$	9.7	2.4	0.3
$k = 1000$	1	0.2	0.03

Table 2. Area of ambiguity search space in case the ionospheric delay is assumed absent.

$A(I = 0)$	$\sigma_P = 60\,\mathrm{cm}$	$\sigma_P = 30\,\mathrm{cm}$	$\sigma_P = 10\,\mathrm{cm}$
$k = 1$	4.87	2.43	0.81
$k = 5$	0.97	0.49	0.16
$k = 10$	0.49	0.24	0.08

4.   STATISTICS OF THE DECORRELATED AMBIGUITIES

With the LAMBDA-method the original ambiguities are transformed to new ambiguities that are more precise and less correlated. The results in terms of precision, elongation and correlation for both the original and transformed ambiguities are shown in the following table for the case the ionosphere is assumed present ($\sigma_\Phi = 0.3$ cm and $k = 1$). The indices "N" and "z" denote respectively the original and transformed ambiguities.

Table 3. Precision (σ), elongation (e) and correlation (ρ) of original and transformed ambiguities, for the case the ionosphere is present.

$I \neq 0$	σ_{N1}	σ_{N2}	σ_{z1}	σ_{z2}
$\sigma_p = 60\,\text{cm}$	32.2	32.1	31.2	0.99
$\sigma_p = 30\,\text{cm}$	16.2	16.0	15.6	0.50
$\sigma_p = 10\,\text{cm}$	5.4	5.3	5.2	0.17
	e_N	e_z	ρ_N	ρ_z
$\sigma_p = 60\,\text{cm}$	67	31	.9995	.0135
$\sigma_p = 30\,\text{cm}$	67	31	.9995	.0123
$\sigma_p = 10\,\text{cm}$	67	31	.9995	.0002

The two transformed ambiguities, z_1 and z_2 are almost statistically independent. The ambiguity transformation also reduces e. This change however, is not as spectacular as the change in ρ. The consequence can be seen in the precision of the transformed ambiguities. Both transformed ambiguities have a better precision than the original L_1 and L_2 ambiguities, but only one of the two transformed ambiguities has a precision which is drastically better than the precision of the original ambiguities. The ambiguity transformations are automatically provided for by the LAMBDA-method. They vary for varying levels of code and phase precision, but they are independent of the number of samples used. For the case $\sigma_\Phi = 0.3\,\text{cm}$ and $\sigma_P = 60\,\text{cm}$ the decorrelating ambiguity transformation reads

$$z_1 = -7N_1 + 8N_2, \quad z_2 = 1N_1 - 1N_2 \tag{6}$$

For the case the ionosphere is assumed absent, the results are given in table 4. (with $\sigma_\Phi = 0.3\,\text{cm}$, and $k = 1$). Again a drastic decrease in ρ is achieved. Also e is reduced in size, and this change is now far more spectacular. With reference to the arithmetic and geometric mean of the ambiguity variances, the drastic decrease in both ρ and e therefore also results in a major improvement of ambiguity precision. Again the ambiguity transformations vary for varying levels of code and phase precision. For the case $\sigma_\Phi = 0.3\,\text{cm}$ and $\sigma_P = 60\,\text{cm}$ the optimal transformation is given as

$$z_1 = -7N_1 + 9N_2, \quad z_2 = -4N_1 + 5N_2 \tag{7}$$

Table 4. Precision (σ), elongation (e) and correlation (ρ) of original and transformed ambiguities, for the case the ionosphere is absent.

$I = 0$	σ_{N1}	σ_{N2}	σ_{z1}	σ_{z2}
$\sigma_p = 60\,\text{cm}$	4.46	3.46	0.32	0.50
$\sigma_p = 30\,\text{cm}$	2.23	1.74	0.29	0.29
$\sigma_p = 10\,\text{cm}$	0.74	0.58	0.16	0.17
	e_N	e_z	ρ_N	ρ_z
$\sigma_p = 60\,\text{cm}$	206	1.6	.9995	.18
$\sigma_p = 30\,\text{cm}$	103	1.6	.998	$-.42$
$\sigma_p = 10\,\text{cm}$	34	1.4	.98	.32

REFERENCES

Jonge, P.J. de, C.C.J.M. Tiberius (1995): Integer Ambiguity Estimation with the LAMBDA-Method. In: *Proceedings of Symposium "GPS Trends in Precise Terrestrial, Airborne, and Spaceborne Applications"*, XXI General Assembly of IUGG, July 2-14, 1995, Boulder, Colorado, USA, 5 p.

Teunissen, P.J.G. (1993): *Least-Squares Estimation of the Integer GPS Ambiguities*. Invited Lecture, Section IV Theory and Methodology, IAG General Meeting, August 1993, Beijing, China. Also in LGR-Series No.6, 16 p.

Teunissen, P.J.G. (1994a): A new Method for Fast Carrier Phase Ambiguity Estimation. *Proceedings IEEE Position Location and Navigation Symposium PLANS'94*, pp. 562-273

Teunissen, P.J.G. (1994b): *The Invertible GPS Ambiguity Transformations*. In: Publications of the Delft Geodetic Computing Centre, LGR-Series, No. 9.

Tiberius, C.C.J.M., P.J. de Jonge (1995): Fast Positioning using the LAMBDA-Method. In: *Proceedings DSNS95*, April 24-28, Bergen, Norway.

INTEGER AMBIGUITY ESTIMATION WITH THE LAMBDA METHOD

Paul de Jonge and Christian Tiberius[1]
Delft Geodetic Computing Centre (LGR), Faculty of Geodetic Engineering
Delft University of Technology, Delft, The Netherlands

INTRODUCTION

High precision relative GPS positioning is based on the very precise carrier phase measurements. In order to achieve high precision results within a short observation time span, the integer nature of the GPS double difference ambiguities has to be exploited. In this contribution we concentrate on the integer ambiguity estimation, which is one of the steps in the procedure for parameter estimation, see section 2 in [2].

The integer ambiguity estimation will be carried out with the LAMBDA method. LAMBDA stands for Least-squares AMBiguity Decorrelation Adjustment. After applying a decorrelating transformation, a sequential conditional adjustment is made upon the ambiguities. As a result, integer least-squares estimates for the ambiguities are obtained.

INTEGER MINIMIZATION

The second step of the estimation procedure [ibid] consists of

$$\min_a \|\hat{a} - a\|^2_{Q_{\hat{a}}^{-1}} \text{ with } a \in Z^n, \tag{1}$$

where $\hat{a}$ is the real valued least-squares estimate for the n-vector of double difference ambiguities, and $Q_{\hat{a}}$ the variance-covariance matrix. This minimization yields the integer least-squares estimate, $\check{a}$, for the vector of ambiguities.

For the computation of the integer least-squares estimate, we use the LAMBDA method, see [1] and [4]. The LAMBDA method consists of

1. the decorrelation of the ambiguities by a reparametrization Z^* of the original ambiguities a to new ambiguities: $z = Z^*a$, and

2. the actual ambiguity estimation

The efficiency of the method comes from the decorrelation step and has been explained in detail by analysis of the precision and correlation of the GPS double difference ambiguities in [3]-[5]. The actual integer minimization is done for the transformed ambiguities.

In this paper we will concentrate on the actual integer ambiguity estimation. We assume to have the vector with real valued estimates (for either original or transformed ambiguities) from the float solution, $\hat{a}$, and the variance-covariance matrix $Q_{\hat{a}}$. Decorrelating the

---

[1] Supported by the Lely foundation of Rijkswaterstaat

ambiguities is not a prerequisite for the integer ambiguity estimation, it is, however, largely beneficial for the efficiency of the search.

INTEGER AMBIGUITY ESTIMATION

We start with the LDL^* decomposition, see [6], of the inverse of the variance-covariance matrix $Q_{\hat{a}}^{-1}$ where L is a unit lower triangular matrix, and D a diagonal matrix, $D = diag(d_1, \ldots, d_n)$. It is easily constructed once the Cholesky decomposition $Q_{\hat{a}}^{-1} = GG^*$ is available.

As discussed in [1], no standard techniques are available for solving (1). A discrete search is employed instead. An ellipsoidal region in R^n is taken, on the basis of which a search is performed for the minimizer of (1).

$$(\hat{a} - a)^* Q_{\hat{a}}^{-1} (\hat{a} - a) \leq \chi^2 \tag{2}$$

(2) is referred to as (ambiguity) search ellipsoid. Integer vectors a that satisfy (2) are called candidates; they are on or inside the search ellipsoid. The candidate with the least distance to $\hat{a}$ is the integer least squares estimate $\check{a}$. For a discussion on the value for χ^2, the constant that controls the size of the ellipsoid, the reader is referred to [1] and the section on the volume of the ellipsoid in this paper.

With the LDL^*-decomposition of matrix $Q_{\hat{a}}^{-1}$, expanding (2) yields

$$\sum_{i=1}^{n} d_i \left[(a_i - \hat{a}_i) + \sum_{j=i+1}^{n} l_{ji} (a_j - \hat{a}_j) \right]^2 \leq \chi^2 \tag{3}$$

Equation (3) is the point of departure for the construction of bounds per ambiguity parameter, see also [8].

SEQUENTIAL CONDITIONAL ADJUSTMENT

Equation (3) is just an algebraic development of (2). The search can also be given a statistical interpretation: the sequential conditional adjustment, see section 5 of [1]. The conditional least-squares estimate for ambiguity i reads

$$\hat{a}_{i|i+1,\ldots,n} = \hat{a}_i - \sum_{j=i+1}^{n} l_{ji}(a_j - \hat{a}_j) \tag{4}$$

and the elements of the inverse of matrix D in the LDL^* decomposition are the corresponding conditional variances.

$$d_i^{-1} = \sigma_{\hat{a}_{i|i+1,\ldots,n}}^2 \tag{5}$$

The conditional estimate $\hat{a}_{i|i+1,\ldots,n}$ is the estimate for a_i conditioned on a_j with $j = i+1, \ldots, n$. Equation (4) clearly shows that conditioning on a_j for $j = i + 1, \ldots, n$ affects the estimate for a_i due to the correlation between the ambiguities. Only in case there is no correlation, $L = I$, we have

$$\hat{a}_{i|i+1,\ldots,n} = \hat{a}_i \tag{6}$$

The term between the square brackets in (3) is the difference of a_i and $\hat{a}_{i|i+1,...,n}$ and together with (5), it can be rewritten in

$$\sum_{i=1}^{n} \frac{(a_i - \hat{a}_{i|i+1,...,n})^2}{\sigma_{\hat{a}_{i|i+1,...,n}}^2} \leq \chi^2 \tag{7}$$

COMPUTATION OF THE BOUNDS

By means of a sequential conditional adjustment, the full ellipsoid (2) will be searched through for candidates for the vector of ambiguities. The adjustment starts with ambiguity a_n, and when a_n through a_{i+1} have been conditioned, we can construct from (3) bounds for ambiguity a_i. The interval for ambiguity a_i is defined by

$$-\sqrt{bound_i} \leq (a_i - \hat{a}_{i|i+1,...,n}) \leq \sqrt{bound_i} \tag{8}$$

It will be searched through in a straightforward manner from left to right, i.e. from the lower to the upper bound. With a valid integer in the interval (8), one proceeds with the next ambiguity a_{i-1}. If for a certain ambiguity a_l no valid integers can be found, one returns to the previous ambiguity a_{l+1} and takes the next valid integer for this ambiguity. Once an integer is encountered, that satisfies interval (8) for ambiguity a_1, a full candidate vector is found. The search terminates when all valid integers have been treated and one is back at the last ambiguity a_n. The full ellipsoid has been searched through. This is the so-called depth-first search.

In terms of the sequential adjustment, we start with a conditioning on a_n and end with a conditioning on a_1. In this way the bounds for the ambiguities a_n through a_1 are constructed in a recursive way. It starts with

$$bound_n = \frac{\chi^2}{d_n} \tag{9}$$

When the sequential adjustment is at ambiguity i, the interval is given by (8), showing that the interval for a_i is centered at the conditional estimate $\hat{a}_{i|i+1,...,n}$.

From (7) it can be seen that the conditional variances play a decisive role in the bounds for the ambiguities. The smaller the conditional variance, the smaller the interval [5].

THE VOLUME OF THE ELLIPSOID

The volume, expressed in [cycles^n], of the ellipsoidal region (2) is given by

$$E_n = \chi^n \sqrt{|Q_{\hat{a}}|} V_n \tag{10}$$

see [7]. In (10) the volume function is

$$V_n = \frac{2}{n} \frac{\pi^{\frac{n}{2}}}{\Gamma(\frac{n}{2})} \tag{11}$$

where Γ is the gamma function. With the LDL^* decomposition and the conditional variances being available from matrix D (5), the determinant becomes

$$|Q_{\hat{a}}| = \prod_{i=1}^{n} \sigma_{\hat{a}_{i|i+1,...,n}}^2 \tag{12}$$

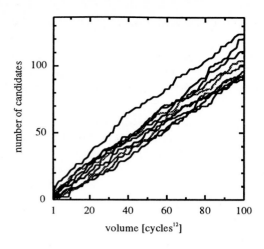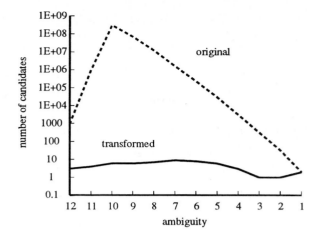

Figure 1: Volume E_n versus number of candidates k in ellipsoid, (left), number of candidates per ambiguity (right).

This shows that the volume E_n can easily be computed.

The volume E_n turns out to be a good indicator for the number of candidates (grid points) contained in the ellipsoid. A volume of E_n cycles^n corresponds to approximately $k = nint(E_n)$ candidates. The mismatch is caused by the discrete nature of the gridpoints. Centering the same ellipsoid at a different location, may result in a different number of candidates, $k \in Z$, while the volume, $E_n \in R$, remains unchanged. The value χ^2 can be taken such that a certain number of candidates will be inside the ellipsoidal region. In this way the size of the ellipsoid is controlled prior to the search. A straightforward search can then be performed to obtain the requested number of candidates. A list with the best k candidates, ordered after their norms, can be set up and updated during the sequential adjustment.

The purpose of estimation is to find the integer least-squares estimate $\check{a}$. Current validation procedures require also the second best candidate $\check{a}'$. The above procedure easily allows both candidates to be output by the integer least-squares algorithm.

EXAMPLE

Figure 1 (left) shows ten experiments, each with two epochs (sampling interval 1 second) of dual frequency phase data to seven satellites (12 ambiguities) on a 2.2 km baseline. The actual number of candidates contained in the ambiguity search ellipsoid is given as function of the volume of the ellipsoid [cycles^{12}]. The volume ranges from 1 to 100. All lines run under 45 degrees approximately. It shows that the volume is a good indicator for the actual number of candidates in the ambiguity search ellipsoid.

For the first of the above ten experiments, we compare the search on the original ambiguities with the search on the transformed ambiguities. The transformation is realized by the decorrelating Z^* reparametrization of the LAMBDA method. In figure 1 (right) we give the number of valid integers per ambiguity, encountered during the full search. The volume of the ambiguity search ellipsoid is $E_{12} = 2.8$ [cycles^{12}].

When a baseline (three coordinate unknowns) is observed for a short time span, three conditional variances, of the original ambiguities, are very large and the remaining ones are very small. From figure 1 (right) we see that for ambiguity a_{10} (the third ambiguity in the search) there are over 3.10^8 candidates. After having proceded to ambiguity a_1, there are only 2 full candidate vectors left, which implies that there are very many so called 'dead ends'. By the Z^*-transformation, the spectrum of conditional variances is flattened, as is the graph on the number of candidates per ambiguity. The search on the transformed ambiguities can be performed very efficiently. There are only very few dead ends left.

The computation time, on a 486 PC 66 MHz, for searching the original ellipsoid is 1 hour and 12 minutes; the total time for constructing the Z^*-transformation and subsequently searching the ellipsoid of the transformed ambiguities is only 20 milliseconds.

CONCLUDING REMARKS

The LAMBDA method for integer estimation of the GPS double difference ambiguities consists of firstly a decorrelation kf the ambiguities and secondly a sequential conditional adjustment of the ambiguities. In this paper the second feature has been discussed. The integer minimization problem is attacked by a discrete search over an ellipsoidal region, the ambiguity search ellipsoid.

The shape and orientation of the ellipsoid are governed by the variance covariance matrix of the ambiguities. The size of the ellipsoid can be controlled prior to the search using the volume function. The volume gives an indication of the number of candidates contained in the ellipsoid. Making a request for a few candidates and carrying out the search on the transformed ambiguities, makes that a straightforward implementation of the search can be used. A limited number of candidates will be gathered of which one is the integer least squares estimate for the vector of ambiguities.

Acknowledgement. Professor P.J.G. Teunissen is acknowledged for his comments and suggestions.

REFERENCES

[1] Teunissen, P.J.G. (1993). Least-squares estimation of the integer GPS ambiguities. Invited lecture. Section IV Theory and Methodology, IAG General Meeting. Beijing, China. (16 p.) Also in *Delft Geodetic Computing Centre LGR series* No. 6.

[2] Teunissen, P.J.G. (1994) The least-squares ambiguity decorrelation adjustment: A method for fast GPS integer ambiguity estimation. *Manuscripta Geodaetica*. (18 p.) Received: June 30, 1994, accepted for publication: November 14, 1994.

[3] Teunissen, P.J.G. (1994). On the GPS double difference ambiguities and their partial search spaces. Paper presented at the III Hotine-Marussi symposium on Mathematical Geodesy. L'Aquila, Italy. May 29 - June 3, 1994. (10 p.)

[4] Teunissen, P.J.G. and C.C.J.M. Tiberius (1994). Integer least-squares estimation of the GPS phase ambiguities. *Proc. of KIS'94*. Banff, Canada. pp. 221-231.

[5] Teunissen, P.J.G., P.J. de Jonge and C.C.J.M. Tiberius (1994). On the spectrum of the GPS DD-Ambiguities. *Proc. of ION GPS-94* Salt Lake City, USA. pp. 115-124.

[6] Golub, G.H. and C.F. van Loan (1989). *Matrix computations*. Second edition. The Johns Hopkins University Press, Baltimore, Maryland, USA.

[7] Apostol, T.M. (1969). *Multi-variable calculus and linear algebra, with applications to differential equations and probability*. Calculus Vol. 2, Wiley, New York.

[8] Wübbena, G. (1991). *Zur Modellierung von GPS-Beobachtungen für die hochgenaue Positionsbestimmung*. Universität Hannover.

THE IMPACT OF AMBIGUITY RESOLUTION ON GPS ORBIT DETERMINATION AND ON GLOBAL GEODYNAMICS STUDIES

L. Mervart, G. Beutler, M. Rothacher, S. Schaer
Astronomical Institute, University of Berne,
CH-3012 Bern, Switzerland

ABSTRACT

The *Center for Orbit Determination in Europe (CODE)* is one of the Analysis Centers of the *International GPS Service for Geodynamics (IGS)*. Currently (mid 1995), the data from about 60 globally distributed permanent GPS tracking sites are processed at CODE day after day. Various parameter types, in particular site coordinates, orbit parameters and earth orientation parameters, are estimated. Since the beginning of the 1992 IGS Test Campaign on 21 June 1992 the Bernese GPS Software Version 3 used by CODE was subject to a significant and continuous development. In particular *ambiguity resolution techniques* suitable for long baselines were developed, their impact on orbit determination, earth rotation parameters and station coordinates was studied.

Our new ambiguity resolution strategies allow us to find the correct integer values for the initial phase ambiguity parameters in the baseline mode even without having access to precise code measurements. This aspect is of special interest when analyzing data of the IGS network under the regime of *Anti-Spoofing (AS)*. The new methods make consequent use of the IGS products. They proved to be powerful up to baseline lengths of about 2000 km. The principles of the new algorithms are outlined, the expected accuracy improvements for the parameters of interest are discussed and initial results based a test implementation at the CODE processing center are presented.

THE EFFECT OF AMBIGUITY RESOLUTION IN GLOBAL GPS NETWORKS

It is known that ambiguity fixing is of great importance for short *occupation times* (shorter than about 1 hour). If global networks are processed considerably longer sessions are used. [Mervart, 1995] studied the improvement in the accuracy of the coordinates achieved by the ambiguity fixing for various session lengths. Figure 1 shows the RMS error of the Helmert transformation between the *"true"* coordinates (stemming from the processing of very long series of data) and the coordinates estimated using various session lengths (in this case the European part of the IGS Core Network was involved, details are given in [Mervart, 1995]).

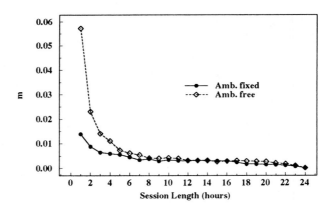

Figure 1: Rms of 7-parameter Helmert transformations of solutions of a specified length with respect to the *"true"* coordinate set

According to Figure 1 we cannot expect dramatic quality differences between the coordinates stemming from ambiguities fixed solutions resp. free solutions if sessions longer than about 6 hours are used. However, if the RMS errors of the site coordinates are inspected in more detail we observe that the fixing of the ambiguities improves the accuracy of the west–east component by about 50 % even for 24 hours session. The quality of the north-south and the height components remains more or less the same.

In [Mervart, 1995] the impact of ambiguity resolution on the other estimated parameters was discussed. Studies of the formal RMS errors of the orbital parameters gave the following results:

Table 1: Formal rms errors of the orbital elements from a 3-days solution (from [Mervart, 1995])

	a	e	i	Ω	ω	u_0
	m	10^{-10}	\multicolumn{4}{c}{10^{-3} ″}
float	0.0038	3.16	0.095	0.16	2.15	0.377
fixed	0.0024	2.85	0.068	0.11	1.99	0.241
improv.	37 %	10 %	28 %	31 %	7 %	36 %

In our daily processing the earth orientation parameters (the coordinates of the ephemeris pole and the UT1 – UTC drift) are estimated, too. According to the formal RMS errors of these parameters we may expect the following improvement:

Table 2: Formal rms errors of the earth orientation parameters (from [Mervart, 1995])

	X-pole	Y-pole	UT1-UTC drift
	10^{-5} ''		10^{-6} s/day
float	4.4	4.3	5.7
fixed	4.1	4.1	4.9
improv.	7 %	5 %	14 %

The results achieved by [Mervart, 1995] were encouraging enough to set up the ambiguity resolution procedure for the CODE routine IGS processing. We had to cope with the following problems:

(a) Due to the poor quality of the code measurement under the Anti-Spoofing (AS) we had to develop a code independent ambiguity resolution strategy.

(b) The ambiguity resolution algorithm has to be very efficient in order to allow to process the entire IGS Core Network daily.

The result of the development of the ambiguity fixing algorithm is the so-called Quasi-Ionosphere-Free (QIF) ambiguity resolution strategy to be outlined now.

QUASI-IONOSPHERE-FREE (QIF) AMBIGUITY RESOLUTION STRATEGY

Considering a dual-band GPS receiver, we may write the following two phase observation equations (pertaining to one epoch):

$$L_1 = \varrho + \Delta_{ion} + \lambda_1\, n_1 \,, \tag{1a}$$

$$L_2 = \varrho + \frac{f_1^2}{f_2^2}\, \Delta_{ion} + \lambda_2\, n_2 \,, \tag{1b}$$

where L_1, L_2 are the biased carrier-phase measurements, ϱ is the non-dispersive delay, lumping together the effects of geometric delay, tropospheric delay, and clock signatures, Δ_{ion} is the ionospheric delay for L_1 carrier, f_1, f_2 are the frequencies of the L_1 and L_2 carriers, λ_1, λ_2 the corresponding wavelengths, and n_1, n_2 are the initial phase ambiguities.

Due to the effect of ionospheric refraction it is mandatory to use the ionosphere-free linear combination L_3 if a global network is processed:

$$L_3 = \varrho + \tilde{B}_3 \,. \tag{2}$$

Denoting by b_1 and b_2 the (real valued) estimates of the (unknown) integer ambiguities n_1 and n_2 corresponding to the value $\tilde{B}_3$ we may express the estimated ionosphere-free bias $\tilde{B}_3$ in narrow-lane cycles (one cycle corresponding to a wavelength of $\lambda_3 =$

$c/(f_1 + f_2) \approx 11$ cm, $c =$ velocity of light):

$$\begin{aligned}
\tilde{b}_3 &= \frac{\tilde{B}_3}{\lambda_3} = \tilde{B}_3 \cdot \frac{f_1 + f_2}{c} = \frac{f_1}{f_1 - f_2} b_1 - \frac{f_2}{f_1 - f_2} b_2 \\
&= \beta_1 b_1 + \beta_2 b_2 .
\end{aligned} \tag{3}$$

Introducing the L_3-bias associated with (resolved) ambiguities n_1 and n_2

$$b_3 = \beta_1 n_1 + \beta_2 n_2 \tag{4}$$

we may use the difference

$$d_3 = |\tilde{b}_3 - b_3| \tag{5}$$

as a *test criterion* for accepting or selecting the pair n_1, n_2 of integer ambiguities as the correct solution. However, many pairs n_1, n_2 give small differences d_3. These pairs lie on a narrow band in the (n_1, n_2) space. A unique solution only results *if it is possible to limit the search range.*

For baselines longer than about 10 km the processing of the two frequencies L_1 and L_2 separately does not give sufficiently good initial real valued estimates b_1 and b_2 due to the influence of the ionospheric refraction. Therefore it is necessary to estimate the ionospheric delay too. Due to space restrictions we refer to [Mervart, 1995] and [Schaer, 1994] for details of the implementation of the QIF ambiguity resolution strategy and the modeling of the ionosphere.

INITIAL RESULTS

To assess the quality of our estimations we compared the satellite positions stemming from subsequent 3-day orbits. The principle is shown in Figure 2.

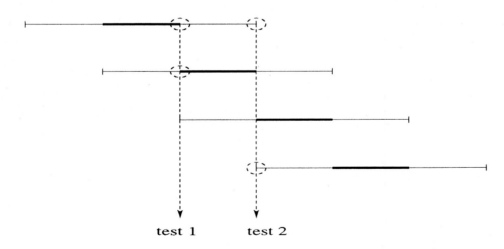

Figure 2: Consistency tests of 3-day orbits

Figure 3 shows the mean differences in the satellite positions. Because the modeling of satellite orbits is more difficult during eclipsing seasons, the eclipsing and non eclipsing satellites are treated separately.

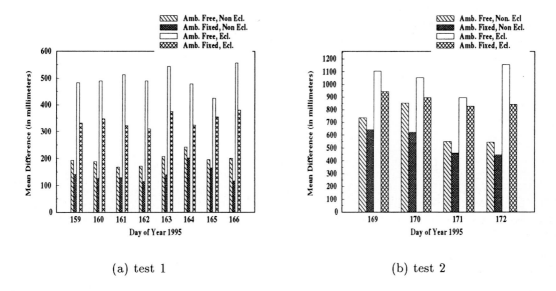

(a) test 1   (b) test 2

Figure 3: Orbit quality estimated from discontinuities at day boundaries according to Figure 2. (tests 1 and 2)

CONCLUSIONS

According to the results shown in Figure 3 we conclude, that the fixing of the ambiguities considerably improves the consistency between subsequent arcs. It should be mentioned that the consistency itself is not necessarily equal to the quality of the orbits. However having no "ground truth" for the orbits, it seems to be relevant to use the discontinuities at day boundaries as a quality criterion. We may expect that with the growing number of permanent stations in the IGS Core Network the improvement achieved due to the fixing of the ambiguities will increase. The ambiguity-fixed solution is the official CODE solution released to the scientific community since GPS week 807 (starting June 25, 1995).

REFERENCES

Mervart, L. (1995): *Ambiguity Resolution Techniques in Geodetic and Geodynamic Applications of the Global Positioning System.* Ph.D. Thesis, Printing office, University of Berne.

Schaer, S. (1994): *Stochastische Ionosphärenmodellierung beim Rapid Static Positioning mit GPS.* Lizentiatsarbeit. Astronomisches Institut, Druckerei Universität Bern.

ON-THE-FLY AMBIGUITY RESOLUTION FOR LONG RANGE GPS KINEMATIC POSITIONING

Shaowei Han and **Chris Rizos**
School of Geomatic Engineering
The University of New South Wales
Sydney NSW 2052, AUSTRALIA

ABSTRACT

Precise long range GPS kinematic positioning to decimetre accuracy requires that the cycle ambiguities are fixed. In this paper we describe an 'on-the-fly' ambiguity resolution method based on the condition that the initial integer ambiguities are first assumed known, using a static ambiguity resolution technique at the beginning of a survey session, and the ambiguities are *recovered* when cycle slips occur 'on-the-fly'. The cycle slip detection and repair procedure uses a variety of data including one, or two, precise pseudo-range measurements to determine possible cycle slip candidates on two combinations of carrier phase data (the standard 'widelane' combination with an effective wavelength of 0.8619 metres, and the $-7 \cdot \varphi_1 + 9 \cdot \varphi_2$ combination with an effective wavelength of 14.6526 metres). Three validation and rejection criteria are applied to determine the correct cycle slips on the L1 and/or L2 carrier phase observations.

GPS LONG RANGE KINEMATIC POSITIONING

In the last few years, interest in long range GPS kinematic positioning has been growing for airborne applications where GPS reference receivers cannot be set up near the survey area, such as out on the continental shelf areas, and remote and inaccessible land areas, etc. The distances between receivers may range from hundreds to thousands of kilometres and the accuracy requirement is generally at the decimetre level or higher. How to determine the initial integer ambiguity *and* repair cycle slips over such long baselines when the antenna is in motion? This question will be briefly discussed in this paper.

The observation equations and ambiguity resolution procedures for high precision long range *static* positioning were described in, for example, Blewitt (1989), Dong & Bock (1989), Goad (1992), and are based on the following equations:

$$R_1 = \rho + \frac{I}{f_1^2} + \varepsilon_{R_1} \qquad (1a) \qquad\qquad R_2 = \rho + \frac{I}{f_2^2} + \varepsilon_{R_2} \qquad (1c)$$

$$\varphi_1 \lambda_1 = \rho - \frac{I}{f_1^2} + N_1 \lambda_1 + \varepsilon_{\varphi_1} \qquad (1b) \qquad\qquad \varphi_2 \lambda_2 = \rho - \frac{I}{f_2^2} + N_2 \lambda_2 + \varepsilon_{\varphi_2} \qquad (1d)$$

where R_1 and R_2 are the one-way precise pseudo-ranges; φ_1 and φ_2 are the one-way carrier phase observations in units of cycles; ρ is the geometric range from station to satellite; I is a function of the Total Electron Content; f_1, f_2 and λ_1, λ_2 are the frequencies and wavelengths of the L1 and L2 carrier waves respectively; N_1 and N_2 are the integer cycle ambiguities of the L1 and L2 phase observations; and ε is the observation noise with respect to the observation type indicated by its subscript. Based on eqn (1), the real-valued ambiguity ($N_{i,j}$) estimation formula for a carrier phase combination $\varphi_{i,j}$ can be written as:

$$N_{i,j} = \varphi_{i,j} - \frac{9240(i+j)+289\cdot i}{2329\cdot\lambda_1}\cdot R_1 + \frac{9240(i+j)+289\cdot j}{2329\cdot\lambda_2}\cdot R_2 \qquad (2)$$

The double-differenced ionosphere-free observable combination can be represented by the formula:

$$\Delta\nabla\varphi_{77,-60}\lambda_{77,-60} = \Delta\nabla\rho + \Delta\nabla N_{77,-60}\lambda_{77,-60} + \varepsilon_{\Delta\nabla\varphi_{77,-60}\lambda_{77,-60}} \qquad (3)$$

Because the wavelength of the ionosphere-free combination is very small (0.63cm), $\Delta\nabla N_{77,-60}$ will be difficult to resolve. However, after the widelane ambiguities $\Delta\nabla N_{1,-1}$ are fixed and $\Delta\nabla N_{77,-60}$ is replaced by $\Delta\nabla N_1$ and $\Delta\nabla N_{1,-1}$, eqn (3) can be written as:

$$\Delta\nabla\varphi_{77,-60}\lambda_{77,-60} = \Delta\nabla\rho + \Delta\nabla N_1\cdot(17\lambda_{77,-60}) + \Delta\nabla N_{1,-1}\cdot(60\lambda_{77,-60}) + \varepsilon_{\Delta\nabla\varphi_{77,-60}\lambda_{77,-60}} \qquad (4)$$

The third term on the right hand side is therefore a known quantity. Any search technique can be used to resolve $\Delta\nabla N_1$. Comparing this equation with the L1 phase observation equation, the wavelength of $\Delta\nabla N_1$ in eqn (4) is equivalent to 10.7cm ($17\lambda_{77,-60} = 10.7$cm). After determination of the initial ambiguities $\Delta\nabla N_{77,-60}$, using $\Delta\nabla N_{1,-1}$ and $\Delta\nabla N_1$, the ionosphere-free combination (eqn (3)) can be used for the subsequent GPS long range kinematic positioning to decimetre accuracy within a Kalman filter algorithm, as described by Colombo et al. (1995) for an experiment with baselines greater than 1000km.

AMBIGUITY RESOLUTION ON-THE-FLY

After the initial integer ambiguities are determined, ambiguity resolution 'on-the-fly' will become an ambiguity *recovery* problem if cycle slips appear. Cycle slip detection and repair using one-way data was first suggested by Blewitt (1990) for the static environment. Han & Rizos (1995) proposed the widelane and $\varphi_{-7,9}$ combinations for detection and repair of cycle slips for the kinematic positioning mode.

Real-Valued Cycle Slip Estimation

Eqn (2) is suitable for the computation of the real-valued widelane ambiguity at every epoch, but not precise enough to compute the ambiguities for other phase combinations (Ibid, 1995), hence the ionosphere-biased formula is used instead (mod. of eqn (1)):

$$N_{i,j} = \varphi_{i,j} - \frac{R}{\lambda_{i,j}} + \gamma\frac{I}{f_1^2} \qquad (5)$$

where

$$\gamma = \frac{\alpha_{i,j}+\beta}{\lambda_{i,j}} \qquad (6)$$

$$\alpha_{i,j} = \frac{4620\cdot i+5929\cdot j}{4620\cdot i+3600\cdot j}; \qquad \beta = \begin{cases} 1 & \text{for} & R = R_1 \\ 1.647 & \text{for} & R = R_2 \\ 1.323 & \text{for} & R = (R_1+R_2)/2 \end{cases} \qquad (7)$$

R, the precise pseudo-range, in eqn (5) can be chosen as being either R_1 or R_2, or the mean value of R_1 and R_2, which minimise the P-code pseudo-range noise (depending on the data available). The larger the wavelength of $\varphi_{i,j}$, the smaller the noise of $N_{i,j}$, if the ionospheric delay can be obtained with high precision. Therefore, the widelane and $\varphi_{-7,9}$ combinations are selected for the cycle slip detection and repair process.

The biases caused by the ionospheric delay have a very strong correlation between epochs. If a few epochs are used to define a linear function, and to subsequently predict the value at the next epoch, the differences between the predicted value $N^-_{1,-1}(k)$ (or $N^-_{-7,9}(k)$) using the previous epochs and the computed value $N_{1,-1}(k)$ (or $N_{-7,9}(k)$), using observations at this epoch from eqns (2) and (5), can be obtained:

$$DN_{1,-1}(k) = N_{1,-1}(k) - N^-_{1,-1}(k) \qquad\qquad DN_{-7,9}(k) = N_{-7,9}(k) - N^-_{-7,9}(k) \qquad (8)$$

If no cycle slip or multipath disturbance is present, the noises of $DN_{1,-1}(k)$ and $DN_{-7,9}(k)$ are dependent on the correlations of the ionospheric delay and the noise of the observations. If the ionospheric delay changes rapidly, the noises will be larger. Figs. 1 and 2 illustrate their noise level. On the other hand, discrete jumps in these quantities may indicate cycle slips $CS_{1,-1}$ and/or $CS_{-7,9}$. These are subject to the odd-even relations:

if $CS_{1,-1}$ is even $\rightarrow$ $CS_{-7,9}$ has to be even;   if $CS_{1,-1}$ is odd $\rightarrow$ $CS_{-7,9}$ has to be odd

Obviously it is easy to determine $CS_{1,-1}$ when $CS_{-7,9}$ is known and vice versa. The cycle slips on φ_1 and φ_2 can then be determined. For a static receiver, this step can detect and repair almost all cycle slips. However, for a receiver in motion, especially where the satellite elevation angle is relatively low ($< 40^\circ$), or when a few tens of seconds of data is missing, the cycle slip(s) cannot be determined as one unique set. Hence, several cycle slip candidate sets can be formed, and the following tests will be necessary.

Validation Criteria

Using eqn (8), the real-valued cycle slip estimates, and their standard devations can be computed. The search regions for $DN_{1,-1}$ and $DN_{-7,9}$ can be formed in the one-way case and then used to create the search region for the double-differenced observable. For each double-differenced cycle slip, the following tests should be applied.

Test 1: Tests on the innovation sequences of $\Delta\nabla\varphi_{1,-1}$ *and* $\Delta\nabla\varphi_{-7,9}$. The ionospheric delay value for each double-differenced observable can be predicted using the ionospheric delay values computed at the previous epochs, and the ionosphere-corrected observable of $\Delta\nabla\hat{\varphi}_{1,-1}$ and $\Delta\nabla\hat{\varphi}_{-7,9}$ can be obtained. The innovation values can be computed:

$$\Delta\nabla L_{1,-1} = (\Delta\nabla\hat{\varphi}_{1,-1} - \Delta\nabla CS_{1,-1})\lambda_{1,-1} - HX(-) \qquad (9)$$

$$\Delta\nabla L_{-7,9} = (\Delta\nabla\hat{\varphi}_{-7,9} - \Delta\nabla CS_{-7,9})\lambda_{-7,9} - HX(-) \qquad (10)$$

If $\Delta\nabla L_{1,-1} \sim N(0, D_{1,-1} + HP(-)H^T)$ or $\Delta\nabla L_{-7,9} \sim N(0, D_{-7,9} + HP(-)H^T)$, $\Delta\nabla CS_{1,-1}$ or $\Delta\nabla CS_{-7,9}$ will pass the test, otherwise the $\Delta\nabla CS_{1,-1}$ or $\Delta\nabla CS_{-7,9}$ candidate set should be rejected.

X(−) and P(−) are the Kalman predicted position and its variance matrix; H is the design matrix of the Kalman filter; $D_{1,-1}$ and $D_{-7,9}$ are the variances of $\Delta\nabla\hat{\varphi}_{1,-1}$ and $\Delta\nabla\hat{\varphi}_{-7,9}$.

Test 2: Test on the quadratic form of the residuals of $\Delta\nabla\hat{\varphi}_{1,-1}$. Using the ionosphere-corrected double-differenced widelane observable and the Kalman filter predicted position, the update position and the quadratic form of the residuals $(QF_{1,-1})_i$ for a cycle slip candidate set i can be computed. The following test can then be applied:

$$\left(QF_{1,-1}\right)_i = V_{1,-1}^T D_{1,-1}^{-1} V_{1,-1} + \delta X_{1,-1}^T P^{-1}(-)\delta X_{1,-1}; \qquad \frac{(QF_{1,-1})_i}{(QF_{1,-1})_{min}} < \xi_{F_{m,m};1-\alpha} \tag{11}$$

where $V_{1,-1}$ is the residual vector of the widelane observable at this epoch; $\delta X_{1,-1}$ is the correction to X(−); $\xi_{F_{m,m};1-\alpha}$ is the Fisher percentile for degrees of freedom m and m and confidence level $1-\alpha$; m is the number of double-differenced observations. If a cycle slip candidate set does not pass the test, this set is rejected.

Test 3: Contrast tests on the quadratic form of the residuals of $\Delta\nabla\varphi_{77,-60}$. Using all cycle slip candidate sets that have passed the previous tests, the cycle slip candidates of the ionosphere-free observable can be formed. Fixing one set of cycle slips, the current epoch data is used within the Kalman filter and the quadratic form of the residuals $QF_{77,-60}$ should be computed for cycle slip candidate set. If the smallest $(QF_{77,-60})_{min}$ and $(QF_{77,-60})_i$ with respect to a cycle slip candidate set i are not consistent with the relation:

$$\frac{(QF_{77,-60})_i}{(QF_{77,-60})_{min}} < \xi_{F_{f,f};1-\alpha} \tag{12}$$

the cycle slip candidate set i should be rejected. If only the current epoch data are used, f=m. If all other sets, except the one that has the smallest $(QF_{77,-60})_{min}$, are rejected, the cycle slip set corresponding to $(QF_{77,-60})_{min}$ is selected as the correct set. Otherwise all cycle slip sets that have passed the test will be treated as candidates for the next epoch.

EXPERIMENTS

This experiment was carried out on October 6, 1994, using Ashtech Z12 GPS receivers. Two receivers were set up (one on a train), separated by about 272 km at the start. After a period of static tracking, the train receiver moved over a period of about 50 minutes until the maximum distance was about 295km. Eqns (2) and (5) were used to compute $N_{1,-1}(k)$ and $N_{-7,9}(k)$, and then to predict $N_{1,-1}^-(k)$ and $N_{-7,9}^-(k)$ using the previous several epochs (ten epochs are choosen here). The differences between the predicted value and the computed value ($DN_{1-1}(k)$ and $DN_{-7,9}(k)$) are plotted in Figs. 1a and 1b for the train receiver. The noise of $DN_{1-1}(k)$ and $DN_{-7,9}(k)$ is less than about 0.1 cycles. This satellite (PRN 16) has the lowest elevation of the five satellites tracked, and is plotted in Fig. 1a. It can be seen that cycle slip detection and repair is very easy for 1 second data rate observations if there is no data gap. A data gap of 5 minutes was simulated at each epoch and the real-valued cycle slip estimates are plotted in Figs. 2a and 2b. The suggested procedure can therefore successfully determine the correct set of cycle slip values.

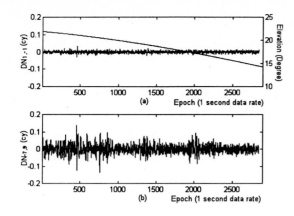

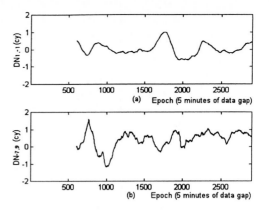

Fig. 1. Real-valued Cycle Slip Estimates
($DN_{1,-1}$ (a) and $DN_{-7,9}$ (b)) for Roving Rcvr,
with Elevation of Satellite

Fig. 2. Real-valued Cycle Slip Estimates
($DN_{1,-1}$ (a) and $DN_{-7,9}$ (b)) for Roving Rcvr,
Assuming 5 Minutes of Data Gap

CONCLUSIONS

If precise pseudo-range data are available, the combined carrier phase observations $\varphi_{1,-1}$ and $\varphi_{-7,9}$ are very useful for cycle slip detection and repair in kinematic data. The suggested procedure can repair data gaps up to 1-5 minutes depending on the receiver type and the ionosphere conditions. However, the procedure requires ambiguity initialisation at the beginning of a session. For short receiver separations at the beginning of a session, the traditional ambiguity resolution method for the static case, or ambiguity resolution 'on-the-fly' for the kinematic case, can be used. However, for large receiver separations at the beginning of a session, static methods are needed to initialise the ambiguities.

REFERENCES

Blewitt, G. (1989). Carrier phase ambiguity resolution for the Global Positioning System applied to geodetic baselines up to 2000km. *J. Geophys. Res.*, **94(B8)**, 10187-10203.

Blewitt, G. (1990). An automatic editing algorithm for GPS data. *Geophys. Res. Letters*, **17(3)**, 199-202.

Colombo, O. L., Rizos, C. and Hirsch, B. (1995). Decimeter-level DGPS navigation over distances of more than 1000km: results of the Sydney Harbor Experiment. Proc. *4th Int. Conf. on Differential Satellite Navigation Systems*, Bergen, Norway, 24-28 April.

Dong, D. N. and Bock, Y. (1989). Global Positioning System network analysis with phase ambiguity resolution applied to crustal deformation studies in California. *J. Geophys. Res.*, **94(B4)**, 3949-3966.

Goad, C. C. (1992). Robust techniques for determining GPS phase ambiguities. Proc. *6th Int. Geodetic Symp. on Satellite Positioning*, Columbus, Ohio, 17-20 March, 245-254.

Han, S. W. and Rizos, C. (1995). A suggested procedure for on-the-fly ambiguity resolution for long range kinematic positioning. Proc. *4th Int. Conf. on Differential Navigation Systems*, Bergen, Norway, 24-28 April.

GPS PHASES: SINGLE EPOCH AMBIGUITY
AND SLIP RESOLUTION

W Pachelski, Space Research Center,
Bartycka 18A, 00-716 Warsaw, Poland;
phone +48.22.403766; fax +48.39.121273;
e-mail: wp@cbk.waw.pl

ABSTRACT

An observational equation of the GPS carrier phase contains the pair station - satellite specific and epoch independent bias $\gamma_r^s = \psi^s(t_0) - \psi_r(t_0) + N_r^s$, in which N_r^s is an integer ambiguity and $\psi^s(t_0)$, $\psi_r(t_0)$ are transmitter and receiver initial phases. Through sequential processing of phases we update in each epoch, among other unknowns, the γ_r^s estimates, provided specific *minimal configurations* of satellites, stations and already processed epochs are satisfied. All second differences of the phases, e.g. with respect to a given reference satellite and reference receiver, $\nabla\Delta\gamma_r^s = \nabla\Delta N_r^s$, should be then integers on each L1 and L2 band. These conditions can be solved for all N_r^s's (thus implying new γ_r^s-values) about current estimates of the γ_r^s's as soon as the integer values are found by means of a proper search procedure.

Cycle slips come into view as outliers of observations produced by rapid changes of particular γ_r^s values. In that case a new observation sequence is created, for which new γ-parameters are estimated, and then consequently constrained for ambiguities.

INTEGER AMBIGUITIES IN GPS PHASE OBSERVATION EQUATION

According to (Landau, 1988; Lindlohr, 1988) an **udifferenced GPS carrier phase observation equation**, corrected for the ionosphere and troposphere:

$$u_r^s(t) = -\frac{f}{c}\left(1 - \frac{\dot{\rho}_r^s(t)}{c}\right)\rho_r^s(t) + \left(1 - \frac{\dot{\rho}_r^s(t)}{c}\right)\alpha_r(t) + \beta^s(t) + \gamma_r^s - \psi_r^s(t) - d_{ion} + d_{trop} \qquad (1)$$

contains the receiver r and satellite s specific biases (mainly clock offsets), $\alpha_r(t)$, $\beta^s(t)$, as well as the pair satellite-receiver specific and epoch independent bias γ_r^s defined as:

$$\gamma_r^s = \psi^s(t_0) - \psi_r(t_0) + N_r^s \qquad (2)$$

where $\psi^s(t_0)$ and $\psi_r(t_0)$ are initial phases at some epoch t_0, and N_r^s is the phase integer ambiguity (ρ_r^s is the transmitter-to-receiver path length and contains desired parameters, i.e. station and - possibly - satellite coordinates).

As shown by Lindlohr and Wells (1985), and by Schaffrin and Grafarend (1986), the **equation (1) is rank deficient** with respect to biases $\alpha_r(t)$, $\beta^s(t)$ and γ_r^s with the rank defect equal to:

$$R + S + T - 1 \tag{3}$$

for R receivers (stations), S satellites (transmitters) and T observation epochs. A possible complete set of constraints to resolve this deficiency can be the following (Pachelski, 1992a, b, c):

1) one of clocks, e.g. a satellite clock *sref*, is a reference clock, so that:

$$\beta^{sref}(t) = \text{const} \quad \text{for } t = 1, \ldots, T ; \tag{4}$$

2) all other satellite clocks are to be synchronized to the reference clock, i.e.:

$$\beta^s(t) = \beta^{sref}(t) \quad \text{for } s = 1, \ldots, S, \, s \neq sref , \text{ and} \tag{5}$$

3) all receiver clocks are to be synchronized to the reference clock, i.e.:

$$\alpha_r(t) = \beta^{sref}(t) \quad \text{for } r = 1, \ldots, R \tag{6}$$

Similarly, γ_r^s in (1), being usual floating point numbers, are to be constrained according to the following relations resulting from (2):

$$\nabla\Delta\gamma_r^s = \gamma_r^s - \gamma_{rref}^s - \gamma_r^{sref} + \gamma_{rref}^{sref} = I_r^s \quad \text{for} \quad r \neq rref, \quad s \neq sref \tag{7}$$

i.e. double differences of the γ-parameter estimates must be equal to some integer values $I_r^s = \nabla\Delta N_r^s$. These constraints, when applied in the process of the sequential adjustment as soon as the right hand sides have been found, lead to **resolving integer ambiguities** of the observed phases. The values of I_r^s are to be found by means of the proper search proce- dure so, as to fit the floating-point γ-estimates in the best way. Practically, the constraints (7) are strong enough to cause large correlations between γ-parameters, which improve accuracy of the desired parameters. On the other hand, however, this also makes the error hyper-ellipsoid in the space of γ-parameters quite elongated, which allows for a substantial reduction of the search domain for the proper I_r^s. In that case application of the search procedure such as the *LAMBDA method* suggested by Teunissen (e.g. 1993, 1994, 1995), and de Jonge and Tiberius (1995), becomes quite essential.

GROUP SEQUENTIAL ADJUSTMENT OF GPS PHASES

Least-squares estimates of parameters, $\mathbf{x}_k$, of their variance-covariance matrix, $\mathbf{Q}_k$, and of the quadratic form of residuals, s_k, can be **updated with a new group of observations**, composed of new observation equations $\mathbf{v}_m = \mathbf{A}_m\mathbf{x} - \mathbf{l}_m$ and their variance-covariance ma- trix $\mathbf{R}_m$, by means of the following relations (Pachelski, 1980):

$$\left.\begin{array}{l} \mathbf{x}_{k+m} = \mathbf{x}_k + \mathbf{Q}_k\mathbf{A}_m^T\left(\mathbf{A}_m\mathbf{Q}_k\mathbf{A}_m^T + \mathbf{R}_m\right)^{-1}\left(\mathbf{l}_m - \mathbf{A}_m\mathbf{x}_k\right) \\[2mm] \mathbf{Q}_{k+m} = \mathbf{Q}_k - \mathbf{Q}_k\mathbf{A}_m^T\left(\mathbf{A}_m\mathbf{Q}_k\mathbf{A}_m^T + \mathbf{R}_m\right)^{-1}\mathbf{A}_m\mathbf{Q}_k \\[2mm] s_{k+m} = s_k + \left(\mathbf{l}_m - \mathbf{A}_m\mathbf{x}_k\right)^T\left(\mathbf{A}_m\mathbf{Q}_k\mathbf{A}_m^T + \mathbf{R}_m\right)^{-1}\left(\mathbf{l}_m - \mathbf{A}_m\mathbf{x}_k\right) = s_k + \Delta s_m \end{array}\right\} \tag{8}$$

By means of the increment Δs_m we get a „local" measure of accuracy:

$$\sigma_m^2 = \Delta s_m / \Delta n_m \tag{9}$$

where Δn_m is an increment of degrees of freedom.

As soon as a relation between „global" precision and reliability measures becomes optimal for resolving ambiguities (Teunissen, 1995), the $(R-1)(S-1)$ constraints (7) considered as pseudo-observation equations (i.e. with $\mathbf{R}_m=0$) are to be used in the sequential adjustment (8). Each set of the considered I_r^s values can be then tested against its minimal contribution (9) to the being minimized quadratic form s_{k+m} before it is actually used to update the estimates $\mathbf{x}_{k+m}$, $\mathbf{Q}_{k+m}$ and s_{k+m}.

MINIMAL CONFIGURATIONS

Obviously, resolving ambiguities requires all γ-parameters to assess deterministic estimates at a given step of the sequential adjustment. This condition requests a balance among observations, unknowns and datum properties by means of the following basic diophantic inequality:

number of observations $\geq$ number of unknowns – datum rank defect $\tag{10}$

In the case of GPS undifferenced phase observations this leads to the following inequality (Lindlohr and Wells, 1985):

$$RST \geq 3R + RT + ST + RS - (R + S + T - 1) \tag{11}$$

in which the right hand side specifies the number of unknowns in each group: $x_r, \alpha_r(t), \beta^s(t)$ and γ_r^s, as well as the rank defect of the bias design matrix (3).

At the beginning of the sequential adjustment all parameters are undefined, so the so-called „global" configurations necessary to resolve all of them must satisfy (11), hence some resulting minimal configurations of receivers, satellites and observation epochs are the following:

$$
\begin{array}{ccc}
R = 2 & S = 2 & T = 7 \\
2 & 3 & 4 \\
2 & 4 & 3 \\
2 & 7 & 2 \\
4 & 5 & 2 \\
\cdots & \cdots & \cdots
\end{array}
\tag{12}
$$

These are minimal constraints to be satisfied before resolving integer ambiguities.

As soon as station coordinates and γ-parameters become defined in the process of the sequential adjustment, we may also have „**local**" **configurations:**

$$RST \geq 0 + RT + ST + 0 - T \tag{13}$$

needed to resolve specific for a given epoch (group of observations) parameters $\alpha_r(t)$ and $\beta^s(t)$ only. Thus, for a single epoch (T=1) we get the following minimal solution:

$$R = 2, S = 2 \tag{14}$$

Any additional observation allows for a single epoch slip resolution.

SINGLE EPOCH SLIP RESOLUTION

Sequential (epoch-by-epoch) observation and adjustment are parallel processes in which a cycle slip is caused by a rapid change of a signal phase due to an instantaneous loss of lock-on. In the adjustment it comes into view as **an outlier** affecting a particular γ-parameter. It is exposed through an enormous value of a „local" accuracy estimate σ_m in (9), as soon as: 1^0 - a „global" minimal configuration acc. to (11) and (12) has been resolved within the already processed epochs, and 2^0 - an actual epoch contains superfluous observation(s) with respect to those needed for resolving „local" minimal configuration acc. to (13) and (14).

In that case **a new observation sequence** is considered to begin at the given epoch (new initial epoch t_0), so that a new set of γ-parameters is initialized, sequentially estimated and consequently constrained for ambiguities. Based on the preceding observation sequence, the new sequence is assumed to be already resolved for a „global" minimal configuration.

EXAMPLE

```
UPPER ROW:      without ambiguity resolution:
LOWER ROW:      with ambiguity resolution:

NOF: 715    SUM OF SQ. RES.:   0.267279  SIGMA:   0.019
     731                       2.58889            0.060

Final values and mean errors of station positions:
           x                    y                    z
1   3653747.310  0.000  1394811.670  0.000  5021697.910  0.000
         .310    0.000         .670  0.000         .910  0.000
2   3653710.783  0.045  1394992.495  0.030  5021674.673  0.019
         .499    0.002         .530  0.003         .390  0.005
3   3653597.599  0.045  1395552.583  0.030  5021602.621  0.019
         .191    0.002         .559  0.003         .520  0.005

Interstation distances and mean errors:
1   2       185.936   0.023
              .061    0.002
1   3       761.870   0.023
              .939    0.002
2   3       575.934   0.023
              .879    0.002
```

The above example shows, for a trivial case of short baselines and small number (92) of observation epochs, a substantial influence of resolving ambiguities on results of the GPS phase adjustment. Besides of actual influence on final coordinates and baseline lengths, constraining γ-parameters to integers caused the overall SIGMA value to **increase** about 3 times, while the RMS errors of x, y, z-coordinates **decreased** 4 to 20 times, and of baseline lengths more than ten times. This is due to a substantially higher conditioning of the system in the case of fixed ambiguities.

CONCLUSIONS

- An undifferenced approach to processing GPS carrier phase observations in an epoch-by-epoch manner (the sequential adjustment) is useful in an **one-step network adjustment**, directly from raw data to station coordinates.
- It allows for a **detailed analysis of bias parameters and of partial results** in the course of the adjustment.
- In particular, it allows for **integer ambiguity resolution** through proper constraining specific parameters at an arbitrary epoch in the course of the adjustment, as soon as a „global" minimal configuration has been resolved beforehand (i.e. excluding some very initial epochs).
- Data of any epoch (excl. initial ones) can **be inspected against slips** before using them to update parameter estimates.
- Detection of a slip effects in beginning a **new observation sequence** rather than in trying to „repair" the data.

REFERENCES

Gaiovitch I, Pachelski W (1994): Systematic Effects in GPS Positioning, *Artificial Satellites*, Vol. 29, No. 3.

de Jonge P, and Tiberius Ch (1995): Integer Ambiguity Estimation with the LAMBDA-Method. *IUGG XXI Gen. Ass.*, Boulder, Colorado, USA.

Landau H (1988): Zur Nutzung des Global Positioning Systems in Geodäsie und Geodynamik: Modellbildung, Software-Entwicklung und Analyse. *Schriftenreihe Univ. der Bundeswehr München*, Neubiberg, Heft 36.

Lindlohr W (1988): Dynamische Analyse geodätischer Netze auf der Basis von GPS-Phasenbeobachtungen. *DGK, Reihe C*, Heft Nr. 346, München.

Lindlohr W and Wells D (1985): GPS design using undifferenced carrier beat phase observations. *Manuscripta Geodaetica*, 10.

Pachelski W (1980): On the Decomposition in Least Squares (with Examples of its Application in Satellite Geodesy). *DGK, Reihe A*, Heft Nr. 91, München.

Pachelski W (1992a): PHANTASY: The Program for the Sequential Adjustment of Undifferenced GPS Phases. *Proc. 2nd Int. Workshop on High Precision Navigation*, Stuttgart and Freudenstadt, November, 1991, Ferd. Duemmlers Verlag, Bonn.

Pachelski W (1992b): Undifferenced Processing of GPS Phases. *3rd Geodetic Meeting Italia - Polonia*, Trieste, Italy.

Pachelski W (1992c): Sequential Adjustment of Undifferenced GPS Phases. *Proc. Intern. Workshop on Global Positioning Systems in Geosciences*, Tech. Univ. of Crete, Chania, Greece, 8 - 10 June 1992.

Teunissen PJG (1993): Least-Squares Estimation of the Integer GPS Ambiguities. *Invited Lecture, Sec. IV IAG General Meeting*, Beijing, China.

Teunissen PJG (1994): On the GPS double difference ambiguities and their partial search spaces. *III Hotine-Marussi Symposium on Mathematical Geodesy*, L'Aquila, Italy.

Teunissen PJG (1995): The Geometry of the L1/L2 Ambiguity Search Space with and without Ionosphere. *IUGG XXI Gen. Ass.*, Boulder, Colorado, USA.

DIRECT AMBIGUITY RESOLUTION USING INTEGER NONLINEAR PROGRAMMING METHODS

M. Wei and K.P. Schwarz

Department of Geomatics Engineering
The University of Calgary
2500 University Drive, N.W.
Calgary, Alberta, T2N 1N4, CANADA

ABSTRACT

This paper presents a new and efficient strategy for ambiguity resolution on the fly. The new approach is based on the method of Integer Nonlinear Programming (INLP). After discussing some typical methods for INLP, ambiguity resolution of GPS phase measurements based on the Integer Least-Squares Method (ILSM) is formulated as a problem of INLP. A new ambiguity search method is then presented based on a combination of different INLP methods. The new ambiguity search method shows that the search for the optimal integer ambiguities of six to seven satellites can be performed within 0.1 - 0.2 second and thus can be used for real-time applications. Based on this search method, ambiguity resolution on the fly is carried out using a sequential approach which estimates the optimal integer ambiguities at each epoch by using all GPS observations available at that epoch. This approach is very robust because it avoids the critical issue of erroneously rejecting the optimal ambiguities. To validate the estimated integer ambiguities at each epoch, a number of criteria are discussed and tested to ensure the correctness of the estimated integer ambiguities. The method has been successfully tested and has shown robustness as well as reliability.

INTRODUCTION

Without the exact determination of the integer ambiguity precise positioning at the cm level using GPS phase observations cannot be achieved. Thus, ambiguity resolution plays a key role for precise positioning and navigation using GPS carrier phase observations in both the static and kinematic cases. A very good review paper by Hatch and Euler (1994) summarizes many different techniques for ambiguity resolution on the fly (AROF) developed over the last decade. In the following the investigation mainly focuses on single frequency data. The method developed in this paper can also be applied to dual frequency data, e.g. wide lane observations. The principle for both single frequency data and dual frequency data is the same.

The difficulty in ambiguity resolution on the fly is the integer constraint on the ambiguity parameters. Due to this constraint there is no unique analytical solution for the integer ambiguities. Unless very precise code observations are available, the principle of integer ambiguity resolution is mainly based on geometrical constraints which require a significant geometrical change of the GPS constellation. The many uncertainties and errors, such as orbital errors, atmospheric effects, multipath effects and measurement noise of the carrier phase also affect the solution of the integer ambiguity. In this paper, the double differencing technique is applied and the integer ambiguities to be determined are the double differenced integer ambiguities. In this case, the total observation errors are smaller than one cycle of the integer ambiguity for a baseline shorter than a certain distance.

Most methods for ambiguity resolution on the fly are based on the concept of the search space for potential ambiguities. The definition and construction of the search space are different for different search methods. At each epoch the potential ambiguities are tested and identified using different test and validation techniques. The ambiguities which do not pass the test and validation procedure are rejected and the search space is reconstructed for the next epoch. This procedure continues until the optimal integer ambiguities are identified and fixed. In this paper a direct integer ambiguity search (DIAS) method is developed. One major characteristics of this algorithm is that the optimal integer ambiguities are directly searched for without counting the potential solutions, using a quadratic integer programming (QIP) algorithm. Another feature is that the mathematical model used for the direct integer ambiguity search algorithm is based on all GPS data available. Thus the optimal solution of integer ambiguities at each epoch is the global solution for all GPS data available at that epoch which is equivalent to including all geometric constraints of the GPS constellation for the integer ambiguity resolution. The most critical issue for integer ambiguity resolution on the fly is to validate whether the optimal solution is the true solution for the integer ambiguities. Based on the information for the global optimal solution and the second best solution, some very reliable test and validation procedures are applied to the direct ambiguity search algorithm.

MIXED INTEGER LEAST-SQUARES ESTIMATION

The integer ambiguity $\bar{x}$ can be reliably estimated using all observation l_i available at epoch k. The general linear model for the global solution of the integer ambiguities $\bar{x}$ is of the form

$$l_i = A_i x + B_i y_i + n_i \quad \text{for} \quad i = 1, \cdots, k \tag{1}$$

with

$$x = \bar{x} \tag{2}$$

where

$$x \in R^n, \; y_i \in R^m, \; \text{and} \; \bar{x} \in Z^n \tag{3}$$

To estimate the parameters of the mixed integer linear model (1) the following least-squares criterion is applied

$$\Omega_{k/k} = \Omega' + \Omega'' = min \tag{4}$$

with

$$\Omega' = \sum_{i=1}^{k} \{(l_i - A_i \tilde{x} - B_i \tilde{y}_i)^T Q(i)_{ll}^{-1} (l_i - A_i \tilde{x} - B_i \tilde{y}_i)\} = min \tag{5}$$

and

$$\Omega'' = (\tilde{x} - \bar{x})^T Q(k/k)_{\tilde{x}\tilde{x}}^{-1} (\tilde{x} - \bar{x}) = min \tag{6}$$

where Ω' is the quadratic form for the floating ambiguities, Ω'' is the quadratic form for the integer ambiguities, $\tilde{y}$ is the estimate of y, $\tilde{x}$ and $\bar{x}$ are the float and integer estimates of the ambiguity parameter x based on all observations l_i (i=1,$\cdots$,k), and $Q(k/k)_{\tilde{x}\tilde{x}}^{-1}$ is the covariance of the global float estimate $\tilde{x}$.

Equation (5) can be solved using the standard recursive least-squares method. Using the estimated floating ambiguities $\tilde{x}$, the global solution for the integer ambiguities x is obtained by solving equation (6). One can use the quadratic integer programming technique to directly search the optimal integer ambiguity $\bar{x}$.

QUADRATIC INTEGER PROGRAMMING

Quadratic integer programming is of the general form

$$min \; f(x) = -2c^T x + x^T Q x \quad \text{with} \; x \in S \tag{7}$$

where $f(x)$ is the objective function, Q is a n x n symmetric positive matrix, S is the solution space of all feasible solutions defined as follows

$$S = \{ \ \mathbf{x} \ | \ |x_i| < b, \ \mathbf{x} \in \mathbf{Z^n} \ \}$$ (8)

and $|x_i| < b$ defines the search cube.

The Branch and Bound Method

The basic concept of the branch and bound method is to divide and discard. Since the original problem is too difficult to be solved directly, an efficient search can be accomplished by successively dividing the solution space into smaller subsets of feasible solutions. The dividing is done by partitioning the entire space of all feasible solutions into smaller and smaller subsets. The discard (fathoming) is done by evaluating how good the best solution in the subset is to determine whether it should be searched further or discarded. It involves calculating lower bounds f_i^* and upper bound $\hat{f}$ of the objective function. More details can be found in Garfinkel and Nemhauser (1972).

Integer Gradient Direction Method

The k-th step of the direct search can be described as

$$\mathbf{x}_{k+1} = \mathbf{x}_k + \lambda \cdot \mathbf{n}$$ (9)

where $\mathbf{x}_k$ is the base point to start the search at the k-th step, $\mathbf{x}_{k+1}$ is the new point to be searched and $\mathbf{m}$ is an integer direction vector, usually determined by the direction of the steepest descent of the objective functions $f(\mathbf{x}_k)$ at the base point $\mathbf{x}_k$. If the search direction $\mathbf{m}$ is determined the increment scale λ along the direction $\mathbf{m}$ can be determined using a one-dimensional search algorithm.

Discrete Modified Complex Method

In equation (9) the search direction can also be determined without computing the gradient of the objective function. Based on the complex method, a modified simplex method by Box (1965), the discrete modified complex (DMC) method was developed for the integer variable by Beveridge and Schechter (1970) and Fox (1981). The basic idea is to create a complex figure with 2N+1 vertices, with each vertex at a discrete point of the n-dimensional space of points $\mathbf{x}$. Starting from the 2N+1 vertices, the principle of the DMC method is to discard the vertex with the worst objective function value and locate a new vertex along the direction of the line between the poorest vertex and the centroid of the remaining vertices.

IMPLEMENTATION AND VALIDATION

Combining the branch and bound method and the direct discrete search algorithms discussed above, a very efficient direct integer ambiguity search (DIAS) algorithm has been developed at the University of Calgary and implemented in the GPS/INS processing software KINGSPAD (Kinematic Geodetic System for Position and Attitude Determination). The software KINGSPAD can process either GPS observations or INS data as well as the integrated GPS/INS data in both the static and kinematic modes for different applications. For the static and kinematic applications the DIAS algorithm has the following features:

(i) The DIAS algorithm is a direct integer search method, which directly yields the optimal solution of integer ambiguities without counting potential solution at each epoch. This approach is very robust because it avoids the critical issue of erroneously rejecting the optimal ambiguities which is possible using the concept of potential solutions;

(ii) The DIAS algorithm guarantees the optimal solution to be the global optimum of the minimization problem (4);

(iii) The optimal integer ambiguity obtained by DIAS algorithm is the global solution based on all GPS observations.

(iv) The search algorithm is very efficient. At each epoch it typically only takes about 0.1 to 0.2 second for 7 to 8 satellites; it thus can be used for real-time applications;

(v) The search algorithm also gives the second best solution for the integer ambiguities.

(vi) DIAS includes an overall validation test (OVT) procedure to validate the optimal integer ambiguities as the true solution.

Validation Test

Using the optimal and second best solutions one can form very reliable test criteria to validate the obtained optimal solution. The threshold test is given by

$$\text{i)} \qquad \Omega_1'' < \chi \, , \Omega_2'' < \chi, \tag{10}$$

$$\text{ii)} \qquad \Omega_1'' < \chi \, , \Omega_2'' > \chi, \tag{11}$$

$$\text{iii)} \qquad \Omega_1'' > \chi \, , \Omega_2'' > \chi, \tag{12}$$

and the test criterion of the ratio test is given as follows

$$\frac{\Omega_2''}{\Omega_1''} < F, \tag{13}$$

$$\frac{\Omega_2''}{\Omega_1''} > F \tag{14}$$

where Ω_1'' is the quadratic form of the optimal solution and Ω_2'' is the quadratic form of the second best solution.

In principle, using the threshold test and the ratio test, one can decide if the optimal solution is the true solution of the integer ambiguity.

RESULTS AND ANALYSIS

The DIAS technique implemented in KINGSPAD was tested both in the static and kinematic mode. The procedure using DIAS for ambiguity resolution on the fly is shown in Table 1. The first column gives the start time of the DIAS procedure. The second column gives the time when the true solution of integer ambiguities is assumed to be detected by the optimal solution. The third column is the time when the ratio between the second best solution and the optimal solution is larger than 2. At this time the true integer ambiguities are considered to be confirmed. The fourth column is the time when the integer ambiguities are fixed and considered to be the true integer ambiguities. The fifth column is the ratio between the second best solution and the optimal solution at the time when the integer ambiguities are be fixed.

start time	time of detect.	$\Omega_2/\Omega_1 \geq 2$	time of fix	Ω_2/Ω_1	remark
155350	155389	155418	155537	3.9	static, 7 sat.
156800	156935	156965	157084	6.0	15 m/s, 6 sat.
157600	158157	158951	159070	2.6	13 m/s, 6 sat.

Table 1: Ambiguity solution results of the land test, May 8, 1995

As shown in Table 1, the true integer ambiguities can be found by the DIAS algorithm at a very early stage. The time span between the time to start the validation procedure and the time when the true integer ambiguity are fixed varies in each case. This depends largely on the satellite geometry and the size of the observation errors.

CONCLUSIONS

In this paper the direct integer ambiguity search (DIAS) algorithm has been introduced for ambiguity resolution on the fly. The basic principle is the direct search for the global optimal solution of the integer ambiguities using integer nonlinear programming methods.

The major merit of the DIAS method is its robustness, because it always searches the global optimal solution based on all GPS data available at each epoch. The concept of a potential solution is no more necessary. The problem of rejecting the true integer ambiguity can thus be avoided. The critical issue here is to use all GPS data and their covariance information properly. Test results indicate that the DIAS algorithm implemented in KINGSPAD is very efficient. It only takes about 0.1 to 0.2 second for 7 to 8 satellites. The total computation time for the GPS processor in KINGSPAD, including ephemeris computation, navigation algorithm and the DIAS method, is less than 0.3 second using 486/50 PC computer. Thus, real-time applications can be processed.

The question of validating optimal integer ambiguities is the most critical issue for the integer ambiguity resolution procedure. For this purpose the DIAS method uses the second best solution for the integer ambiguities. Based on the optimal solution and the second best solution, a validation procedure using different test algorithms has been developed and tested. Test results show that these algorithms are reliable.

REFERENCES

Beveridge G.S. and R.S. Schechter (1970), *Optimization: Theory and Practice*. McGraw-Hill.

Box, M.J. (1965), A new method of constrained optimization and a comparison with other methods, Comp. J., 8, (1), pp. 42-52, 1965.

Fox, D.B. and J.S. Liebman (1981), A discrete nonlinear simplex method for optimized engineering design. *Engineering Optimization*, Vol. 5, pp. 129-149, 1979.

Garfinkel, R. and G. Nemhauser (1976), *Integer Programming*. John Willey & Sons.

Hatch, R. and H.-J. Euler (1994), Comparison of several AROF kinematic techniques. *Proc. of ION GPS-94*, The Institute of Navigation, Sept., 1994.

ALTERNATIVE FORMS OF GPS ADJUSTMENT MODELS WITH KALMAN GAIN MATRICES

Stanisław Oszczak

Institute of Geodesy and Photogrammetry
Olsztyn University of Agriculture and Technoplogy
Olsztyn, Poland

ABSTRACT

The alternative forms of batch and sequential estimation algorithms with Kalman gain matrices are given. The new approach to the problem of addition of new parameters to observation equations is presented. The generalized Kalman gain symbols (Kalman weighting matrices) introduced by the author and applied to the sequential estimation algorithm give a compact form of final equations for determination of all parametres and their covariance matrix.

INTRODUCTION

The processing procedure of GPS observations requires both batch and sequential estimation algorithms to be applied. For fast, real time GPS applications the sequential Kalman filtering algorithm is used, while for static session or other geodetic determinations the batch algorithm can be more convenient for computations. The duality introduced symbols, in geodesy and Kalman filtering, in both adjustment algorithms can be avoid using the defined Kalman filtering symbols. The paper presents the alternative forms of commonly used adjustment algorithms taking advantage of symbols of the Kalman gain matrices.

BATCH ESTIMATION ALGORITHM (Tapley, 1988)

Given: $\overline{X}_k$, $\overline{P}_k$ and observation y_k together with R_k

where: $\overline{X}_k$ is a priori estimate of the state matrix

and $\overline{P}_k$ is the associated covariance matrix of parameters

R_k is the covariance matrix of observations.

Linearized system of observation equation:

$$y_k = H_k x_k + \varepsilon_k; \quad R_k$$
$$\overline{x}_k = X_k + \eta_k; \quad \overline{P}_k$$

where: $E[\varepsilon_k] = 0, \qquad E[\varepsilon_k \varepsilon_k^T] = R_k$

and $E[\eta_k] = 0, \qquad E[\eta_k \eta_k^T] = \overline{P}_k$

Compute the Least - Squares estimate:

$$\hat{x}_k = (H_k^T R_k^{-1} H_k)^{-1} H_k^T R_k^{-1} y_k =$$

$$= \left[\begin{bmatrix} H_k^T & I \end{bmatrix} \begin{bmatrix} R_k^{-1} & 0 \\ 0 & \bar{P}_{k-1}^{-1} \end{bmatrix} \begin{bmatrix} H_k \\ I \end{bmatrix} \right]^{-1} \left[\begin{bmatrix} H_k^T & I \end{bmatrix} \begin{bmatrix} R_k^{-1} & 0 \\ 0 & \bar{P}_{k-1}^{-1} \end{bmatrix} \begin{bmatrix} y_k \\ x_k \end{bmatrix} \right]$$

$$\hat{x}_k = (H_k^T R_k^{-1} H_k + \bar{P}_k^{-1})^{-1} (H_k^T R_k^{-1} y_k + \bar{P}_k^{-1} \bar{x}_k)$$

$$P_k = (H_k^T R_k^{-1} H_k + \bar{P}_k^{-1})^{-1}$$

Remarks:

- y_k may be one or a batch of observations

- $\bar{x}_k$ may be initial estimate or from processing previous observations

- Charakteristics of P_k:
    - $n \times n$ symetric
    - must be positive definite
    - parameter observability is related to rank of matrix
    - number of observations, m, must be greater than the number of parameters being estimated, n
    - contains a measure of the accuracy estimate, $\hat{x}_k$

- Numerical problems may be encountered in computing $\bar{P}_k^{-1} (n \times n)$

- In the case when $m \gg n$ the batch algorithm may be the best choice for post - processing procedure.

SEQUENTIAL ESTIMATION ALGORITHM (Tapley, 1988)
(with addition of new observations only)
Given: $\bar{x}_k, \bar{P}_k$ and y_k together with R_k

Fundamental Identities:

$$P_k = (H_k^T R_k^{-1} H_k + \bar{P}_k^{-1})^{-1} = \bar{P}_k - \bar{P}_k H_k^T (H_k \bar{P}_k H_k^T + R_k)^{-1} H_k \bar{P}_k$$

$$K_k = \bar{P}_k H_k^T (H_k \bar{P}_k H_k^T + R_k)^{-1} = (H_k^T R_k^{-1} H_k + \bar{P}_k^{-1})^{-1} H_k^T R_k^{-1}$$

$$P_k = \bar{P}_k - K_k H_k \bar{P}_k = (I - K_k H_k) \bar{P}_k$$

Sequential estimate:

$$\hat{x}_k = \bar{x}_k + K_k(y_k - H_k \bar{x}_k)$$

$$P_k = (I - K_kH_k)\bar{P}_k$$

Remarks:
- The Sequential Estimation Algorithm is also referred as the Kalman - Bucy Filter

Advantages of sequential algorithm are as follows:
- Matrix to be inverted R_k^{-1} has the dimensions equal to number of observations $(m \times m)$

- In the case when $n > m$ the sequential procedure can be the best choice for both real time and post - processing GPS procedure
- The presented algorithm can be used also for batch solution. The Fundamental Identities give you a chance to select a proper equation for batch algorithm in the case when $m > n$.

THE SEQUENTIAL ESTIMATION ALIGORITHM EXTENDED FOR NEW PARAMETERS (Oszczak, 1993):
The problem of the two-component adjustment (Vanicek and Krakiwsky, 1982) can be applied to the process of sequential adjustment in the case when new observations as well as new parameters are added to the estimation. In the further part of the work, the same as above denotations are introduced, to be compatibile with those used in the Kalman filtering methods (Kalman and Bucy, 1961, Tapley, 1972).
- The linear model of sequential adjustement extended for new parameters has the form:
(Oszczak, 1986)

$$y = H_1(\bar{x}_1 + \delta x_1) + H_2 x_2 + v$$
$$W = y - H_1\bar{x}_1$$
$$W = H_1\delta x_1 + H_2 x_2 + v$$

$$C_r = \begin{vmatrix} \bar{P}_1 & 0 \\ 0 & R_2 \end{vmatrix}$$

where: $\bar{x}_1$ is a priori estimate of the vector of parameters,

x_2 is a vector of new parameters,

y is the misclosure vector,

H_1, H_2 are the design matrices in the first and the second observation groups, respectively,

δx_1 is the vector of corrections to the parameters determined from the first group,

v is the vector of errors of new observations,

$\bar{P}_1$ is a priori estimate of a covariance matrix of the parameters $\bar{x}_1$,

R_2 is the covariance matrix of observation errors

The solution of the system is:

$$\delta \hat{x}_1 = K_{1,2}W$$

$$\hat{x}_2 = K_2 W$$

$$\hat{v} = K_{V(1,2)}W$$

- Introduced symbols of generalized Kalman weighting matrices (Oszczak, 1986):

$$K_{1,2} = K_1 (I - H_2 K_2)$$
$$K_1 = \bar{P}_1 H_1^T (H_1 \bar{P}_1 H_1^T + R_2)^{-1} = P_1 H_1^T M^{-1}$$
$$K_2 = P_2 H_2^T (H_1 \bar{P}_1 H_1^T + R_2)^{-1} = P_2 H_2^T M^{-1}$$

$$K_{V(1,2)} = K_V (I - H_2 K_2)$$

$$K_V = R_2 I M^{-1}, \quad P_2 = (H_2^T M^{-1} H_2)^{-1}$$

The covariance matrices of the parameters have the forms:

$$P = \begin{vmatrix} (I - K_{1,2}H_1)\bar{P}_1 & -K_1 H_2 P_2 \\ -K_1 H_1 \bar{P}_1 & P_2 \end{vmatrix}$$

- In the case of addition of new observations only the general form of solution of the system can be reduced to the classical form given by Kalman and Bucy (1961):

$$\hat{x}_1 = \bar{x}_1 + K_1 (y - H_1 \bar{x}_1)$$

$$P_1 = (I - K_1 H_1)\bar{P}_1$$

SEQUENTIAL ESTIMATION ALGORITHM (Oszczak, 1993)

Given: $\bar{x}_{k-1}, \bar{P}_{k-1}, X^*_{k-1}$ and y_k, R_k

where: X^*_{k-1} is the vector of nominal values of the parameters at the basic time epoch t_{k-1}

and the observation is taken at the basic time epoch t_k

1. Compute the design matrices H_{k-1} and H_k

2. Compute the misclosure vector y_k

  (the difference between the results of observation and its value obtained from adopted mathematical model)

3. Compute

$$M_k^{-1} = (H_{k-1}\bar{P}_{k-1}H_{k-1}^T + R_k)^{-1}$$
$$K_{k-1} = \bar{P}_{k-1}H_{k-1}^T M^{-1}$$
$$P_k = (H_k^T M_k^{-1} H_k)^{-1}$$

$$K_k = P_k H_k^T M_k^{-1}$$
$$K_{k-1, k} = K_{k-1}(I - H_k K_k)$$
$$W_k = y_k - H_{k-1}\bar{x}_{k-1}$$

$$\delta \hat{x}_{k-1} = K_{k-1,k} W_k$$

$$\hat{x}_{k-1} = \bar{x}_{k-1} + \delta \hat{x}_{k-1}$$

$$\hat{x}_k = K_k W_k$$

4. Replace k with $k+1$ and return to 1

Remarks:

The sequential estimation algorithm has a compact and simple form compatible with the algorithms elaborated for the Kalman filtering methods (Tapley, 1972, 1988).

An advantage of the presented sequential processing algorithm is that the matrix to be inverted will be of the same dimension as the observation error covariance R_k.

In the case of addition of new observations alone, the equations given by the author are reduced to the same form as it was given for Kalman and Bucy filter (Kalman and Bucy, 1961):

$$\hat{x}_1 = \bar{x}_1 + K_1(y_1 - H_1\bar{x}_1)$$

$$P_1 = (I - K_1 H_1)\bar{P}_1$$

REFERENCES

1. Kalman R. E., R. S. Bucy, 1961, New Results in Linear Filtering and Prediction Theory, Trans. ASME, Ser. D., J. Basic Eng. 83: 95-108.
2. Oszczak S., 1986, Application of Filtering Methods to Elaboration of Satellite Data (in Polish), Acta Acad. Agricult. Tech. Olst., Vol. 15, Suppl. A.
3. Oszczak S., 1988, Application of Kalman Weighting Matrices to Sequential Adjustment by Addition of Observations and Parameters, Proceedings of Geodetic Meeting ITALY - POLAND, Bologna.
4. Oszczak S., 1989, Sequential Orbit Improvement with Addition of New Observations and Parameters, Observations of Art. Earth Satellites, Vol. 27, Kraków.
5. Oszczak S., 1993, Compact Form of Sequential Adjustment Algorithm Extended for New Parameters, IAG General Meeting, Beijing.
6. Tapley B. D., 1972, Statistical Orbit Determination Theory, Recent Advances in Dynamical Astronomy, D. Reidel. Publ. Comp. Dodrecht - Holland.
7. Tapley B. D., 1988, International Summer School of Theoretical Geodesy, Assisi, Italy, Lecture Notes.
8. Vanicek P, E. Krakivsky, 1982, Geodesy - the Concept, Noorth - Holland

Status of the GEOSAT software after ten years of development and testing [*]

P. H. Andersen [1]

[1] Norwegian Defence Research Establishment, P.O. Box 25, N-2007 Kjeller, Norway

1 Introduction

Almost ten years ago the GEOSAT software was for the first time presented to the international scientific community in a paper (Andersen, 1986) given at the Fourth International Geodetic Symposium on Satellite Geodesy, held in Austin, Texas. Since then the software has been extended significantly with improved models and the possibility to apply the software for many different space-geodetic measurement techniques. Presently, GEOSAT is the only existing software which can be used for analysis of VLBI data and satellite tracking data from GPS, SLR, PRARE, DORIS, radar altimetry and SST (Satellite-to-Satellite Tracking).

The simultaneous combined analysis of different datatypes, due consideration of the physical interrelations, and presentation of results in a common reference system, are the main ideas behind the developement of the GEOSAT software. As a first step towards this goal, the different techniques have been implemented and validated separately. Furthermore, the first results from a combined VLBI/GPS/SLR analysis are presented.

2 The GEOSAT software

GEOSAT as currently under development at NDRE, is a state-of-the-art software system for high-precision analysis of satellite and radio source tracking data for geodetic and geodynamic applications. GEOSAT is a multi-station, multi-satellite, and multi-measurement technique system, designed to provide a flexible tool for accuracy analysis related to geodetic studies. The software can be applied either in an estimation mode, a simulation mode or in an error analysis mode.

GEOSAT is a library of subroutines written in the FORTRAN language and presently consisting of approximately 100000 statements. The code is highly portable

---

[*]Paper presented at the XXI-th IUGG meeting in Boulder, July, 1995

and can be used on computers running under the UNIX operating system. There are virtually no limits on the maximum numbers of satellites and of tracking stations that can be included; these will be limited only by the external storage size and processing time available on the computer.

GEOSAT has a sophisticated mathematical model (Andersen, 1994) with equations of motion and measurements formulated either in a solar barycentric or a geocentric frame of reference and corrected for relativistic effects. In this software, the most precise reference frames, dynamical models, and measurement models available are used and updated continually as better information comes along.

Sophisticated versions of Bayesian weighted least squares and Kalman filtering techniques are available, and many types of model parameters, including dynamical parameters, can be represented by stochastic models (white or colored noise, or random walk).

3   Software validation

3.1   VLBI data analysis

Geodetic VLBI data have been analyzed using GEOSAT. Station coordinates and velocities, source coordinates, and Earth orientation parameters have been estimated in a global mode, using data from the Extended Research and Development Experiment (ERDE) and the Research and Development series (R&D) within the NASA's CDP (Andersen and Rekkedal, 1995). It is demonstrated using the ERDE data, that the program is capable of calculating station coordinates with an accuracy of a few mm in the local horizontal plane and about 7 mm in the vertical direction. Analysis of the R&D dataset yields errors 2 to 3 times higher in all directions. It has been shown that daily nutation and polar motion parameters can be estimated with an accuracy of typically 0.5 mas, and variation in UT1 with an accuracy of 0.04 ms. The coordinates of primary sources can in most cases be determined with an accuracy of 0.2 to 0.3 mas.

3.2   GPS data analysis

Six days of data from the GIG'91 experiment have been analyzed with a fiducial strategy (Andersen et al., 1993). The results obtained with GEOSAT show daily horizontal and length repeatabilities of 1 part in 10^9 plus 2 mm for baseline lengths up to 4000 km. A direct comparison with results from the GIPSY software shows, with some exceptions, mean differences at the sub-cm level. After transformation to ITRF'90, the rms of the coordinate differences is 14.8 mm. Studies of orbital predictions and comparisons with external high-precision orbits indicate a mean orbit precision of around 35 cm in each cartesian coordinate. Correlations between the GEOSAT and GIPSY solutions indicate some common model deficiencies.

3.3 SLR data analysis

3.3.1 Low-orbit satellites

SLR tracking data for the ERS-1 satellite have been analyzed in order to validate the force model (Andersen et al., 1994). The altitude of 800 km requires a very accurate model in order to recover a high-precision orbit. A realistic surface force model for ERS-1 is used together with the Jacchia 77 atmospheric model, semidaily drag coefficients, a 1-cpr (cycle-per-revolution) sinusoidal along-track acceleration, and the GSFC JGM-2 gravity model. ERS-1 orbits have been derived for 5.5-day arcs of laser tracking data between July 6 and August 12, 1992. Results from overlapping orbits and comparison with precise D-PAF orbits indicate an orbital accuracy of 10-15 cm in the radial direction, $\sim$ 60 cm in the along-track direction and $\sim$ 15 cm in the cross-track direction.

3.3.2 High-orbit satellites

Amplitudes and phases of tidally coherent daily and subdaily variations in the Earth's rotation and geocenter have been estimated from an analysis of twelve months of SLR tracking data from the LAGEOS 1 & 2 and ETALON 1 & 2 satellites, performed between October 1992 and September 1993 (Andersen, 1994). The dataset consisted of 132000 two minute normal points from 37 globally distributed stations. The results show good agreement with results from similar analysis of independent VLBI and SLR data. It was demonstrated that it is possible to determine a one-year continuous high-precision series in UT1 using multi-satellite laser ranging. Polar motion and UT1, with a resolution of 3 days, were determined with a precision of better than 0.2 mas and 0.02 ms. Station coordinates differed from ITRF'91 with 1 - 2 cm for the best stations. The geocenter differed from ITRF'91 by $\sim$ 2 cm. Vertical ocean loading amplitudes and phases were estimated for some selected stations. Zonal gravity coefficients from degree two up to degree six were estimated in addition to $\bar{C}_{21}$, $\bar{S}_{21}$, $\bar{C}_{22}$, $\bar{S}_{22}$.

3.4 PRARE data analysis

51 passes of SLR and 10 passes of METEOR 3 PRARE range and range-rate tracking data in the period February 16-19, 1994, have been analyzed with the GEOSAT software (Andersen, 1995). The results indicate that the noise level of the PRARE data is very small but the data contain some range biases still to be explained. The total *a posteriori* rms of residuals for the SLR, PRARE range and PRARE range-rate data were 8.4 cm, 4.5 cm and 0.3 mm/s correspondingly. The average PRARE range bias for all ten passes was estimated to be 4 cm with a one-sigma uncertainty of 16 cm.

3.5 Crossover satellite altimetry data analysis

ERS-1 satellite altimetry data and SLR data (39 passes from 13 stations) from the period September 19-24, 1992, have been analyzed in a combined mode. 50 globally

COMBINED VLBI/GPS/SLR : REPEATABILITY				
$\sigma = a + bL$	South	East	Height	Length
a(mm)	1.7	2.1	4.2	2.3
b(ppb)	0	0	0	0.18

Table 1: Coordinate repeatability of collocated stations with high-precision eccentricity vectors. Two of the stations, Ft. Davis (MDO1) and Wettzell (WETT), had also some passes of SLR data.

GPS ECCENTRICITY CHECK				
Station	ΔX(mm)	ΔY(mm)	ΔZ(mm)	σ (mm)
ONSA	0.8 ± 1.5	3.6 ± 0.5	0.5 ± 1.4	3
WETT	-2.8 ± 1.3	-3.8 ± 1.7	-1.3 ± 0.6	3
KOKB	0.8 ± 1.5	0.6 ± 1.3	-0.5 ± 2.0	3
MDO1	-0.5 ± 1.9	2.2 ± 2.4	-0.6 ± 2.2	12

Table 2: The difference between VLBI to GPS eccentricity vectors estimated by GEOSAT and the vectors applied by IERS in the derivation of ITRF'93. The last column shows the precision of the ground tie.

distributed altimetry crossover points (XA) were generated from the raw data and applied in the analysis. The *a posteriori* rms of residuals was 10 cm for SLR and 15 cm for XA.

4 Combined VLBI - GPS - SLR data analysis

The GEOSAT combined analysis included VLBI (the R& D network) and GPS data (the IGS network), and SLR data for LAGEOS 1 & 2, from January 13-15 and from January 20, 1994. Data from 2 collocated VLBI/GPS/SLR stations, 4 VLBI/GPS stations, 28 GPS-alone stations and 12 SLR-alone stations were applied in the analysis.

No steps were taken in the analysis to optimize the relative data weighting between the three techniques. Typically 40000 GPS pseudorange measurements with a weight based on $\sigma = 1$ m, 40000 GPS phase measurements with a weight based on $\sigma = 7$ mm, 6000 VLBI group delay measurements with a weight based on $\sigma = 7$ mm, and 300 SLR measurements with a weight based on $\sigma = 5$ cm, were analyzed for each day.

All parameters were estimated simultaneously in a free network mode. One common set of station coordinates were derived for each collocated station with contributing data from all techniques when available. Table 1 presents the coordinate repeatability of collocated stations with high-precision eccentricity vectors. It can be seen that the repeatability in baseline length is almost independent of the distance between the stations.

Eccentricity vectors from VLBI to GPS and/or SLR were also estimated with an *a priori* sigma in accordance with the given precision of the ground tie. The difference between VLBI to GPS eccentricity vectors estimated by GEOSAT and the vectors applied by IERS in the derivation of ITRF'93 is shown in Table 2.

5 Conclusions

A sophisticated high-precision space geodesy software system, GEOSAT, has been developed and validated. GEOSAT is presently the only existing system available for analysis of both VLBI and satellite techniques in common reference frames with consistent models. Some of its capabilities are demonstrated with applications to VLBI, GPS, and SLR, including precise orbit determination for a low-altitude satellite ERS-1 and geodynamical applications of the dense and spherical high-altitude satellites LAGEOS 1 & 2 and ETALON 1 & 2. The quality of the results is comparable with that obtained by the leading international institutions within each technique.

The combined multi-technique analysis strategy has been demonstrated with very promising results.

6 Acknowledgements

The author is grateful to the Norwegian Mapping Authority for the continuous financial support of the project during many years.

7 References

Andersen, P. H. (1986). GEOSAT - a computer program for precise reduction and simulation of satellite tracking data. In proc. of the Fourth Int. Geod. Symp. on Satellite Geodesy, Austin, Texas, April 28 - May 2.

Andersen, P. H. (1994). High-precision Station Positioning and Satellite Orbit Determination, PhD. thesis, the University of Oslo, NDRE/PUBL-95/01094, The Norwegian Defence Research Establishment.

Andersen, P. H. (1994). Measuring rapid variations in Earth orientation, geocenter and crust with satellite laser ranging. Accepted for publication in Bulletin Geodesique.

Andersen, P. H. (1995). METEOR 3 PRARE data analysis at NDRE, EUCLID RTP9.1 Research note, July.

Andersen, P. H., S. Hauge and O. Kristiansen (1993). GPS relative positioning at the level of one part per billion. Bulletin Geodesique, 1993, Vol 67, No. 1, pp. 91-106.

Andersen, P. H. and S. Rekkedal (1995). VLBI data analysis with the GEOSAT software, Bulletin Geodesique, Vol 69, No. 3, pp. 125-134.

Analysis of Data from the VLBI-GPS Collocation Experiment CONT94 [*]

P. H. Andersen [1], O. Kristiansen [2] and N. Zarraoa [2]

[1] Norwegian Defence Research Establishment, P.O. Box 25, N-2007 Kjeller, Norway
[2] Geodetic Institute, Norwegian Mapping Authority, 3500 Hønefoss, Norway

1   Introduction

During January 1994, a special VLBI experiment took place. For the first time in VLBI history, three different networks were scheduled to operate in continuous mode for almost two weeks. This experiment will be referred to as "CONT94". The main purpose of this setup was to generate a data base of the best quality for a thorough comparison of VLBI and GPS capabilities. Therefore, most of the VLBI sites were also occupied with GPS receivers of the best quality.

Complementary to this hardware collocation, we had also access to a solid software collocation, with three different software packages able to analyse each technique or both. The GEOSAT multi-purpose software (Andersen, 1994), developed at the Norwegian Defence Research Establishment, has proven its capacity to process GPS data, VLBI data, and SLR data either individually or simultaneously. We have also applied well known software packages for the analysis of VLBI data (OCCAM, Zarraoa, 1993) and GPS data (GIPSY II, Webb and Zumberge, 1993), in order to compare the estimates of baseline lengths and tropospheric parameters.

2   Data applied in the analysis

We have selected VLBI data from the Research and Development network (R&D), consisting of the seven VLBI stations listed in the first column of Table 1. The R&D stations successfully observed all the scheduled periods. All stations, except one, being collocated with GPS receivers. Both systems were connected to the same Hydrogen-maser at four of the stations. In addition we have applied data from twenty-eight

---

[*]Paper presented at the XXI-th IUGG meeting in Boulder, July, 1995

Station	Instr	Station	Instr	Station	Instr	Station	Instr	Station	Instr
FAIR	VGH	CRO1	G	NLIB	G	PIE1	G	HOBA	G
KOKB	VGHE	MATE	G	FORT	G	ALGO	G	HART	G
WEST	VGH	HERS	G	KOSG	G	GRAZ	G	YAR1	G
LA-V	V	TIDB	G	NYAL	G	TROM	G	USUD	G
MDO1	VGHE	TSKB	G	TAIW	G	MCMU	G	MASP	G
ONSA	VGHE	SANT	G	YELL	G	PAMA	G	MADR	G
WETT	VGHE	METS	G	RIC5	G	GODE	G	GOLD	G

Table 1: The table shows the stations applied in the analysis. The instrumental codes are defined as follows: V: VLBI, G: GPS, H: Collocated Hydrogen-maser clock, E: Eccentricity vector between VLBI and GPS is known with a precision of 12 mm or better (1-σ).

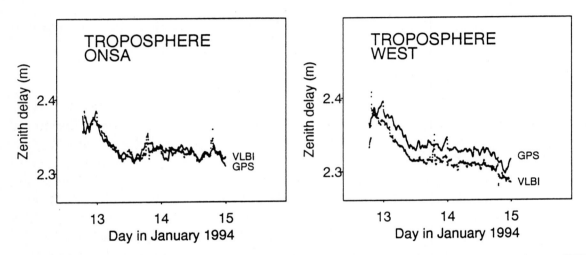

Figure 1: Total zenith delay estimated from VLBI data with OCCAM and from GPS data with GIPSY for the Onsala and Westford stations.

other GPS receivers participating in the International GPS Service for Geodynamics (IGS) network.

The R&D VLBI data were processed using both the GEOSAT and the OCCAM 3.3 software packages. The data from the two other VLBI networks were also analyzed, but only with OCCAM. For the GPS data set, data from all stations in Table 1 were analyzed for all days, except January 20, using GIPSY and only for January 13-15 and January 20 using GEOSAT.

3 Estimation of tropospheric parameters

A comparison of independent GPS and VLBI estimates of the tropospheric and clock parameters can provide an important *a priori* knowledge that can be used in the proper combination of the two data types. We will concentrate on the tropospheric

ECCENTRICITY CHECK			
Station	Component	GEOSAT (mm)	ITRF/IGS (mm)
ONSA	X	52626 ± 4	52631 ± 3
	Y	-40470 ± 3	-40464 ± 3
	Z	-43874 ± 6	-43865 ± 3
WETT	X	38705 ± 3	38697 ± 3
	Y	117424 ± 3	117417 ± 3
	Z	-59315 ± 2	-59322 ± 3
KOKB	X	-506 ± 11	-512 ± 3
	Y	-19410 ± 7	-19409 ± 3
	Z	-42239 ± 3	-42235 ± 3
MDO1	X	-5989580 ± 13	-5989575 ± 12
	Y	3788656 ± 7	3788650 ± 12
	Z	4541696 ± 6	4541694 ± 12

Table 2: Comparison of eccentricity vectors estimated using the GEOSAT software and vectors applied by IERS in the derivation of ITRF'93.

solutions here and return to the clock problem in a later report.

With the exception of a small bias, due to different antenna heights, both techniques are affected by the same tropospheric effects. GPS offers a better horizontal distribution of data and a higher data rate. VLBI offers an improved elevation-dependent range of data. In Fig. 1 we present the estimated tropospheric zenith delays for Onsala and Westford obtained individually from VLBI and GPS data and after accounting for the GPS and VLBI antenna height differences. The VLBI estimates are the filtered values without any smoothing. In most cases the VLBI-GPS tropospheric biases were a few mm except for the Westford station with a bias of 18 mm. After removing the tropospheric bias, the rms between the VLBI and GPS estimates is typically around 10 mm or less. Kokee was an exception with a rms of 21 mm. The dicrepancies might be due to the application of two different programs and to the use of unsmoothed tropospheric VLBI estimates. It is interesting to observe that even with the use of unsmoothed results, the VLBI estimates seem to be less noisy than the smoothed GPS estimates.

4   Separate and combined data analysis

The GEOSAT analysis included GPS data from January 13-15 and from January 20, and VLBI data from January 12-24. First, the VLBI and the GPS data were analyzed separately. From the mean values of the estimated coordinates for each technique, a set of VLBI/GPS eccentricity vectors, given in Table 2, were calculated for Onsala (ONSA), Wettzell (WETT), Kokee (KOKB) and Ft. Davis (MDO1). For these stations we also list the high-precision eccentricity vectors applied by IERS in their derivation of the ITRF'93 reference frame. The precision given for the GEOSAT

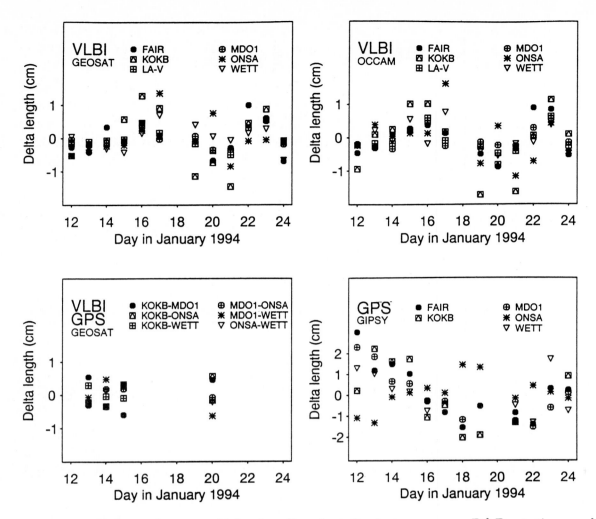

Figure 2: Daily estimates of the baseline lengths between any R&D station and Westford obtained from VLBI analysis with the GEOSAT software (upper left) and from the OCCAM software (upper right). The combined solutions with GEOSAT are given in the lower left plot while the daily GPS solutions with GIPSY are in the lower right plot.

eccentricities are based on the daily repeatabilities of the GPS and VLBI solutions. The average eccentricity residual difference is below 6 mm which is remarkably low remembering that both the local tie errors, errors in the GPS solutions and errors in the VLBI solutions contribute. This high consistency between the GPS and VLBI results can be the consequence of using very consistent models for the analysis of the two data types within the same program.

In a series of combined GPS/VLBI analyses, common coordinate estimates for four of the collocated stations were obtained using GEOSAT. All other parameters had individual estimates for each of the two techniques. The relative data weight between the two data types were selected on the basis of results from a series of runs with simulated data and with variable relative data weighting. The daily variations of the baseline lengths between the stations with precise eccentricity vectors is shown in the lower left plot of Fig. 2. The daily solutions for the baseline lengths obtained from either VLBI or GPS data is also shown. It can be seen from the lower right plot that the GPS-alone solutions with GIPSY give more noisy estimates than the VLBI-alone solutions with GEOSAT or OCCAM. The agreement between the results obtained from VLBI analysis with the GEOSAT and the OCCAM programs is excellent, once a small systematic scale difference is removed. The repeatability level achieved for the baseline length using the CONT94 data set is around 0.7 parts per billion, yielding the best results ever obtained from VLBI analysis. However, we have observed a systematic behaviour in the estimates of the baseline lengths that is clearly seen from day to day. We still do not have an explanation for this behaviour, which is common to all baselines. If it could be removed, the rms would decrease to the level of a few parts in 10^{-10}.

5   Conclusions

The agreement in the results obtained from VLBI analysis with the GEOSAT and the OCCAM programs is excellent, once a small systematic scale difference is removed.

Baseline lengths and tropospheric parameters seem to be more precisely estimated with VLBI data than with GPS data.

A combined analysis of VLBI and GPS data with consistent models has for the first time been demonstrated with very promising results. Such a combination of data will be especially advantageous in analysis of VLBI sessions with poor network geometry.

6   References

Andersen, P. H. (1994). High-precision Station Positioning and Satellite Orbit Determination, PhD. thesis, the University of Oslo, NDRE/PUBL-95/01094, The Norwegian Defence Research Establishment.

Webb, F. H. and J. F. Zumberge (1993). An introduction to GIPSY/OASIS II, Jet Propulsion Laboratory, 4800 Oak Grove Drive, Pasadena, CA 91109.

Zarraoa, N. (1993). OCCAM 3.2. Status report. In proceedings of the IX European VLBI Meeting, Univ. Bonn Geod. series, 81, pp. 25-26. Ed. J. Campbell and A. Nothnagel.

A NEW SOFTWARE FOR GPS DATA PROCESSING.
Work in progress and preliminary results.

B. Betti[1], M. Crespi[2], B. Marana[3], G. Venuti[3]
[1] DIGET, Politecnico di Torino - Italy
[2] DITS, Universita' di Roma - Italy
[3] DIIAR, Politecnico di Milano - Italy

INTRODUCTION

BA.M.BA. is a new software for GPS data processing. It works following a multibase approach to process observations collected in static mode. The structure of the program is modular, basically constituted by three parts devoted to perform the following operations:
preprocessing
first processing
final processing
The software is written in standard FORTRAN77 both for VAX and PC platforms.
It requires, as basic input, GPS data in RINEX v.2 format, i.e. ASCII files generated by the standard public domain Bernese software.
The strategy of solution works via classical double differences approach; however it is important to underline that there are some innovative modules applied to preprocessing, to cycle-slips detect and repair technique and to the solution of initial phase ambiguity problem. By now only the preprocessing module is working and it has been partially tested; the implementation of the first processing step module is in progress, while the theoretical background supporting the last module of the program has been completely analyzed and successfully applied to the solution of a small simulated example. We plan to reach our final goal in summer 1996.
Now let us analyze in details the three different modules of the program.

PREPROCESSING

It works through a preliminary screening of raw data RINEX files and then it provides the least squares approximated estimates of phase-smoothed ranges at each epoch and real initial phase ambiguities; further it is able to detect and repair gross cycle-slips and to evaluate the evolution of ionospheric delay. This approximated solution is achieved starting from already existing methodologies (Euler-Goad, 1991), but following a completely new procedure based on a simple least squares approach applied to the following observation equations for each satellite and epoch:

$$C_1 = \rho + I/f_1^2$$
$$\lambda_1\Phi_1 = \rho - I/f_1^2 + N_1\lambda_1$$
$$Y_2 = \rho + I/f_2^2$$
$$\lambda_2\Phi_2 = \rho - I/f_2^2 + N_2\lambda_2$$

Common to all four equations is ρ which represents the sum of geometrical contributes due to distance between receiver and satellite, clock errors, S/A effects, nondispersive contributions such as tropospheric refraction and other smaller effects (relativistic).

The least squares estimate provides:

$$\hat{N}_1 = E(N_1(t))$$

$$\hat{N}_2 = E(N_2(t))$$

$$\hat{J} = E(J(t)) \qquad\qquad (J = I/f_1^2)$$

$(\bullet(t) = $ value of parameter $\bullet$ at epoch t$)$

The last preprocessing step provides least squares estimates of approximated double differences real phase ambiguities.

FIRST PROCESSING

It provides, via a least squares solution technique, the approximated estimates of station coordinates, based on phase-smoothed ranges and ephemeris information (broadcast or precise), and baseline vectors (lengths and cartesian components), based on double difference observations (codes and phases) processing.

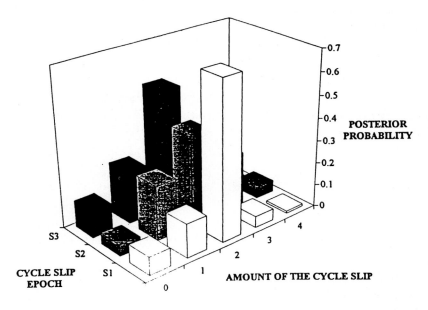

Fig. 1. Posterior distribution of the finite and discrete variable "cycle-slip amount" at each epoch.

321

Then it works through least squares triple difference phase observations to refine cycle-slips detection and repair and to provide baseline vectors estimates.

The procedure of detection and repair of residual cycle-slips is then refined following two different strategies: via a classical least squares double difference processing and residual data snooping and via a new Bayesian approach to determine the posterior distribution of the finite and discrete variable "cycle-slip amount" at each epoch and the most probable cycle-slip epoch (Fig. 1).

Finally, through double differences, the last step provides the estimate of real initial phase ambiguities, baseline lengths and cartesian components.

FINAL PROCESSING

It processes double difference phases following a new Bayesian method (Betti, Crespi, Sanso', 1993) leading to the final estimate of the baseline vectors.

The solution is obtained by suitably averaging the different (fixed) solutions, corresponding to different combinations of integer ambiguities, weighing them with the probability (density) of getting each particular combination.

This procedure has the advantage to overcome the problem of strongly underestimating the variance of the point coordinates, which occurs when ambiguities are just fixed to integer values without taking into account the error on the estimates due to a wrong choice of the combination of integer values (Fig. 2).

The Bayesian estimates are:

$$\hat{\xi}_B = \sum_\beta \xi_\beta p(\beta) \qquad \text{(weighed average)}$$

$$C_{\hat{\xi}_B \hat{\xi}_B} = \sigma_0^2 N_{xx}^{-1} + \sum_\beta \xi_\beta \xi_\beta^T p(\beta) - \sum_{\beta\beta'} \xi_\beta \xi_\beta^T p(\beta) p(\beta') \qquad \text{(covariance matrix)}$$

$$p(\beta) = \frac{e^{-\frac{1}{2\sigma_0^2}\bar{\psi}(\beta)}}{\sum_\beta e^{-\frac{1}{2\sigma_0^2}\bar{\psi}(\beta)}} \qquad \text{(weights)}$$

where:

$\hat{\xi}_B$ represents the least squares solution for geometric (baseline vectors) parameters;

β is a real value originated by the difference between the floating estimate of the ambiguities and the chosen combination of integers;

N_{xx} is the normal matrix corresponding to the ξ parameters;

$\bar{\psi}(\beta)$ is the reduced quadratic form which has to be minimized to provide the particular solution β.

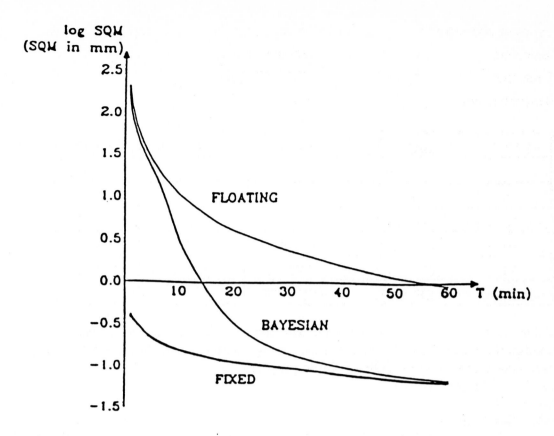

Fig. 2. Mean square estimation error (Bayesian) of the bayesian estimation, compared to the floating and the fixed ambiguity solutions.

ORBIT IMPROVEMENT

The program finally allows to get both a least squares definitive estimate of phase-smoothed ranges at each epoch, based on cycle-slips free phase data, and corrected ionospheric delays.

Then, starting from station positions and from the information provided by the new estimate of phase-smoothed ranges, it is also possible to get a least squares estimate of satellite orbits.

PRELIMINARY RESULTS

We made a comparison between approximated real initial double differences L1 phase ambiguities, estimated by the preprocessing module, and the corresponding integer parameters estimated by TOPAS Turbo program.

Table 1 shows the results, we obtained, versus those provided by TOPAS Turbo for a base of about 7.5 Km.

Types of observations: L1, L2, C/A, Y1-Y2
Receivers: Trimble 4000 SSE
Time span: ~ 1^h 25^m
Baseline length: ~ 7.5 Km

SAT.#	PHSRNG (P)	TOPAS(T)	EPOCH #	(P-T)
	[cycles]	[cycles]		[cycles]
16	-320989.9	-320993	170	3.1
18	1284795.2	1284787	99	8.2
19	891991.7	893987	170	4.7
24	1070143.2	1070133	31	10.2
27	REF. SAT.	REF. SAT.	170	--

Table 1. Comparison between approximated real initial double differences L1 phase ambiguities, estimated by the preprocessing module, and the corresponding integer parameters estimated by TOPAS Turbo program.

REFERENCES

Betti B., Crespi M., Sanso' F. : "A geometric illustration of ambuguity resolution in GPS theory and a Bayesian approach" - Manuscripta Geodetica, volume 18, number 5, 1993.

Crespi M. : "GPS data processing: advantages and limits of traditional approaches" - PhD thesis (in italian), 1992.

Euler H.J., Goad C.C. : "On optimal filtering of GPS dual frequency observations without using orbit information" - Bulletin Geodesique, 65: 130-143, 1991.

Landau H. : "TOPAS Turbo Software version 3.3 user's manual" - terraSat GmbH, 1993.

Landau H. : "On the use of the Global Positioning System in Geodesy and Geodynamics: Model Formulation, Software Development, and Analysis" - English translation of the author's PhD thesis, 1989.

Marana B. : "GPS signals preprocessing" - GNGTS symposium, 1994.

Teunissen P.J. : "The least-squares ambiguity decorrelation adjustment: a method for fast GPS integer ambiguity estimation" - Publications of the Delft Geodetic Computing Centre, No. 9, 1994.

Low-cost GPS Time Synchronization: The "Totally Accurate Clock"

Thomas A. Clark
Laboratory for Terrestrial Physics
NASA Goddard Space Flight Center
Greenbelt, MD, 20771
clark@tomcat.gsfc.nasa.gov

The "Totally Accurate Clock" ("*TAC*") has been developed as a simple, low-cost way to provide accurate and precise epoch timing for VLBI and SLR stations. The *TAC* provides a precise one pulse/second (1PPS) signal for external use which has been shown to have 30-50 nsec RMS scatter (the range being dependent on the measurement averaging interval) when compared to Hydrogen Maser frequency standards with no operator intervention. This precision is achieved despite the effects of S/A (Selective Availability, wherein the US Military intentionally degrades the stability of the GPS satellite clock signals). S/A effects are reduced by averaging over many GPS satellites since the S/A modulation is incoherent between the GPS satellites. The overall Allan Variance has been measured to be ~ $(50 \text{ nsec}/T)$ for time intervals T from a few seconds to 3 days. The accuracy of the timing pulse has been shown to be <20 nsec compared to the USNO master clocks. The 1PPS timing epoch can be set (under software control) to any arbitrary offset from the UTC second with 1 nsec resolution.

The *TAC* is a small, stand-alone device based on a commercial single-frequency GPS receiver module plus a small circuit board that provides RF amplification of the GPS signals, provides buffered 1 PPS outputs, and makes an RS-232 1PPS signal available for computer use on the DCD handshaking line. In our realization, the two circuit boards are mounted in a 18.5 x 13 x 5.5 cm box along with a NiCd battery and all I/O connections. The reproduction cost of the *TAC* hardware is about $1100 US including power supply and antenna. Following our preliminary announcement of the *TAC* implementation in early 1995, we have been contacted to produce more than 50 copies the *TAC* for Space Geodesy and Radio Astronomy Observatories around the world.

The *TAC* operations are controlled by an external computer through an RS-232 interface; the 1PPS timing signal is sent to the computer on the RS-232 interface for synchronization. A software program called *SHOWTIME* provides support for the *TAC* and runs on IBM-PC clones; *SHOWTIME* assists in initial setup and displays time-of-day, precise timing and GPS status information. We plan to include many of the *SHOWTIME* functions for use in the *LINUX*-based Mark-4 VLBI Field System control software in the future. Other groups have implemented an *UNIX*-based `timed` daemon for computer network time synchronization.

Information on the *TAC* design and the *SHOWTIME* software are available by anonymous ftp on the Internet from:

```
ftp://aleph.gsfc.nasa.gov/GPS/totally.accurate.clock
```

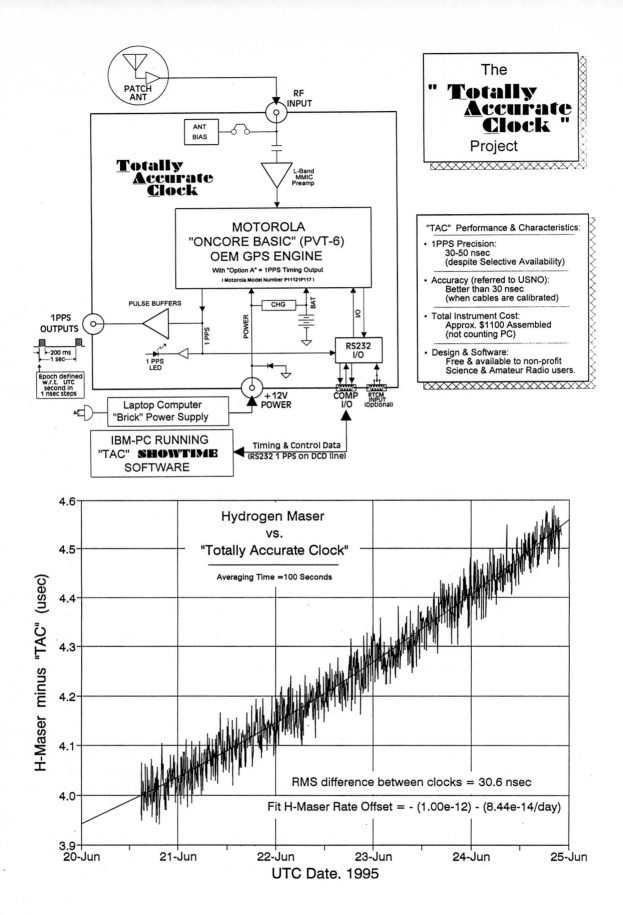

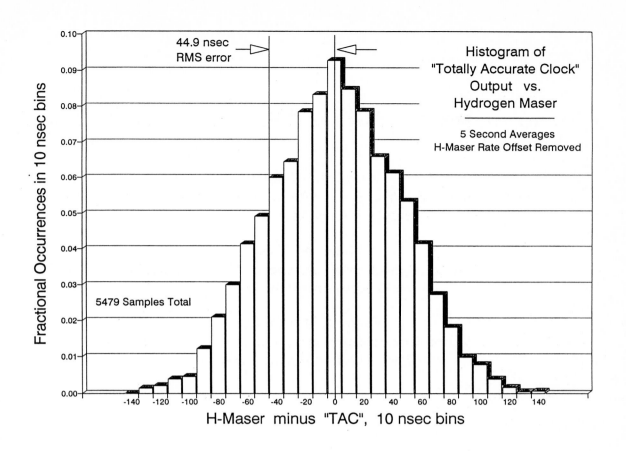

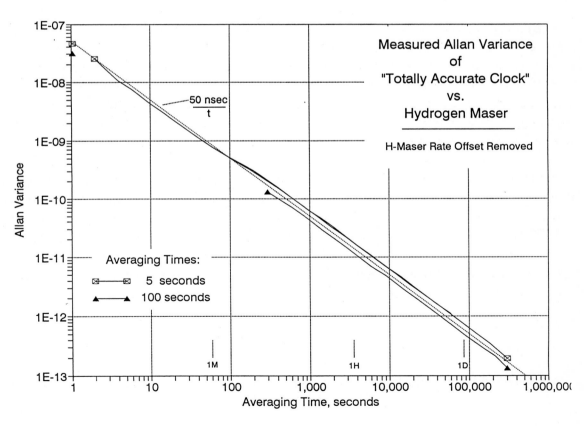

Characterizations of GPS User Antennas: Reanalysis and New Results

Bruce R. Schupler
AlliedSignal Technical Services Corporation, Columbia, Maryland 21045 USA
(schuplb@thorin.atsc.allied.com)

Thomas A. Clark
Goddard Space Flight Center, Greenbelt, Maryland 20771 USA
(clark@tomcat.gsfc.nasa.gov)

Roger L. Allshouse
AlliedSignal Technical Services Corporation, Columbia, Maryland 21045 USA
(allshor@clmmp003.atsc.allied.com)

INTRODUCTION

For several years we have been measuring the RF characteristics of a variety of GPS user antennas at both L1 and L2 using the Goddard Space Flight Center anechoic chamber. The properties that we have measured have included the location of the antenna phase center, and the amplitude, phase, and axial ratio response of the antenna as a function of zenith distance and azimuth. The use of our phase patterns and associated phase centers in the processing of GPS data has proven to be useful. However, a close examination of the phase patterns that we have measured has shown that the locations of the phase centers that we have published may not be optimal.

As we have discussed previously [*Schupler and Clark, 1991*], ignoring the location of the phase center or the effect of the phase pattern can have a serious effect on the quality of the recovered geodetic product. It can produce a systematic error in the baseline between two sites in the cases of a single occupation or of multiple occupations that have exactly the same configuration at each time at each site (antennas, antenna orientation, observing strategy, etc.). If multiple occupations that do not have exactly the same configuration at each time at each site are involved, ignoring the antenna related effects can increase the scatter in the recovered baselines.

ANALYSIS PROCEDURE

The phase center locations and phase patterns that we have previously published [*Schupler et. al., 1994*] and distributed were those obtained by the personnel who

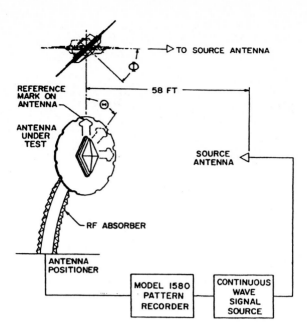

Fig. 1. Measurement geometry in the GSFC anechoic chamber.

operated the anechoic chamber and did not take into account the possible effects of miscentering of the true antenna phase center with respect to the center of rotation of the antenna positioner. The effects of this miscentering can be modeled as follows:

$$a_O + a_U \sin(\Phi) + a_E \cos(\Phi)\sin(\Theta) + a_N \cos(\Phi)\cos(\Theta) \tag{1}$$

where a_U is the component of miscentering in the up direction and a_E and a_N are the miscentering components in the East and North directions. (See Figure 1 for the definitions of Φ and Θ.)

An error in the vertical position of the phase center manifests itself as an increased amplitude of the phase pattern. An error in the horizontal position of the phase pattern manifests itself as a tilt in the phase pattern.

When we attempted to fit the miscentering correction shown in Equation (1) to the phase pattern data that we had collected for the various antennas, it quickly became apparent that the phase center location which we recovered was quite sensitive to the zenith angle range over which the fit was performed. This motion at 10 degree increments of the cutoff angle from 10 to 110 degrees along with the recovered position of the phase center for the Trimble 2202000 antenna at L1 and the location of our original phase center determination are shown in Figure 2. (Note that the tics on the North and East axes are spaced at 0.2 mm while those on the vertical axis are spaced at 25 mm.)

It also was apparent that the phase pattern generated by applying the displacement coefficients calculated for various cutoff angles was well constrained within the cutoff angle range which determined the coefficients, but was not well behaved outside of this range. (Figure 3 shows an example of this behavior for the Trimble 2202000 antenna at L1.) This observation bears directly on the choice of the zenith distance cutoff angle

329

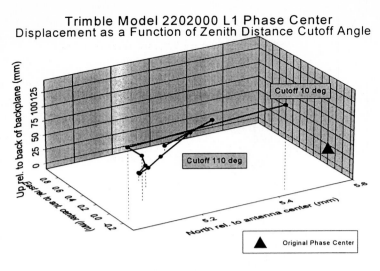

Fig. 2. Phase center motion as a function of data cutoff angle for the Trimble 2202000 antenna at L1.

which should be used to determine the phase center for each antenna. This canonical zenith distance cutoff angle should be large enough that all plausible GPS observations fall within the well constrained portion of the pattern, but not so large that the fitted location of the phase center does not closely reflect the effective phase center location for a typical observing scenario.

Based on a consideration of these two factors, we have chosen to define the phase centers and phase patterns for the various antennas using fits to data taken within 80 degrees of the zenith. This cutoff angle is greater then the zenith distance cutoff typically used for geodetic GPS observations at the present (75 degrees), yet is close

Fig. 3. Adjusted phase patterns are well constrained only within the range of data used in the adjustment.

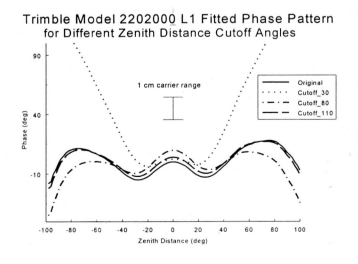

| Antenna | Original Phase Center | | | | | |
| | L1 | | | L2 | | |
	Up (mm)	North (mm)	East (mm)	Up (mm)	North (mm)	East (mm)
Dorne-Margolin T	1.59	3.97	1.94	65.79	3.59	1.16
Trimble 2202000	19.4	5.6	0.0	2.6	2.82	-0.55
Ashtech L1/L2	29.2	0.62	5.06	-20.3	0.62	5.06
Dorne-Margolin DMC 146.6	75.1	0.0	0.0	96.7	0.0	0.0
Micropulse 90LL12300	39.62	0.78	-0.78	68.84	1.57	0.0
TI	221.6	0.0	0.0	197.7	0.0	0.0
Ashtech Geodetic	37.08	0.88	-1.66	37.08	2.39	-1.55
Macrometer	107.1	-0.3	-0.3	91.69	-0.97	-0.37
Trimble 14532-00	25.7	-1.68	-3.94	-1.32	-2.25	-1.13

Table 1. Phase centers as originally determined in the GSFC anechoic chamber.

enough to this value that the effective phase center is well represented by our adjusted position. The selection of a cutoff angle to define the phase centers and phase patterns somewhat greater then the presently used processing cutoff angle also allows the processing cutoff to be lowered in response to hardware or analysis improvements without the necessity of redefining the phase characteristics of the antenna.

RESULTS

The locations of our originally determined phase centers are shown in Table 1 while the locations of our phase centers fitted to data out to an 80 degree zenith distance cutoff are shown in Table 2. In all cases the North and East offsets are referred to the physical center of the antenna and the Up offset is referenced to the rear surface of the antenna backplane (in general, the surface of the antenna that would rest on the ground if the signal amplifier were removed).

The phase pattern labeled Cutoff_80 in Figure 3 indicates the typical flattening of the phase pattern that we obtain by removing the miscentering of the antenna as compared to what we originally obtained.

An examination of Tables 2 and 3 shows that the differences between the original and adjusted phase center locations are almost completely in the antenna vertical. This is a result of the limited zenith distance range that the chamber operators used in determining the phase center as well as the difficulty in adjusting the position of the anechoic chamber antenna positioner along this axis.

The adjusted phase patterns of the various antennas that we have measured along with the phase center positions are available in machine readable form via anonymous FTP from bfecvlbi.atsc.allied.com.

Antenna	Phase Center Fitted With 80 Degree Zenith Distance Cutoff					
	L1			L2		
	Up (mm)	North (mm)	East (mm)	Up (mm)	North (mm)	East (mm)
Dorne-Margolin T	53.88	4.71	2.08	82.99	3.50	1.41
Trimble 2202000	2.71	5.27	0.65	7.03	3.12	-0.25
Ashtech L1/L2	16.23	1.37	5.44	3.22	1.72	5.69
Dorne-Margolin DMC	85.2	0.39	1.32	104.8	0.34	0.28
146.6						
Micropulse 90LL12300	58.64	1.04	-0.83	83.35	1.54	0.11
TI	222.5	-0.64	-0.20	198.4	-1.13	0.22
Ashtech Geodetic	46.66	1.22	-2.23	44.34	3.84	-3.73
Macrometer	111.4	-0.24	-0.39	94.94	-1.44	-0.37
Trimble 14532-00	19.25	-1.48	-5.58	-3.48	-2.62	-2.45

Table 2 - Phase centers after adjustment for miscentering using data out to a zenith distance cutoff of 80 degrees.

CONCLUSIONS

The location of the best fit phase center for a given antenna is quite sensitive to the zenith distance range of the data that is used to determine the phase center. Furthermore, the phase patterns calculated for a particular fitted phase center are well constrained only within the zenith distance range used to determine the phase center.

It is crucial that the phase center and phase pattern which are used in the data analysis process are self consistent. Figures 2 and 3 show the great range in phase patterns which are obtained for a variety of assumed phase center locations. Clearly it would not be appropriate to pair a phase center location with a phase pattern that corresponds to a different phase center. GPS data analysts must take care to assure that the phase patterns and centers which they apply pass this consistency test.

Acknowledgment. This work was supported by contract NAS5-31742 between NASA and the AlliedSignal Technical Services Corporation. We would like to thank Bob King for pointing out an error in an earlier version of this paper.

REFERENCES

Schupler, B. R., and Clark, T. A., How Different Antennas Affect the GPS Observable, *GPS World*, 2, 32-36, 1991

Schupler, B. R., Allshouse, R. L., and Clark, T. A., Signal Characteristics of GPS User Antennas, *Navigation*, 41(3), 277-295, 1994

AZIMUTH- AND ELEVATION-DEPENDENT PHASE CENTER CORRECTIONS FOR GEODETIC GPS ANTENNAS ESTIMATED FROM GPS CALIBRATION CAMPAIGNS

M. Rothacher, W. Gurtner, S. Schaer, R. Weber
Astronomical Institute, University of Berne, Switzerland

W. Schlüter, H.O. Hase
Institut of Applied Geodesy, Wettzell, Germany

ABSTRACT

Using the GPS data of a special antenna calibration campaign involving most of the current high-precision geodetic receivers, mean antenna offsets and elevation- and azimuth-dependent phase center variations were estimated. The phase center variations were modeled using spherical harmonics. Due to the fact that the antennas were rotated by 180° and also interchanged between sessions, it was possible to estimate site coordinates *and* horizontal antenna phase center offsets at the same time. Rotating the antennas also made it possible to distinguish between multipath and real antenna phase patterns in the azimuth and to solve the problem posed by the "northern hole" in the satellite distribution.

INTRODUCTION

It is well-known that biases caused by phase center variations are one of the accuracy-limiting factors in high-precision GPS processing, in particular when combining different antenna types. Corrections obtained from anechoic chamber measurements (one of the possibilities to get information on phase center variations) did not find a wide user group so far. When applying these corrections to the processing of short baselines, some inconsistencies still remain. The estimation of antenna phase center variations from GPS phase data — discussed in this paper — represents an alternative method to obtain phase center corrections that is (apart from the absolute calibration) independent of anechoic chamber results and allows to check and complement those. Only the differences between antenna phase center variations are, however, accessible to GPS.

THE WETTZELL ANTENNA CALIBRATION CAMPAIGN

In spring 1995, March 20–24, an antenna calibration campaign was organized by the *Institut für Angewandte Geodäsie (IfAG)* in Wettzell. IfAG was supported by the *Deutsches Geodätisches Forschungsinstitut München* supplying two Wild GPS Systems 200 and the *Institut für Erdvermessung, Universität Hannover*, contributing two Ashtech Z-12 receivers. Observations were taken during 4 sessions of 24 hours each. Between day 079 and 080 (first and second session) all antennas — with the exception of the permanent Rogue and Turborogue antennas (points L and M) — were rotated by 180°. For days 081 and 082 the antennas were switched between the pillars. The details of the observation scenario are summarized in Table 1. Table 1 also lists the various antenna types involved.

| | Day 079 | | Day 080 | | Day 081 | | Day 082 | |
	Antenna	N/S	Antenna	N/S	Antenna	N/S	Antenna	N/S
A	Trimble Comp. 1	N	Trimble Comp. 1	S	Trimble Geod. 2	N	Trimble Geod. 2	S
B	Turborogue 1	N	Turborogue 1	S	Trimble Comp. 2	N	Trimble Comp. 2	S
C	Ashtech 1	N	Ashtech 1	S	Leica Inter. 2	N	Leica Inter. 2	S
D	Leica Inter. 1	N	Leica Inter. 1	S	Turborogue 2	N	Turborogue 2	S
E	Trimble Geod. 1	N	Trimble Geod. 1	S	Ashtech 2	N	Ashtech 2	S
F	Leica Inter. 2	N	Leica Inter. 2	S	Turborogue 1	N	Turborogue 1	N
G	Ashtech 2	N	Ashtech 2	S	Leica Inter. 1	N	Leica Inter. 1	N
H	Turborogue 2	N	Turborogue 2	S	Trimble Comp. 1	N	Trimble Comp. 1	N
I	Trimble Comp. 2	N	Trimble Comp. 2	S	Trimble Geod. 1	N	Trimble Geod. 1	N
K	Trimble Geod. 2	N	Trimble Geod. 2	S	Ashtech 1	N	Ashtech 1	N
L	Rogue IGS	N	Rogue IGS	N	Rogue IGS	N	Rogue IGS	N
M	Turborogue IGS	N	Turborogue IGS	N	Turborogue IGS	N	Turborogue IGS	N

Table 1: Observation Scenario of the Wettzell Antenna Calibration Campaign, March 20–24, 1995. The column "N/S" gives the orientation of the antenna.

ESTIMATION STRATEGIES

The processing of the GPS data was performed in two steps: (1) the estimation of mean antenna phase center offsets and (2) the estimation of elevation- and azimuth-dependent phase center variations.

In the first step the antenna phase center *offsets* were estimated *together* with site coordinates. This was possible because most of the antennas were rotated by 180° between the sessions. The *Bernese GPS Software* was modified to enable the estimation of horizontal and vertical antenna offsets allowing different antenna orientations and grouping of antennas. To prevent the normal equation system from becoming singular all site coordinates were constrained in all components within 3 mm to the terrestrial ground truth values. *Horizontal* antenna offsets were freely estimated except those of the two IGS sites where the antennas were not rotated (see Table 1) and where therefore 3-mm constraints were applied. All *vertical* offsets were constrained with 5 cm. To avoid a deterioration of the results due to multipath an elevation cut-off angle of 20° was used in this step.

The second step consisted in estimating the coefficients of a *spherical harmonics expansion* of the elevation- and azimuth-dependent phase center variations (see also (Rothacher et al., 1995) for details on this modeling technique). When only elevation-dependent coefficients were determined, the *station heights* had to be fixed to their ground truth values. Coefficients up to degree 10 and order 0 (11 parameters) were estimated. To compute both the elevation- and azimuth-dependence *all three components* of the site coordinates had to be fixed. Here coefficients up to degree 10 and order 5 (91 parameters) were set up. For all computations of the second step the *Trimble Geod. 1 antenna* was used as a reference antenna — only differences between antenna phase variations may be determined by GPS — and the antenna offsets determined in the first step were introduced. As we will see later, the multipath environment is certainly not optimal around the Wettzell facilities (radio telescopes and buildings). Therefore phase center variations were estimated down to 15° elevation only.

RESULTS

Antenna Phase Center Offsets

The *vertical* antenna phase center offsets computed from a combined solution of all the sessions are given in Figure 1.

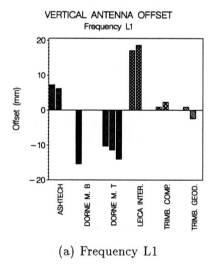

(a) Frequency L1

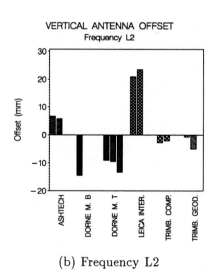

(b) Frequency L2

Figure 1: Vertical Antenna Offsets Determined from GPS Data with Respect to the Official IGS Values for Both Frequencies

The zero line is referring to a "mean" offset of all antennas and the values are corrections with respect to the official IGS antenna offsets (see files `ANTENNA.GRA` and `RCVR_ANT.TAB` at the *IGS Central Bureau Information System* described in (Gurtner et al., 1995)). We see that the differences between antenna types may be up to about 3 cm (e.g. Dorne Margolin – Leica Inter.). These offsets are in good agreement with values obtained from the calibration campaign published in (Rothacher et al., 1995). The site coordinates estimated in separate L1 and L2 solutions were compared to each other and to the "ground truth" available allowing for a translation between the individual coordinate sets. The L1 and L2 solutions agreed with each other on the 1 mm level (rms) in all components, whereas an rms of about 2 mm resulted when comparing the L1-solution with the terrestrial solution.

Elevation- and Azimuth-Dependent Phase Center Corrections

The results of the estimation of elevation-dependent variations using all four sessions are given in Figure 2 for the first frequency — corrections obtained with respect to the vertical antenna offsets shown in Figure 1.

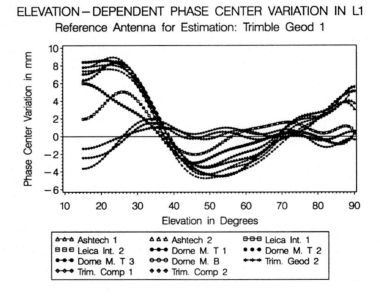

Figure 2: Elevation-Dependent Phase Center Variations in L1 for all Antennas Taking Part in the Wettzell Calibration Campaign

The phase center variations of antennas of the same type are in general consistently determined within about 1 mm. It is also evident that the Ashtech and Dorne Margolin antennas exhibit quite a different elevation-dependence compared to the Trimble Geod. antennas (used as reference here). This differences are responsible for the problems reported by several groups when combining Trimble with Ashtech or Dorne Margolin antennas (see e.g. (Rocken et al., 1992), (Mader et al., 1994), or (Breuer et al., 1995)).

The advantages gained by rotating the antennas during the campaign are becoming very evident when looking at the azimuth-dependence. Figure 3a shows the phase center variations in L1 in elevation and azimuth for the Ashtech 1 antenna as determined using the data of day 079. The antennas were oriented towards the north for this day. The region in the north, where no satellites can ever be tracked, is surrounded by dashed lines. Figure 3b shows the results for the same antenna pair on day 080, the antennas now being oriented towards the south.

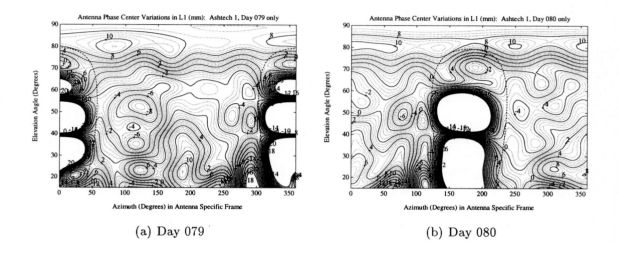

(a) Day 079   (b) Day 080

Figure 3: Elevation/Azimuth-Dependent Phase Center Variations in L1 as determined from GPS Data with respect to the Trimble Geod. 1 antenna

The "azimuth" in Figures 3a and 3b is referring to the azimuth of the antenna alignment (measured relative to the north direction indicated on the antenna itself). Therefore the "northern hole" is at an azimuth angle of $180°$ in Figure 3b. It is obvious then, that when combining the GPS data of all days into one solution the "northern hole" is no problem any more, so that the azimuth-dependent corrections of the combined solution may also be used for data taken at different geographical latitudes. We would like to point out two different features visible in Figure 3. The values obtained for day 079 (Figure 3a) indicate a large "anomaly" at an elevation of $20°$ and an azimuth of $290°$. The same feature can be detected on the Figure 3b at the same elevation angle but at an azimuth of about $110°$ (in both cases to the west of the "northern hole"). Therefore this anomaly cannot be part of the antenna phase center behaviour but is probably a multipath effect. We see that this anomaly is extended up to more than $30°$ elevation, indicating that much care has to be taken not to interpret multipath effects as antenna phase center variations. The combined solution (not shown here) reduces the impact of multipath considerably. The feature at an elevation of about 50° and an azimuth of 120° for day 079, however, reappears at the same location for day 080, as it should happen in the case of a *real* antenna pattern. These examples show the clear advantage of rotating the antennas, when trying to obtain azimuth-dependent corrections.

CONCLUSIONS

Using the GPS data of the antenna calibration campaign organized in Wettzell, by the *Institute for Applied Geodesy*, it could be shown that phase center offsets and elevation-dependent phase center variations may be derived from GPS data. The interchange of antennas between sessions allowed us to estimate not only the antenna offsets but also the horizontal positions of the points.

Because the antennas were pointing towards the north for some sessions and to the south for other sessions azimuth-dependent corrections could be estimated that are not affected by the "depopulated" region of the satellite coverage. In addition it is an interesting method to identify and detect multipath effects. As already pointed out in (Rothacher et al., 1995) phase center variations from chamber tests are necessary to (a) check the results obtained using GPS data only and (b) to get the absolute calibration of the variations not accessible to GPS as an interferometric technique. The combination of both types of results should finally lead to a consistent, reliable set of antenna corrections for all the major geodetic antenna types. There is still some way to go before reaching this goal.

REFERENCES

Breuer, B., J. Campbell, B. Görres, J. Hawig, R. Wohlleben, 1995, Kalibrierung von GPS-Antennen für hochgenaue geodätische Anwendungen, Paper accepted for publication in *Zeitschrift für satellitengestützte Positionierung, Navigation und Kommunikation*, April 1995.

Gurtner, W., R. Liu, 1995, The Central Bureau Information System, *International GPS Service for Geodynamics, Annual Report 1994*, ed. J.Z. Zumberge, R. Liu, and R. Neilan, JPL, California, in preparation.

Mader, G.L., J.R. MacKay, 1994, *Calibration of GPS Antennas*, draft publication.

Rocken, C., 1992, *GPS Antenna Mixing Problems*, UNAVCO Memo, November 12, 1992, Boulder, Colorado.

Rothacher, M., S. Schaer, S., L. Mervart, G. Beutler, 1995, Determination of Antenna Phase Center Variations Using GPS Data, *Proceedings of the IGS Workshop*, Potsdam, Germany, May 15-17, 1995, in press.